Geoinformatics: Methods and Applications

Editor: Noel Lane

R CALLISTO REFERENCE

www.callistoreference.com

Callisto Reference,
118-35 Queens Blvd., Suite 400,
Forest Hills, NY 11375, USA

Visit us on the World Wide Web at:
www.callistoreference.com

ISBN: 978-1-64116-577-8 (Hardback)

Cataloging-in-publication Data

Geoinformatics : methods and applications / edited by Noel Lane.
    p. cm.
Includes bibliographical references and index.
ISBN 978-1-64116-577-8
1. Geoinformatics. 2. Geoinformatics--Methodology. I. Lane, Noel.
G70.212 .G46 2022
910.285--dc23

Table of Contents

Preface

Geoinformatics is a field of science and technology that is concerned with the use and development of information science infrastructure for developing solutions in cartography, geosciences and geography. At its core, the field encompasses technologies that enable the acquisition, analysis and visualization of spatial data. Geospatial analysis and modelling, information system design, geospatial databases, human-computer interaction, etc. are applied to geoinformatics studies. Geoinformatics also uses geovisualization and geocomputation for analysis of geoinformation. The applications of geoinformatics are diverse, from public health, urban planning and land use management to meteorology, climate change, and environmental modelling and analysis. There has been rapid progress in the field of geoinformatics and its applications are finding their way across multiple industries. From theories to research to practical applications, case studies related to all contemporary topics of relevance to this field have been included herein. This book aims to equip students and experts with the advanced topics and upcoming concepts in this area.

The information shared in this book is based on empirical researches made by veterans in this field of study. The elaborative information provided in this book will help the readers further their scope of knowledge leading to advancements in this field.

Finally, I would like to thank my fellow researchers who gave constructive feedback and my family members who supported me at every step of my research.

Editor

Remote sensing of vegetation cover changes in the humid tropical rainforests of Southeastern Nigeria (1984–2014)

Friday Uchenna Ochege[1,2]* and Chukwunonyelum Okpala-Okaka[2]

*Corresponding author: Friday Uchenna Ochege, Laboratory for Cartography and GIS, Department of Geography & Environmental Management, University of Port Harcourt, Choba, Rivers State, Nigeria; Faculty of Environmental Studies, Department of Surveying and Geoinformatics, University of Nigeria, Enugu, Nigeria
E-mail: uchenna.ochege@uniport.edu.ng

Reviewing editor: Louis-Noel Moresi, University of Melbourne, Australia

Abstract: This study demonstrates a 30-year multi-temporal variations in vegetation cover changes as a means of filling the vegetation knowledge gap in the humid tropical forests of southeastern Nigeria. Landsats 4TM, 5TM and 7ETM+ data-sets were accessed and analysed using the Maximum Likelihood Classification algorithm to discriminate and geovisualize the spatiotemporal variations in the general vegetation and other land cover types, from 1984 to 2014. This was supported with detailed field surveys in dry and rainy seasons of 2011 and 2014 to ascertain the status of wide-ranging vegetation cover stands. A 44% vegetation decline was recorded given the reduction in dense vegetation spatial extent from 330.63 km^2 in 1984 to 170.87 km^2 in 2014. Sparse vegetation equally increased in spatial extent by 25% given the variations registered from 6.86 km^2 in 1984 to 97.16 km^2 in 2014. The reduction in vegetation cover was found to have been replaced by increase in other land cover types—residential (18.97 km^2) and industrial areas (39.87 km^2). Suggesting that, heterogeneity in the spatial distribution of land resources, in addition to weak concerns towards preserving the accruing benefits of vegetation resources attracted anthropogenic phenomenon (e.g. urbanization) to vegetated

ABOUT THE AUTHORS

Friday Uchenna Ochege is a PhD candidate with special interest in the application of Cartographic, GIS and Remote Sensing techniques to environment related issues. He has been involved in internationally and nationally funded research projects. We are currently working on the application of geospatial techniques to modeling climate change impacts on agriculture and watershed resilience and resilient landscapes in the Humid Tropics of Southeastern Nigeria. We aim to develop frameworks for non-technical policy makers that will guide in sustainable land use.

Chukwunonyelum Okpala-Okaka is a professor in the Department of Surveying and Geoinformatics, Faculty of Environmental studies at the University of Nigeria. His area of research interest is Cartography, Remote Sensing and GIS modelling. He currently teaches both undergraduate and postgraduate courses in Cartography and GIS and its environmental applications, within Nigeria. He has over 29 years of teaching experience with several scientific publications to his credit.

PUBLIC INTEREST STATEMENT

This study provides baseline information on the spatial variations and temporal patterns of vegetation cover and other land-use changes in the humid tropical rainforest of Nigeria using satellite data-sets. Key factors responsible for the changes have been identified and documented. The validated results helped to reveal interclass confusion especially as earlier report assert that the region lacks adequate monitoring framework and information on vegetation cover changes. With this research, appropriate place-specific reinvestigations and or decisions can be made, such as rapid identification of alternatives where necessary, and to incorporate same in the formulation of sound policies required in the sustainable governance and management of natural resources in the humid tropical rainforest.

areas. As such, strengthening institutional monitoring and urban planning frame-works would help to improve sustainable governance of the tropical rainforests.

Subjects: Vegetation; Environmental Studies & Management; GIS, Remote Sensing & Cartography

Keywords: remote sensing; vegetation cover changes; tropical rainforest; Southeastern Nigeria

1. Introduction
Vegetation is the general plant life or the total plant cover forming parts of the biological system and it is the primary producer of any ecosystem (Ochege, 2014b). Local and regional wide-ranging natural vegetations are important component on earth and they govern all forms of life (Millennium Ecosystem Assessment, 2005). They provide food, oxygen, fertility and finally enhance survival of all living beings (Matlack, 1994; Mckey, Waterman, Gartlan, & Struhsaker, 1978; Newbold et al., 2014). For the earth's environment, natural vegetation constitutes the biologically richest ecosystems and play vital roles in regional hydrology, carbon storage and the global climate dynamics (Du et al., 2015; Forkel et al., 2013; Igbawua, Zhang, Chang, & Yao, 2016).

The benefits accruing to humankind from the natural functioning of a healthy and productive vegetation system cannot be over emphasized. Because, the natural vegetation provides not only the basic needs of life (i.e. food, clothing and shelter), but enables purification of water bodies (Friberg et al., 2011), manages diseases (Fisher & Turner, 2008), regulates climate and the functioning of biosphere (Millennium Ecosystem Assessment, 2005) and provides mankind with spiritual fulfilment that contributes to improving quality of life (Food & Agriculture Organisation, 2013). As such, vegetation remains the lifeblood of human societies around the world (United Nations Environment Programme & the International Institute for Sustainable Development, 2004).

Nevertheless, human being has long distinguished himself from other species by shaping ecosystem forms and processes using fire, tools and technologies that are beyond the capacity of other organisms including natural vegetation stands (Smith, 2007). As a result, man is inextricably and unavoidably attached to the environment, which often time gives him the nerve to impact negatively on forests and natural vegetations resulting in other land uses.

Based on the considerations of the Intergovernmental Panel on Climate Change (IPCC), land use and land cover are not technically synonymous. Land cover is the observed physical and biological cover of the Earth's land, such as vegetation or man-made features, whilst the total arrangements, activities and inputs undertaken in a certain land cover type (e.g. a set of human actions) for both social and economic purposes (e.g. grazing, timber extraction, conservation) are referred to as land use (Watson et al., 2000). Since vegetation is continuously changing—alteration in the surface components of the vegetation cover, the rate of change can either be dramatic and or abrupt—as exemplified by fire, subtle and or gradual (Lund, 1983; Milne, 1988). This is a common global scenario, especially, with regard to forest cover and biomass accumulation or mass deforestation (Hansen et al., 2013).

Africa has been observed to be experiencing the fastest rate of vegetation change and most of its segments are already evidently been impacted and plagued with diverse ecological problems (Erika et al., 2015; Ofori, Owusu, & Attuquayefio, 2014; Yeshaneh, Wagner, Exner-Kittridge, Legesse, & Blöschl, 2013). Especially, as land use remains a significant driver of habitat degradation and removal and associated losses in biodiversity and vegetation resources (Lucas et al., 2015). According to Odjugo and Ikhuoria (2003), these problems, though anthropogenic in context, do have a direct link with the ongoing global climatic and environmental change. Nevertheless, natural vegetation equally suffers from these consequences, as they are profoundly been altered by human activities, and so, few natural stands remain (FAO, 2012).

In Nigeria, a wide range of vegetation types exist, and they reflect past and present climatic variations (Federal Republic of Nigeria, 1977; Igbawua et al., 2016; Odjugo, 2010). Generally, the southern part of the country is flanked by maritime stretches where the sandy uniformity is occasionally broken by mangrove ecosystem, bushes, lush vegetation, hardy trees and putrid water, while the northern part displays segments of croplands and rangelands that are heavily populated by grasses of varied species (Igbozurike, 1975). The global land cover map by DIVA-GIS (2016) reveals Nigeria's current vegetation cover belts (see Figure 1(A)). From top to bottom indicate Sahel Savannah, Sudan Savannah, Guinea Savannah and Forest vegetation (Igbawua et al., 2016).

Based on the obvious reasons of regional vegetation dynamics which has been influenced by global environmental change (Igbawua et al., 2016), the natural vegetation of Nigeria has ever since been under considerable threat like those of most other parts of tropical Africa (Adekunle, Olagoke, & Ogundare, 2013; FAO, 2012; Federal Republic of Nigeria, 1977). The World Resources Institute (WRI) did identify Nigeria's humid tropical rainforests as one of the most ecologically vibrant places on the planet because it is home to over 4,850 different plants and tree species, 1,340 species of animals, among which are 274 mammals, 860 birds (Meduna, Ogunjimi, & Onadeko, 2009; World resources, 1987). Despite these rich and abundant vegetation resources, the country is highly rated for unsustainable exploitations, deforestation and forest degradation among others (Momoh, 2014). Yet, the country lack adequate monitoring framework (Erika et al., 2015) especially, in those steadily urbanizing locations of the humid tropical rainforest belts (Ofori et al., 2014) including southeastern Nigeria i.e. Umuahia.

Prior to Umuahia been declared a headquarter city in Nigeria by the national government on the 27 August 1991, Umuahia was previously known to be home to ecological parks, protected forests, unprotected and undeveloped forest areas and trees growing amidst scattered residential settlements. These natural green covers provide different benefits, ecosystem services and support to well-being, ecosystem health, urban livelihoods and other advantages to the society at large. Notwithstanding these recognized benefits to sustainable development and environmental sustainability to the area, spatial quantification of local vegetation changes has been lacking for this part of Nigeria.

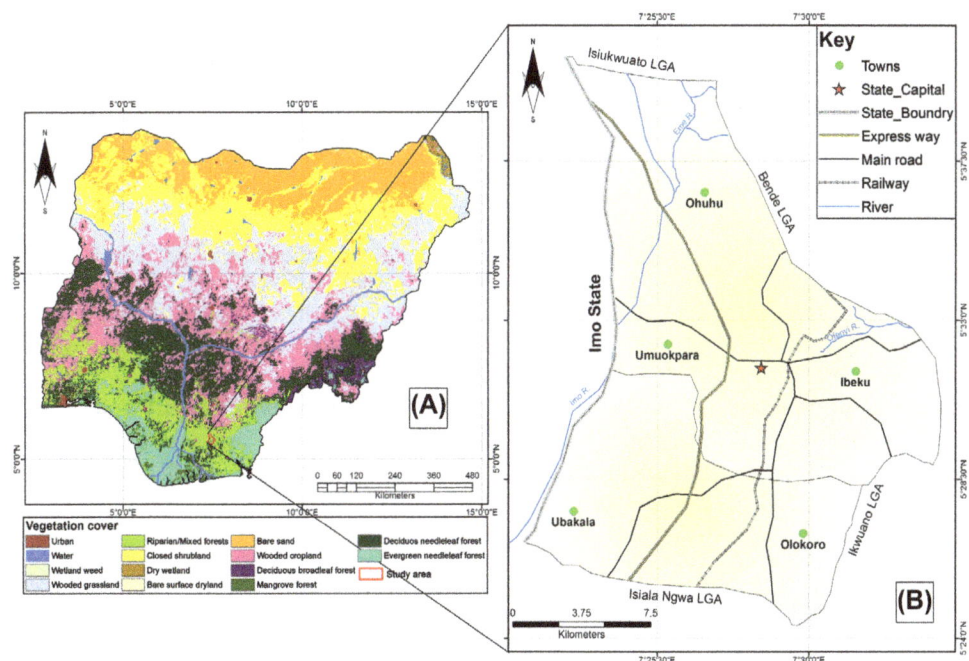

Figure 1. (A) Nigeria's vegetation cover (DIVA-GIS, 2016) and (B) the study area.

Remote sensing technology has proven essentially relevant in establishing land use and land cover monitoring frameworks at different scales (Hansen et al., 2013; Loveland & Dwyer, 2012). It is the science by which information about an object is obtained from electromagnetic radiation reflected from the surface of that object (Jensen, 2014; Lillesand, Kiefer, & Chipman, 2015). Its mechanisms have revolutionized traditional mapping methods (Ariti, van Vliet, & Verburg, 2015; Newbold et al., 2014; Okpala-Okaka & Igbokwe, 2010; Scull et al., 2016; White & Oates, 1999), by advancing and characterizing the spatiotemporal distribution of environmental phenomena using air-borne sensor platforms and image processing and interpretation techniques or packages (Chen, Lu, Luo, & Huang, 2015; Dubula, Tesfamichael, & Rampedi, 2016; Forkel et al., 2013; Hansen et al., 2013; Loveland & Dwyer, 2012; Luo, Zhou, Chen, & Li, 2008). The advent of remote sensing paved way for improved methodological mapping of vegetation cover changes since the establishment of Landsat mission in 1972 (Coppin & Bauer, 1996; Coppin, Jonckheere, Nackaerts, & Muys, 2004; Loveland & Dwyer, 2012).

Several researchers have comprehensively used Landsat imageries, with its extensive data archive at no cost and suitable spectral and spatial resolutions to detect and quantify vegetation cover changes (Chen et al., 2015; Luo & Dai, 2012; Luo et al., 2008; Omo-Irabor et al., 2011; Zhu, Liu, & Chen, 2012). But literature search, with focus on the Umuahia segment of the tropical rainforest belt of Nigeria shows a dearth of information about the dynamics of vegetation cover changes in the area, probably due to limited accessible monitoring data and a lack of appropriate research methods.

This paper, therefore, initiated a remote sensing-based vegetation baseline assessment that is nonexistent in the study area (Erika et al., 2015), as a strategy for informing the non-technical and expert policy makers involved in the sustainable governance and development of the region.

This is necessary to understand the dynamic nature of emerging regional and local vegetation and land cover types that may have been impacted over time. The information will be useful in streamlining policy efforts towards sustainable urban growth in the long-term, land cover recovery, agricultural resilience and carbon storage/sequestration in the face of increasing changing climate. The following specific objectives have been addressed:

(1) to estimate the spatial extent of the vegetation cover changes using remotely sensed satellite data;

(2) to quantify the temporal changes based on spatial extent of the vegetation cover, before and after 1991 (the year the study area was officially created) to a near present date—2014; and,

(3) to identify and document responsible factors for the changes in terms of areal extent of the study area.

The results generated in this study would, thus, show the spatial variations and pattern of vegetation cover distribution of different vegetation classes. While their validation helped to reveal inter-class confusion than could be resolved with the use of other biased information (Foody, 2002), especially as it had been established that the region lacked adequate vegetation monitoring data (see, for example, Erika et al., 2015). In this way, appropriate place-specific reinvestigations and or decisions can be achieved, such as rapid identification of alternatives where necessary, and to incorporate same in the formulation of sound policies required in the governance and natural resources management framework of the region (see, for example, Larcom, van Gevelt, & Zabala, 2016; Obydenkova, Nazarov, & Salahodjaev, 2016).

2. Materials and method

2.1. Case study area

Geographically referred to as Umuahia, the study area (latitude 5°26′06.00″N to 05°36′04.00″N and longitude 07°21′50.00″E to 7°34′03.00″E) covered the administrative boundary of the present day Umuahia North and Umuahia South local government areas in Abia State. It lies entirely in the southeastern segment of the humid tropical rainforest in Nigeria (Figure 1(B)). It covers an area of about 363.13 km² amidst a constellation of scattered villages and towns within a 15 km radius, while the total area covered by the metropolis is about 71 km² (Ejenma, 2013).

The climate is humid tropical rainforest—Koppen Af (Kottek, Grieser, Beck, Rudolf, & Rubel, 2006). Daily average insolation is generally low—4.8 h, nevertheless the area experiences mean annual maximum temperature of 31°C with little daily variations (Iloeje, 2007). Meteorologically, Umuahia experiences an annual mean rainfall of 2,278 mm, with eight months of precipitation, which extends from early March to late October (Figure 2). Meaning, that, the study area witnesses two major seasons—dry season and rainy season. The dry season is dominated by a period of short spell of dry/cool season referred to as *harmattan*. Usually, the heaviest of monthly total of rains → 363 mm (Figure 2), is experienced more in September, while the month with the lowest rainfall fluctuates between January and December, except for recent climatic anomalies with random variance of precipitation in other locations (e.g. Nsukka & Ogbudu) (Campling, Gobin, & Feyen, 2001).

The soils of Umuahia have been greatly influenced by different ancient geologic formations, climate, vegetation, and general topographic configuration of the area (Onu, Opara, & Ehirim, 2012). The major rivers that drain the catchment area of Umuahia are the Imo river on the west axis, Kwa-Ibo river draining southwardly and Enyong creek on the east axis. River Eme with it various tributaries in Ohuhu flows into the Imo river while the Ofenyi river tributaries transverse the Ibeku landscape to deposit into the Enyong creek. These rivers and streams are ephemeral and dry up after the cessation of rainfall. The study area currently serves as one of the nation's commercial hubs for economic development, with high potentials for growth and development.

2.2. Field observations

Validation of remotely sensed data through field surveys, documented information and discussions with the local people has proven useful in several land use/land cover (LU/LC) studies (see, for example, Ariti et al., 2015; Scull et al., 2016; Yeshaneh et al., 2013). In this study, vegetation of the study area was carefully observed by extensive field surveys conducted during rainy season, and the early dry season in 2011 and 2014 respectively. Ground truth of relevant land cover types in the study area were collected alongside plant species and other landscape features. A Garmin 62s Global Positioning System (GPS) with 3 m accuracy was used to identify infrastructural facilities that were relocated to

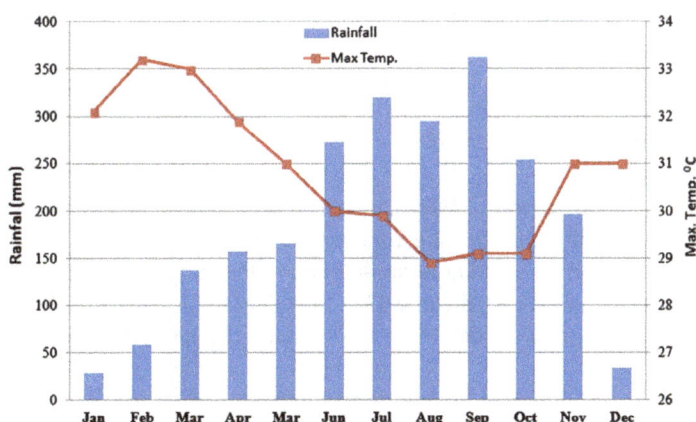

Figure 2. Monthly distribution of rainfall and temperature over Umuahia.

previously vegetated areas. Likewise, certain geographical features, ranging from landscape configurations to vegetation types, were equally identified. This was necessary, to complement and affirm the relationship and variations that may be there, would have occurred and the rationale behind the changes.

As such, the study area lies in the lowland rainforest vegetation belt typified by almost continuous cover of *riparian* forest along stream valleys. Tilling, in most cases, for agriculture and road construction, has resulted in an ecological situation where the normal development of vegetation is markedly retarded (Onu et al., 2012). Although, most of these locations have been drastically affected by human activity so much, large sections of the upper rain forest zone may be called an "oil palm bush". However, the observed vegetation of the area is characterized by an abundance of seed-bearing plants species, such as *Brachustegia eurycoma, Carya* spp., *Steroulia, Canarium, Cassia siame, Triplochitan scleroxylong, Meliaceae, Mitragyna ciliate, Khaya* spp., *Entandrophrayma* spp., *Lovia trichiloides, Nauclea diderrichii, Gmelina arborea, Neliaceae, Acacia nilotica* and *Gum arabi.* Others include *Cauarea and Tarminalia.* Many times, they occur in about 100 species per hectare. It is this great abundance of species that makes the area rich in terms of biomass productivity of all terrestrial ecosystems, therefore, requiring a periodic assessment.

2.3. Satellite remote sensing data

Remote sensing offers priceless source of geospatial information for ecological and vegetation resource management (Hagenlocher, Lang, & Tiede, 2012; Hansen et al., 2013; Lillesand et al., 2015; Luo & Dai, 2012). With reference to complex terrain, data unavailability, area coverage, data obtained by remote sensing can be timely, cost-effective, and objectively presented and demonstrated for either specific or post assessment—spatiotemporal investigations (Hagenlocher et al., 2012; Omo-Irabor et al., 2011). In this study, vegetation variation in Umuahia was mapped and visualized from Landsat data covering the total extent occupied by the study area.

Given the different software algorithm-based platforms available for ecological studies (Coppin et al., 2004), ERDAS Imagine 2014 and ArcGIS 10.1 were used for image pre-processing so as to maintain data spatial reference consistency to the georectified format—the Universal Transverse Mercator (UTM) zone 32 North and the World Geodetic System 1984 datum, at the various stages. As such, classification challenges (Huang, Lu, Zhang, & Plaza, 2014; Lillesand et al., 2015) were minimized by selecting suitable dates—1984, 1991 and 2014. This was necessary for maintaining seasonal uniformity as an a priori for increased ground vegetation and cloud cover during the wet season (Churches, Wampler, Sun, & Smith, 2014; Schwartz, 2003). More importantly, the 1991 data equally served as a second baseline date for the reason of the study area's official pronouncement as headquarter to a major state in southeastern Nigeria. A scan line corrector (SLC) error experienced by the Landsat 7/ETM+ was corrected with a gap fill function (Zhu et al., 2012) in ERDAS Imagine 2014.

Datasets of 1984, 1991 and 2014 (path 188, row 56 with a spatial resolution of 30 × 30 m) were obtained from the United States Geological Survey (USGS) archives at www.earthexplorer.usgs.gov. After image acquisition, bands 5 (mid-IR): 1.55–1.75 μm, 4 (NIR): 0.76–0.90 μm and 3 (red): 0.63–0.69 μm in each of the image scene were stacked together to form a single multispectral image data-set using the "layer stack" function in ERDAS Imagine.

The United Nations Fund for Population Activities (UNPFA) had recognized that population growth and its resultant human influence constitute serious pressures on global natural resources especially on local and regional forests ecosystems and natural vegetation (United Nations Population Fund, 1991). Using the annual population growth rate of 2.83%, (Abia State of Nigeria, 2005), the population of the area was projected to present to determine the influence of population growth on vegetation resources.

So, the resulting multispectral images based on the layer-stacked data-sets were used as the baseline data—before impact imagery (1984–1991—*marked year of population influx in the study*

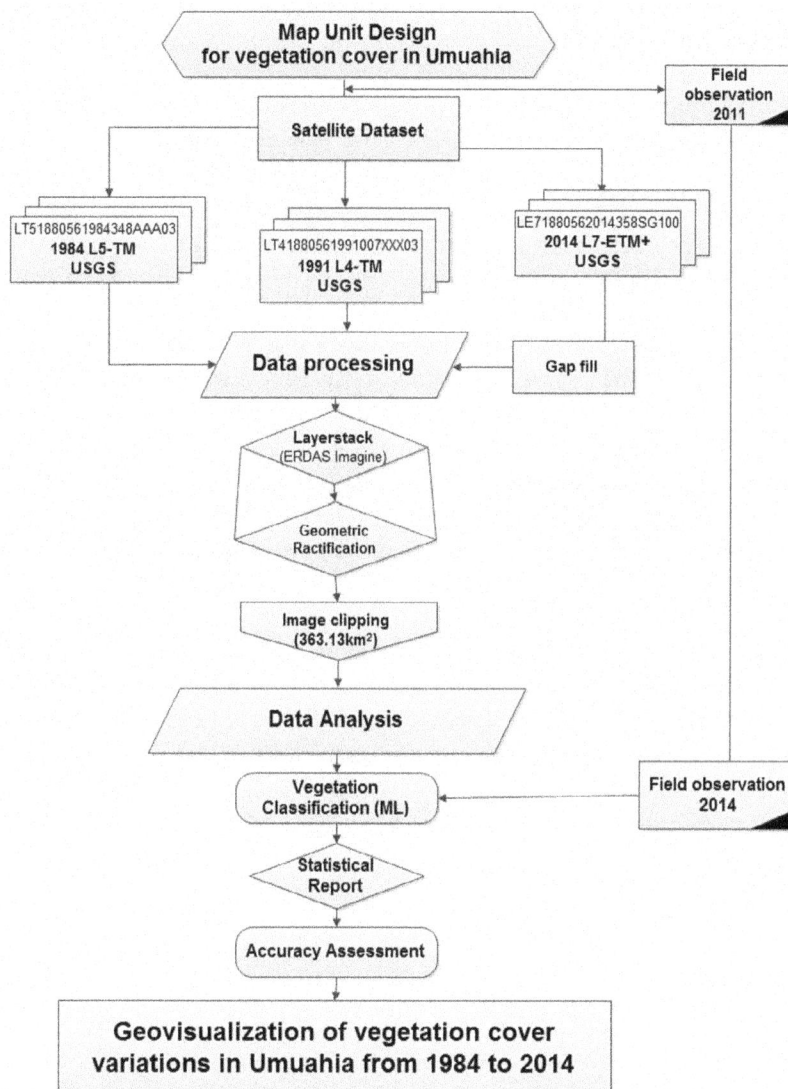

Figure 3. Vegetation cover mapping workflow.

area). While the analysed data-sets served as the post impact assessment imagery (1991–2014). The process allowed for straightforward detection of changes and human-induced impacts on the landscape/vegetation cover over the period of study—30 years. The vegetation classification workflow for the study consists of several other steps and stages, as illustrated in Figure 3.

3. Data analysis

3.1. Image classification
Since land use cover change is a continues issues (Watson et al., 2000), detecting spatial and temporal patterns of vegetation cover changes with satellite data depends largely on pixel-based spectral signatures or vegetation indices (Chen et al., 2015). Based on their spectral signatures or vegetation intensity values, the Landsat image pixels covering the study area were organized into a finite set of classes that represents surface types. This can be done in two ways; supervised and unsupervised classification (Lillesand et al., 2015). This study used the maximum likelihood classification (MLC) method of the set of the supervised classification algorithms in grouping vegetation cover changes. Several other methods do exist (Minimum distance technique, Mahalanobis distance technique and Parallelepiped classification methods (Soofi, 2005), but the maximum likelihood

method is preferred because it required field observation to aid the classification procedure which tends towards accurate data analysis. The analysis was performed with ERDAS Imagine.

Using the (MLC) on ERDAS Imagine (Congalton & Green, 1999), the enhanced false colour composite bands 5, 4 and 3 of the different years depicting the vegetation image pixels were trained and categorized into appropriate classes. A total of five classes were discriminated (i.e. densely vegetated and sparsely vegetated areas, water bodies, congested residential and built-up industrial areas). The same five classes have been adopted in this work. This classification pattern is in accordance with the 2010 updated version of the global ecological zones for forest reporting by FAO (2012).

The supervised classification maps, showing the spatial extent and variations in forest cover across the study area, are presented in Figure 5. For effective visual interpretation, suitable colour patterns have been used to identify and show the various classes. A class name is assigned to a colour. Thick green represents dense vegetation, light green is for sparse vegetation (Ochege, 2014a), ox-blood is used to show congested sections in the study area, light brown represents built-up areas while blue is used to show the presence of any kind of water body in the area. This pattern is acceptable according to the vegetation classification standard as modified by Anderson, Hardy, Roach, and Witmer (1976).

3.2. Accuracy assessment

It is usually very important to ensure the correctness of the classification analyses because it seeks to measure the quality of results shown on the classification maps (Banko, 1998; Congalton & Green, 1999). The accuracy assessment can either be quantitative or qualitative. In this study, we evaluated the accuracy process of vegetation classification by visual inspection of the classified image in ERDAS Imagine using Kappa statistics function on a scale range of −0.1 to 1 (Congalton, 1991). Kappa statistics is calculated as in Equation (1):

...

$$K = \frac{N \sum_{i=1}^{x} x_{ii} - \sum_{i=1}^{r} (x_{i+} \times x_{+i})}{N^2 - \sum_{i=1}^{r} (x_{i+} \times x_{+i})} \qquad (1)$$

where N = is the total number of samples in the matrix, r = corresponds to the number of rows in the matrix, x_{ii} = is the number in row i and column i, x_{+i} = is the total for row i, and x_{i+} = is the total for column i.

We had generated 150 (30 for each class) reference sites (Figure 4) which were based on the simple random sampling technique (Lins & Kleckner, 1996).

Each sample point was assigned the ideal class value during field observations, and was used to enhance geovisual inspections. These points were further superimposed on the classified images. Features of the representing land use pixels (class) that correspond with each point were compared with the features that existed on the ground by also using the federal government approved national vegetation atlas base map of Nigeria (Federal Republic of Nigeria, 1977), Google Earth pro web-based GIS, in addition to other reference information obtained from field observations. Then, the following accuracy assessment estimators: the error matrix, overall accuracy, producer's accuracy, user's accuracy and the kappa coefficient, were computed in ERDAS for each of the year understudied—1984, 1991, 2014 (see Section 4.5).

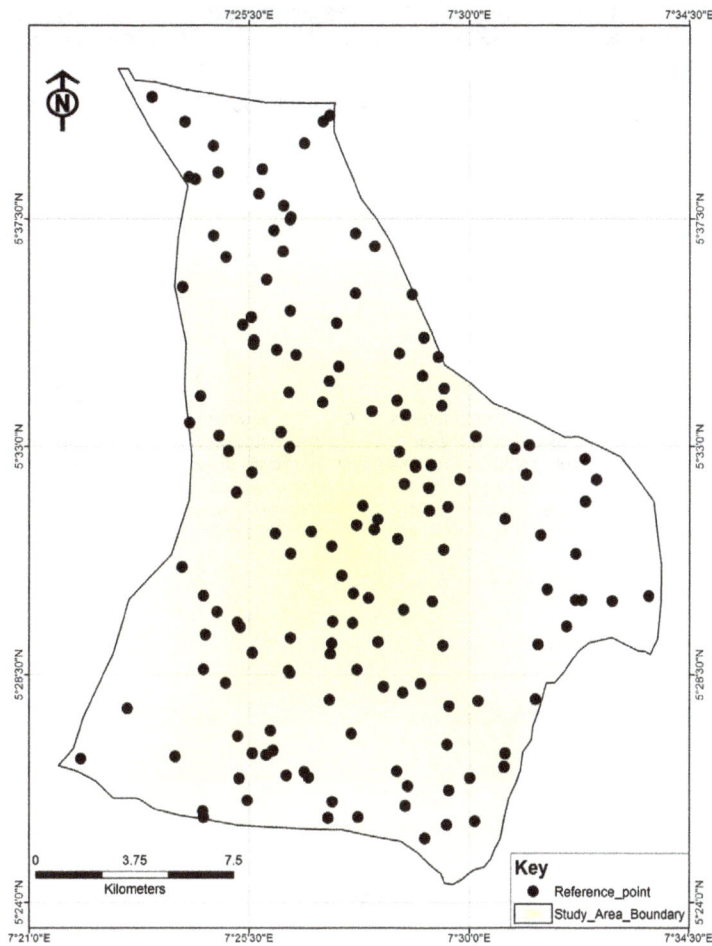

Figure 4. Simple random points generated for vegetation classification validation.

4. Results and discussion

4.1. Spatio-temporal dynamics of vegetation cover changes

Results from this study include a vegetation cover statistics and land-cover classification maps of the years understudied—1984, 1991 and 2014. From 1984 to 1991, dense vegetation (healthy vegetation) had reduced in percentage by 21.2%. Between 1991 and 2014, it further reduced by 22.8%. While sparse vegetation (disturbed or unhealthy vegetation cover) increased by 18.32% (1984–1991) and 6.64 (1991–2014). Likewise, residential and built up areas increased by 0.09% and 2.76% in 1991, and 5.13 and 8.94% in 2014, respectively (Table 1).

4.2. Vegetation and other land cover stands as at 1984

The 1984 classification map clearly shows that dense vegetation covered most of the fragments in the study area followed by patches of other cover types—sparse vegetation, water bodies, congested and built-up areas (Figure 5(A)). As such, vegetation resources in the study area may not be adjudged to be entirely untouched or referred to as pristine because of the scattered pattern of sparse vegetation which indicates disturbances and stress on vegetation canopies and phenology (Ochege, 2014a). In this regard, there are chances that most vegetation fragments in the humid tropical rainforests of southeastern Nigeria have experienced one form of human or natural interruptions.

Often times, fragments of tropical forests are considered primary or virgin, whereas, in the actual sense, they have passed through a number of stages to become secondary forests (Aubreville, 1938;

Table 1. Changes in vegetation and other land cover types in Umuahia

Land cover	Total area covered			% change		
	1984 (%)	1991 (%)	2014 (%)	1984–1991	1991–2014	1984–2014
Dense vegetation	330.63 (91.05)	253.65 (69.85)	170.87 (47.05)	−21.2	−22.8	−44
Sparse vegetation	6.86 (1.8)	73.05 (20.12)	97.16 (26.76)	18.32	6.64	24.87
Congested (residential)	9.69 (2.67)	10.04 (2.76)	28.66 (7.89)	0.09	5.13	5.22
Built-up (industrial)	15.07 (4.15)	25.09 (6.91)	54.94 (15.13)	2.76	8.94	10.98
Water body	0.87 (0.24)	1.30 (0.36)	10.46 (2.88)	0.12	2.52	2.64

Notes: Total area occupied by each cover type was calculated as follows: (CT/SA) × 100%; where CT = area occupied by each cover type of year under consideration, SA = 363.13 km² is the area covered by the study area; while the percentage change (1984–2014) for each cover type was derived by subtracting the percentage change of each cover type in 1984 from those in 2014, likewise for time interval of 1984–1991 and 1991–2014. In furtherance of this exploratory analysis, Table 1 is summarized as follows: dense vegetation and sparse vegetation covers represent vegetation resources, congested (residential) and Built-up(industrial) areas represent urban encroachment, while water body represents Water (see Appendix 2).

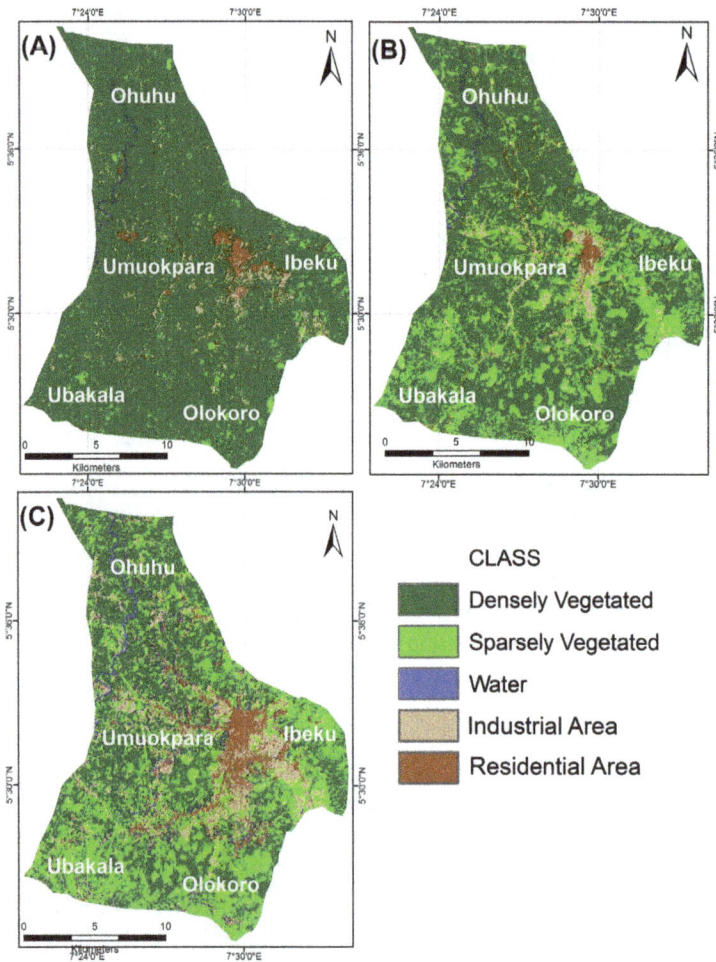

Figure 5. Vegetation classification of the study area (A) in 1984, (B) in 1991, and (C) in 2014.

Figure 6. Harvested Woodstock within the study area.

Budowski, 1970; Bush & Colinvaux, 1994). For instance, the Okomu Forest Reserve in southwest Nigeria was considered to be a primary forest by Richards (1939), but later studies by Jones (1956) revealed extensive charcoal and pottery deposits and a tree population structure reflecting "second growth" vegetation. Recent studies now provide evidence that the forests of Okomu can be traced back to a period soon after 700 years ago, following a period of intensive human use (White & Oates, 1999).

Conducting interviews or discussions with the local people is one way of ascertaining age long anthropogenic impacts on past vegetation status, vegetation loss and their root causes (Ariti et al., 2015; Ofori et al., 2014; Yeshaneh et al., 2013). Critical information obtained through informal discussions with some indigenous people of the area during the field observations, suggests that: *most rural dwellers in the study area depend to a large extent on wood fuel energy for cooking and timber logging for furniture and other kinds of woodwork.* Figure 6 show harvested wood fuel from the Ibeku axis of the study area, intended for domestic energy consumption. This practice is neither new nor exclusive to the Umuahia ecological zone of the humid tropical rainforest, but is obtainable in most parts of the country (Anyiro, Ezeh, Osondu, & Nduka, 2013; Food & Agriculture Organization of the United Nations, 1981; Jones, 1956; Tee, Ancha, & Asue, 2009; White & Oates, 1999).

Likewise, forests and vegetation of the study area may be of similar secondary growth vegetation with those of Okomu forests, as shown by the scattered distribution of stressed vegetation which is an indication of human influence (Figure 5). To say the least, wood fuel harvests are on the increase and constitute a major factor to vegetation cover changes in the humid tropical rainforests of Nigeria and the entire sub-Saharan Africa (Sulaiman, Abdul-Rahim, Mohd-Shahwahid, & Chin, 2017).

Statistical report generated from the 1984 image pixel training show that, dense vegetation dominated a greater portion and occupied a spatial extent of 330.63 km² out of the 385 km² of the entire study area, while sparse vegetation fragments occupied a spatial extent of about 6.86 km². As at 1984, built-up and congested areas occupied 15.07 and 9.69 km², respectively. Given the total study area size of 385 km², the classification analysis accounted for 94.32% of the total land cover. About 5.68% of the study area's land cover was unclassified, and so is unaccounted for. Out of the new land area (i.e. 363.13 km²), dense vegetation totalled 91%, while sparse vegetation, congested and built-up areas occupied 2, 3 and 4%, respectively (Figure 7(A)).

4.3. Vegetation and other land cover changes in 1991
Figure 5(B) shows that increased population among other factors have had significant impact and probably, some negative implication on the spatial extent of vegetation cover changes on the Umuahia segment of the humid tropical rainforest. The statistical report generated from the

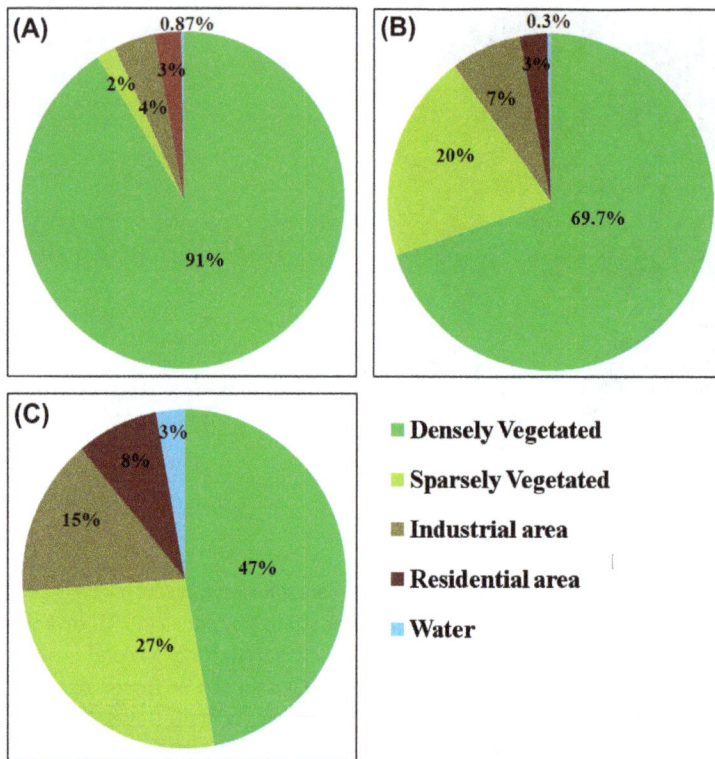

Figure 7. Spatiotemporal variations of vegetation cover of the study area (A) as at 1984, (B) as at 1991, (C) as at 2014.

classification analysis carried out on the 1991 data-set shows that land area covered by dense vegetation decreased from 330.63 km^2 (91%) in 1984 to 253.65 km^2 (69.7%) in 1991, while every other land cover types, i.e. sparse vegetation, congested and built up areas increased in spatial extent (Table 1). As such, 76.98 km^2 which is about 20% of dense vegetation cover was lost between 1984 and 1991. Then fragments occupied by sparse vegetation increased by 18% given the initial 2% as at 1984. Though, congested sections maintained its size, but built-up areas gained by 3% (Figure 7(B)).

It is obvious that population growth is associated with increased demand for land and economic spaces and or exploitation of natural resources—including forest and non-forest products (Obeta, Aujara, Ochege, & Shehu, 2013; United Nations Population Fund, 1991). In this study, data analyses reveal a strong correlation between population growth and vegetation resource depletion. Projected population of Umuahia, which is currently 438,992 rose from 220,104 in 1991 to 359,230 in 2006 (Federal Republic of Nigeria, 2007) show that the area is rapidly urbanizing. Consequently, observations from field surveys revealed serious spread of urban infrastructure and socio-economic activities to previously vegetated sections.

Pieces of information gathered from the indigenous people by preliminary random discussions in 2014 confirm that: (1) urban encroachments (Figure 8) were often not coordinated, (2) have been ongoing since the 1980s, and (3) did significantly accelerate after the study area became state capital in 1991. The people further indicated that: "past methodologies that saw to the sitting and relocation of urban infrastructure to forest-rich zones derided adequate ecological unit accounting and or impact assessment. They maintain that some ecological units of already destroyed forest corridors (e.g. Ibeku and Olokoro) were natural habitats to certain endemic vegetation resources (e.g. Nauclea diderrichii, Myrianthus arboreus P. Beauv., K. Schum., Canarium schweinfurhii, Dialium guineense Wild)". Unfortunately, most of the species are now difficult to come by even within other sections of the study area (See, for example, Meregini, 2005).

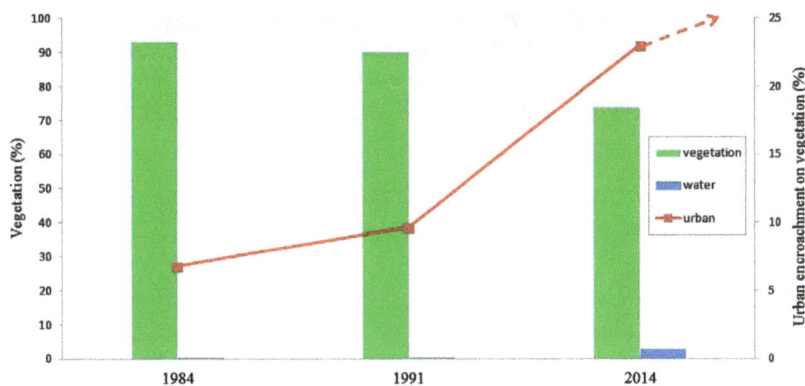

Figure 8. Temporal pattern of urban encroachment on vegetation resources.

These assertions by the local people shows that heterogeneity in the spatial distribution of land resources, in addition to weak concerns towards preserving the accruing benefits of vegetation resources made certain vegetated sites in the study area more attractive for anthropogenic functions of population and urbanization dynamics (Osemeobo, 1988). Thereby, leading to "mixed patterns of land-use and land-cover changes" which are the resultant effect of long-term gaps and or ambiguity in the implementation of urban development plan (Niemelä, 1999; Rojas, Pino, Basnou, & Vivanco, 2013). Suggesting that, before now, it is not unlikely that urban planning actions in the humid tropical rainforest of southeastern Nigeria were perceived as ecologically unsustainable and developmentally biased.

4.4. Vegetation and other land cover changes in 2014
Like the previous result, the 2014 classification map yet shows reduced fragments of dense vegetation on one hand, and on the other hand increased fragments of sparse vegetation, built-up and congested areas and even the water bodies. Indeed, the 2014 classification result is a near-perfect representation of the current vegetation status in the study area as observed during field surveys. It portrays some distinct variations from the previously classified images of 1984, 1991 and 2014, respectively (Figure 5(C)).

The 2014 image classification analysis indicates a serious decline in the phonological quantity and characteristics of general vegetation resources, judging from the spatial dominance of the sparse vegetation. Since 1984, dense vegetation has been decreasing, while other land cover types gain from these losses (Appendices 1 and 2, Table 1, and Figure 8). Initially, dense vegetation occupied about 330.63 km^2, but today it only maintains just about half of that green space i.e. 170.87 km^2. This shows a 44% decrease in its spatial extent. Sparse/unhealthy vegetation on the other hand rose from 2% in 1984 to 27% in 2014; thereby occupying as much as 97.161 km^2 in spatial extent (Figure 7(C)). Built-up and congested areas—*urbanization* have also grown to sustain a steady growth pattern (Figure 8), especially because of the rising population and increasing quest to satisfy human need for urban occupancy in the area.

In 1984, built-up and congested areas occupied 15 and 9.69 km^2, respectively, but the 2014 results show that they currently occupy about 54.94 and 28.66 km^2, in that order. Consequently, the built up area increased by 14% over a period of 30 years while the congested central business district experienced a steady 5% increase over the same period of years. It therefore suggests that Umuahia axis of the humid tropical rainforest is experiencing more of urban diffusion as shown by fragmented patches of other land cover types.

The marked influx of people since 1991 induced urban congestion and increased need for living spaces. Thereby, introducing opportunities for urban planning interventions aimed at sustainable governance of natural resources. The state government did relocate some urban infrastructural facilities e.g. markets, to previously vegetated areas. Yet, the level of increased disturbances as shown

Table 2. Standard summaries of accuracy assessment report							
Vegetation cover	**Water**	**Sparsely vegetated**	**Densely vegetated**	**Industrial**	**Residential**	**Row total**	**User accuracy (%)**
(a) Confusion matrix for 1984 classification							
Water	32	1	0	2	0	35	91
Sparsely vegetated	1	38	1	1	0	41	93
Densely vegetated	0	2	36	1	0	39	92
Industrial	0	1	2	34	0	37	92
Residential	0	0	0	2	21	23	91
Column total	33	42	39	40	21	175	
Producers accuracy (%)	97	90	92	85	100		
(b) Confusion matrix for 1991 classification							
Water	31	0	2	0	0	33	94
Sparsely vegetated	0	37	1	1	0	39	95
Densely vegetated	1	0	38	1	0	40	95
Industrial	0	0	1	34	1	36	94
Residential	1	0	0	1	25	27	93
Column total	33	37	42	37	26	175	
Producers accuracy	94	100	90	92	96		
(c) Confusion matrix for 2014 classification							
Water	35	0	0	0	0	35	100
Sparsely vegetated	0	42	0	0	0	42	100
Densely vegetated	0	0	38	0	0	38	100
Industrial	0	0	1	32	0	33	96
Residential	0	1	0	2	24	27	89
Column total	35	43	39	34	24	175	
Producers accuracy (%)	100	98	97	94	100		

(a) Overall accuracy = 92%; Kappa coefficient = 0.92.
(b) Overall accuracy = 94%; Kappa coefficient = 0 94
(c) Overall accuracy = 97%; Kappa coefficient = 0.97.

by the classification analyses of 1991 and 2014 suggests increased human impacts on the region's ecological landscape. Regrettably, deforestation and habitat loss are the greatest threat to terrestrial biodiversity and ecosystem health which equally leads to species extinction at the time of occurrence and in the future (Millennium Ecosystem Assessment, 2005).

4.5. Accuracy assessment

Using the mapped data against 150 reference data obtained during field observations, the image analysis in this study were subjected to automated quantitative accuracy assessment using the Cohen Kappa's statistics function in ERDAS Imagine, on the scale of 0 to 1 (i.e. Kappa is a value less than or equal to 1, where 1 corresponds to a perfect agreement) (Congalton & Green, 1999). The acceptable standard of overall accuracy for land cover map is set between 80% (Anderson et al., 1976) and 100% (Lins & Kleckner, 1996). In this study, the results of the accuracy assessment obtained for the classified images of 1984, 1991 and 2014 are presented in a standard summaries report in Table 2(a)–(c). The error matrices quantitatively compared the relationship between the classified images with the reference data obtained from field observations. All the classified images—1984, 1991 and 2014 returned high percentage of overall accuracy and Kappa values as follows: 92% (0.92), 94%

(0.94) and 97% (0.97), respectively. This show a high level of conformity with the supervised classification analysis carried out in the study (Table 2).

4.6. Factors responsible for vegetation cover changes in the study area

This study identified four major factors responsible for vegetation cover changes in the humid tropical rainforest segment of Umuahia. (1) Unchecked and increased local demand for forest and non-forest products, (2) Rapidly growing population induced by the reason of the study area's centrality and official recognition as state capital in 1991, (3) Urban diffusion influenced by increased demand for land and economic spaces (i.e. urbanization), (4) Long-term gaps in the implementation of vegetation monitoring frameworks (see, for example, Erika et al., 2015).

Field observations conducted in 2011 and 2014 reveal that the humid tropical rainforest of southeastern Nigeria is a biodiversity rich belt, with abundance of several endemic vegetation resources that provides various benefits to the indigenous people and uphold useful potentials for posterity. Yet, this rich ecologically vibrant area is considered one of the most rapidly urbanizing and endangered areas in Nigeria.

In 1991, the study area was officially declared a headquarter city in southeastern Nigeria. The action trickled-down a ripple cause-effect response of vegetation resources to anthropogenic phenomenon of urbanization. The rising population initiated increased demand for land and economic spaces, which in turn, affected vegetated areas by increased exploitation of natural resources—including forest and non-forest products.

Data extracted from the analyses in this study (Table 1 and Appendix 2) showed increased human impact on vegetation cover. This is indicated by fragmented sparsely vegetated areas, growth of industrial and residential land uses (Figure 8). One of the understudied geographies of population impact on natural resources is the role of institutional governance in monitoring deforestation and vegetation loss in Africa (Erika et al., 2015; Larcom et al., 2016). Nevertheless, like most other regions, the study area continues to experience unrestrained exploitation of natural resources by virtue of domestic need and industrial demand for fuel wood and timber, respectively.

Also, the spatiotemporal extent maps generated in this study show that built-up and residential areas had become fragmented given the accelerated rate of vegetation reduction since 1991 to 2014. Generally, this kind of situation is often attributed to the concentration of built-up patches or new infrastructural developments along emerging economic corridors (Müller, Griffiths, & Hostert, 2016; Simmons et al., 2016).

Based on the forgoing, Umuahia section of the tropical rainforest is currently witnessing its middle phase of urbanization process. This is exemplified by the area experiencing more of urban diffusion, urban growth (Figure 8) and less of vegetation resilience and forest recovery (Table 1, Appendix 1). As such, vegetation reduction in the area is highly correlated with anthropogenic functions from population dynamics.

5. Conclusion and recommendation

This study shows that remote sensing of vegetation is a consistent methodology in ascertaining changes and phenological characteristics in the humid tropical rainforests. The change detection covered a 30-year period, starting from 1984 to 2014, and revealed a reduction in size of healthy vegetation by 159.76 km^2 which indicates a 44% vegetation loss. Composite replacement of healthy vegetation cover in the area is impacted by other land use cover types as follows; 90.3 km^2 (unhealthy vegetation), 18.97 km^2 (congested/residential), 39.87 km^2 (built-up/industrial area), 9.59 km^2 (Water-body). Thereby, indicating a gross encroachment of urban land use of about 23.02% into vegetated areas, and reduced it from 93.03% in 1984 to 73.8% in 2014.

The most significant changes in the spatiotemporal dynamics of land use cover in the study area accelerated after the area became capital city in 1991. Similarly, the much higher percentage of vegetation loses recorded before 1991 (i.e. mid-1980s) is attributed to the persistent unsustainable exploitation of vegetation resources resulting from lack of adequate vegetation monitoring frameworks. Implying that, unsustainable exploitation of vegetation resources, increased economic activities in need of industrial and residential occupancies, in addition to uncoordinated urban expansions constitutes the anthropogenic functions of urbanization currently experienced in the study area. These, therefore suggest that, there is need to strengthen institutional monitoring frameworks that should account for all ecological units in the humid tropical rainforests.

Though limited by data availability, this study provides the baseline information about vegetation depletion in the humid tropical rainforests of Nigeria, and recommends periodic integration of high-resolution satellite images with urban afforestation strategies in natural resources governance through public engagements. Stakeholders and the local people, whom the vegetation resources are domiciled in their communities, can be adequately consulted, sensitized and integrated into every activity that may directly or indirectly affect ecological heritage. This will increase rates of vegetation resilience and forest recovery, such that, developments that disregard ecological losses and urban greening can be monitored effectively.

Acknowledgements
Prof R.N.C. Anyadike (Late) significantly initiated the process that led to this study. We thank our local field guides, and all the reviewers for their helpful comments.

Funding
The authors received no direct funding for this research.

Author details
Friday Uchenna Ochege[1,2]
E-mail: uchenna.ochege@uniport.edu.ng
ORCID ID: http://orcid.org/0000-0002-6661-0966
Chukwunonyelum Okpala-Okaka[2]
E-mail: chukwunonyelum.okpala-okaka@unn.edu.ng
[1] Laboratory for Cartography and GIS, Department of Geography & Environmental Management, University of Port Harcourt, Choba, Rivers State, Nigeria.
[2] Faculty of Environmental Studies, Department of Surveying and Geoinformatics, University of Nigeria, Enugu, Nigeria.

References
Abia State of Nigeria. (2005). Abia state economic empowerment and development strategy. *Abia State Government Publication.* Retrived June 14, 2014, from http://web.ng.undp.org/documents/SEEDS/Abia_State.pdf

Adekunle, V. A. J., Olagoke, A. O., & Ogundare, L. F. (2013). Timber exploitation rate in tropical rainforest ecosystem of southwest nigeria and its implications on sustainable forest management. *Applied Ecology and Environmental Research, 11,* 123–136. http://dx.doi.org/10.15666/aeer

Anderson, J. R., Hardy, E. E., Roach, J. T., & Witmer, R. E. (1976). *A land-use and land cover classification system for use with remote sensor data* (Professional Paper 964). Washington, DC: United State Geological Survey.

Anyiro, C. O., Ezeh, C. I., Osondu, C. K., & Nduka, G. A. (2013). Economic analysis of household energy use: A rural urban case study of Abia State, Nigeria. *Research and Reviews: Journal of Agriculture and Allied Sciences, 2,* 20–27.

Ariti, A. T., van Vliet, J., & Verburg, P. H. (2015). Land-use and land-cover changes in the Central Rift Valley of Ethiopia: Assessment of perception and adaptation of stakeholders. *Applied Geography, 65,* 28–37. http://dx.doi.org/10.1016/j.apgeog.2015.10.002

Aubreville, A. (1938). La forêt coloniale: les forêts de l'Afrique occidentale française [The colonial forest: The forests of French West Africa]. *Annuales Academie des Sciences Coloniales, 9,* 1–245.

Banko, G. (1998). A review of assessing the accuracy of classifications of remotely sensed data and of methods including remote sensing data in forest inventory. In *International Institution for Applied Systems Analysis,* Laxenburg, Austria. IIASSA Interim Report No. IR-98-081. Retrived from http://pure.iiasa.ac.at/5570/1/IR-98-081.pdf

Budowski, G. (1970). The distinction between old secondary and climax species in tropical Central American lowland forests. *Tropical Ecology, 11,* 44–48.

Bush, M. B., & Colinvaux, P. A. (1994). Tropical forest disturbance: Paleoecological records from Darien, Panama. *Ecology, 75,* 1761–1768. http://dx.doi.org/10.2307/1939635

Campling, P., Gobin, A., & Feyen, J. (2001). Temporal and spatial rainfall analysis across a humid tropical catchment. *Hydrological Processes, 15,* 359–375. http://dx.doi.org/10.1002/(ISSN)1099-1085

Chen, Y., Lu, D., Luo, G. P., & Huang, J. (2015). Detection of vegetation abundance change in the alpine tree line using multitemporal Landsat Thematic Mapper imagery. *International Journal of Remote Sensing, 36,* 4683–4701. doi:10.1080/01431161.2015.1088675

Churches, E. C., Wampler, P. J., Sun, W. & Smith, A. J. (2014). Evaluation of forest cover estimates for Haiti using supervised classification of Landsat data. *International Journal of Applied Earth Observation and Geoinformation, 30,* 203–216. doi:10.1016/j.jag.2014.01.020

Congalton, R. (1991). A review of assessing the accuracy of classifications of remotely sensed data. *Remote Sensing of Environment, 37,* 35–46. http://dx.doi.org/10.1016/0034-4257(91)90048-B

Congalton, R. G., & Green, K. (1999). *Assessing the accuracy of remotely sensed data: Principles and practices*. Boca Raton, FL: Lewis Publishers.

Coppin, P., Jonckheere, I., Nackaerts, K., Muys, B., & Lambin, E. (2004). Digital change detection methods in ecosystem monitoring: A review. *International Journal of Remote Sensing, 25*, 1565–1596. http://dx.doi.org/10.1080/0143116031000101675

Coppin, P. R., & Bauer, M. E. (1996). Digital change detection in forest ecosystems with remote sensing imagery. *Remote Sensing Reviews, 13*, 207–234.

DIVA-GIS. (2016). *Land cover map of Nigeria*. Retrieved December, 2016, from http://biogeo.ucdavis.edu/data/diva/cov/NGA_cov.zip

Du, J., Shu, J., Yin, J., Yuan, X., Jiaerheng, A., Xiong, S., ... Liu, W. (2015). Analysis on spatio-temporal trends and drivers in vegetation growth during recent decades in Xinjiang, China. *International Journal of Applied Earth Observation and Geoinformation, 38*, 216–228. doi:10.1016/j.jag.2015.01.006

Dubula, B., Tesfamichael, S. G., & Rampedi, I. T. (2016). Assessing the potential of remote sensing to discriminate invasive *Asparagus laricinus* from adjacent land cover types. *Cogent Geoscience, 2*(1), 1–17.

Ejenma, E. (2013). *Trends and patterns of house rents in Umuahia* (Unpublished PhD Seminar 1). Choba: Department of Geography and Environmental Management, University of Port Harcourt.

Erika, R., Celso, B. L., Martin, H., Erik, L., Robert, O., Arief, W., Daniel, M., & Louis, V. (2015). Assessing change in national forest monitoring capacities of 99 tropical Countries. *Forest Ecology and Management, 352*, 109–123.

FAO. (2012). *State of the world's forests 2012*. Italy: Author.

FAO. (2013). *The latest state of the World's Forest Report of the Food and Agricultural Organisation (FAO) indicate a dangerous cutback in the overall forest reserves in Africa especially between 1990 and 2012*. Italy: Author.

Federal Republic of Nigeria. (1977). *The national atlas of the Federal Republic of Nigeria* (FRN 1st ed.). Lagos: Federal Surveys.

Federal Republic of Nigeria. (2007). *Official gazette: Legal notice on publication of the details of the breakdown of the national and state provisional totals 2006 census*. Lagos: Author.

Fisher, B., & Turner, R. (2008). Ecosystem services: Classification for valuation. *Biological Conservation, 141*, 1167–1169. http://dx.doi.org/10.1016/j.biocon.2008.02.019

Food & Agriculture Organization of the United Nations. (1981). *Map of fuelwood situation on developing countries* (pp. 1–10). Rome: Author.

Foody, G. M. (2002). Status of land cover classification accuracy assessment. *Remote Sensing of Environment, 80*, 185–201. http://dx.doi.org/10.1016/S0034-4257(01)00295-4

Forkel, M., Carvalhais N., Verbesselt J., Mahecha M. D., Neigh S. R. C., & Reichstein M. (2013). Trend change detection in NDVI time series: Effects of inter-annual variability and methodology. *Remote Sensing, 5*, 2113–2144. doi:10.3390/rs5052113

Friberg, N., Bonada, N., Bradley, D. C., Dunbar, M. J., Edwards, F. K., Grey, J., ... Woodward, G. (2011). Biomonitoring of human impacts in freshwater ecosystems: The good, the bad and the ugly. *Advances in Ecological Research*, 1–68. http://dx.doi.org/10.1016/B978-0-12-374794-5.00001-8

Hagenlocher, M., Lang, S., & Tiede, D. (2012). Integrated assessment of the environmental impact of an IDP camp in Sudan based on very high resolution multi-temporal satellite imagery. *Remote Sensing of Environment, 126*, 27–38. http://dx.doi.org/10.1016/j.rse.2012.08.010

Hansen, M. C., Potapov, P. V., Moore, R., Hancher, M., Turubanova, S. A., Tyukavina, A., ... Townshend, J. R. G. (2013). High-resolution global maps of 21st-century forest cover change. *Science, 342*, 850–853.

Huang, X., Lu, Q., Zhang, L., & Plaza, A. (2014). New postprocessing methods for remote sensing image classification: A systematic study. *IEEE Transactions on Geoscience and Remote Sensing, 52*, 7140–7159. http://dx.doi.org/10.1109/TGRS.2014.2308192

Igbawua, T., Zhang, J., Chang, Q., & Yao, F. (2016). Vegetation dynamics in relation with climate over Nigeria from 1982 to 2011. *Environmental Earth Sciences, 75*, 1–16.

Igbozurike, M. U. (1975). Vegetation types. In G. E. K. Ofomata (Ed.), *Nigeria in maps* (pp. 30–32). Benin City: Ethiope Publishing House.

Iloeje, N. P. (2007). *A new geography of Nigeria*. Ikeja: Longman Nigerian PLC.

Jensen, J. R. (2014). *Remote sensing of the environment: An earth resource perspective* (2nd ed.). Harlow: Pearson.

Jones, E. W. (1956). Ecological studies of the rain forest of southern Nigeria IV. The plateau forest of the Okomu Forest Reserve, Part 2. The reproduction and history of the forest. *The Journal of Ecology, 44*, 83–117. http://dx.doi.org/10.2307/2257155

Kottek, M., Grieser, J., Beck, C., Rudolf, B., & Rubel, F. (2006). World map of the Köppen-Geiger climate classification updated. *Meteorologische Zeitschrift, 15*, 259–263.

Larcom, S., van Gevelt, T., & Zabala, A. (2016). Precolonial institutions and deforestation in Africa. *Land Use Policy, 51*, 150–161. http://dx.doi.org/10.1016/j.landusepol.2015.10.030

Lillesand, T., Kiefer, R., & Chipman, J. (Eds.). (2015). *Remote sensing and image interpretation* (7th ed.). Hoboken, NJ: John Wiley & Sons.

Lins, K. S., & Kleckner, R. L. (1996). Land cover mapping: An overview and history of the concepts. In J. M. Scott, T. H. Tear, & F. Davis (Eds.), *Gap analysis: A landscape approach to biodiversity planning* (pp. 57–65). Bethesda, MD: American Society for Photogrammetry and Remote Sensing.

Loveland, T. R., & Dwyer, J. L. (2012). Landsat: Building a strong future. *Remote Sensing of Environment, 122*, 22–29. http://dx.doi.org/10.1016/j.rse.2011.09.022

Lucas, R., Blonda, P., Bunting, P., Jones, G., Inglada, J., Arias, M., ... Charnock, R. (2015). The earth observation data for habitat monitoring (EODHaM) system. *International Journal of Applied Earth Observation and Geoinformation, 37*, 17–28. http://dx.doi.org/10.1016/j.jag.2014.10.011

Lund, H. G. (1983). Change: Now you see it—now you don't! In *Proceedings of the international conference on renewable resource inventories for monitoring changes and trends* (pp. 211–213). Corvallis, OR: Oregon State University.

Luo, G. P., & Dai, L. (2012). Detection of alpine tree line change with high spatial resolution sensed data. *Journal of Applied Remote Sensing, 2*. doi:10.1117/1.JRS.7.073520

Luo, G. P., Zhou, C. H., Chen, X., & Li, Y. (2008). A methodology of characterizing status and trend of land changes in oases: A case study of Sangong River watershed, Xinjiang, China. *Journal of Environmental Management, 88*, 775–783.

Matlack, G. R. (1994). Vegetation dynamics of the forest edge—Trends in space and successional time. *The Journal of Ecology*, 113–123. http://dx.doi.org/10.2307/2261391

Mckey, D., Waterman, P. G., Gartlan, J. S., & Struhsaker, T. T. (1978). Phenolic content of vegetation in two African rain forests: Ecological implications. *Science (Washington, DC); (United States), 202*.

Meduna, A. J., Ogunjimi, A. A., & Onadeko, A. (2009). Biodiversity conservation problems and their implications on Kainji Lake National Park, Nigeria. *Journal of Sustainable Development in Africa, 10*, 59–73.

Meregini, A. O. (2005). Some endangered plants producing edible fruits and seeds in Southeastern Nigeria. *Fruits, 60*, 211–220. http://dx.doi.org/10.1051/fruits:2005028

Millennium Ecosystem Assessment. (2005). *Ecosystems and human well-being: Synthesis*. Washington, DC: Island Press. Retrieved November 10, 2011, from http://www.maweb.org/documents/document.356.aspx.pdf

Milne, A. K. (1988). Change direction analysis using Landsat imagery: A review of methodology. In *Proceedings of the IGARSS'88 Symposium Edinburgh, Scotland, ESASP-284* (pp. 541–544). Noordwijk: ESA.

Momoh, S. (2014). Endangered timber business. *Vanguard, 2*, 4.

Müller, H., Griffiths, P., & Hostert, P. (2016). Long-term deforestation dynamics in the Brazilian Amazon—Uncovering historic frontier development along the Cuiabá-Santarém highway. *International Journal of Applied Earth Observation and Geoinformation, 44*, 61–69. http://dx.doi.org/10.1016/j.jag.2015.07.005

Newbold, T., Hudson, L. N., Phillips, H. R., Hill, S. L., Contu, S., Lysenko, I., ... Purvis, A. (2014). A global model of the response of tropical and sub-tropical forest biodiversity to anthropogenic pressures. *Proceedings of the Royal Society B: Biological Sciences, 281*, 20141371. http://dx.doi.org/10.1098/rspb.2014.1371

Niemelä, J. (1999). Ecology and urban planning. *Biodiversity & Conservation, 8*, 119–131. http://dx.doi.org/10.1023/A:1008817325994

Obeta, M. C., Ochege, F. U., Aujara, I. Y., & Shehu, S. M. (2013). Socio-economic significance of woodstocks and non-timber forest products in Jigawa State, Nigeria. In L. Popoola, F. O. Idumah, O. Y. Ogunsanwo, & I. O. Azeez, (Eds.). *Forest industry in a dynamic global environment. Proceeding of the 35th Annual conference of the forestry Association of Nigeria* (pp. 727–735). Sokoto: Forestry Association of Nigeria.

Obydenkova, A., Nazarov, Z., & Salahodjaev, R. (2016). The process of deforestation in weak democracies and the role of Intelligence. *Environmental Research, 148*, 484–490. http://dx.doi.org/10.1016/j.envres.2016.03.039

Ochege, F. U. (2014a). *Spatio-temporal variations of vegetation cover over Umuahia, Abia state Nigeria from 1984 to 2014* (MSc thesis). University of Nigeria, Nsukka.

Ochege, F. U. (2014b). *Geospatial assessment of forest degradation in Sagbama, Niger delta region of Nigeria* (MSc thesis). University of Edinburgh, Scotland.

Odjugo, P. A. O. (2010). General overview of climate change impacts in Nigeria. *Journal of Human Ecology, 29*, 47–55.

Odjugo, P. A. O., & Ikhuoria, A. I. (2003). The impact of climate change and anthropogenic factors on desertification in the semi-arid region of Nigeria. *Global Journal of Environmental Science, 2*, 118–126.

Ofori, B. Y., Owusu, E. H., & Attuquayefio, D. K. (2014). Ecological status of the Mount Afadjato-Agumatsa range in Ghana after a decade of local community management. *African Journal Ecology, 53*, 116–120.

Okpala-Okaka, C., & Igbokwe, J. I. (2010). Revision of Nsukka N.E Topographic Map Sheet 287 1:50,000 (1964) using Nigeria SAT-1 Imagery. *Nigerian Journal of Space Research, 7*, 13–24.

Omo-Irabor, O., Olobaniyi, S. B., Akunna, J., Venus, V., Maina, J. M., & Paradzayi, C. (2011). Mangrove vulnerability modelling in parts of Western Niger Delta, Nigeria using satellite images, GIS techniques and Spatial Multi-Criteria Analysis (SMCA). *Environmental Monitoring and Assessment, 178*, 39–51. http://dx.doi.org/10.1007/s10661-010-1669-z

Onu, N. N., Opara, A. I., & Ehirim, C. N. (2012). Delineation of Active fractures in a gully erosion area using geophysical methods: Case study of the Okigwe—Umuahia Erosion Belt, Southeastern Nigeria. *International Journal of Science and Technology, 2*(4). ISSN 2224-3577.

Osemeobo, G. J. (1988). The Human causes of forest depletion in Nigeria. *Environmental Conservation, 15*, 17–28. http://dx.doi.org/10.1017/S0376892900028411

Richards, P. W. (1939). Ecological studies on the rain forest of southern Nigeria. 1. The structure and floristic composition of primary forest. *The Journal of Ecology, 27*, 1–61. http://dx.doi.org/10.2307/2256298

Rojas, C., Pino, J., Basnou, C., & Vivanco, M. (2013). Assessing land-use and-cover changes in relation to geographic factors and urban planning in the metropolitan area of Concepción (Chile). *Implications for biodiversity conservation. Applied Geography, 39*, 93–103.

Schwartz, M. D. (2003). *Phenology: An integrative environmental science* (vol. 39). Netherlands: Springer.

Scull, P., Cardelús, C. L., Klepeis, P., Woods, C. L., Frankl, A., & Nyssen, J. (2016). The resilience of Ethiopian church forests: Interpreting aerial photographs, 1938–2015. *Land Degradation & Development*.

Simmons, C., Walker, R., Perz, S., Arima, E., Aldrich, S., & Caldas, M. (2016). Spatial patterns of frontier settlement: Balancing conservation and development. *Journal of Latin American Geography, 15*, 33–58. http://dx.doi.org/10.1353/lag.2016.0011

Smith, B. D. (2007). The ultimate ecosystem engineers. *Science, 315*, 1797–1798. http://dx.doi.org/10.1126/science.1137740

Soofi, K. (2005). *Remote sensing lecture: Image classification. UCSD: Satellite remote sensing - SIO 135/SIO 236*. Retrieved from http://topex.ucsd.edu/rs/classification.pdf

Sulaiman, C., Abdul-Rahim, A. S., Mohd-Shahwahid, H. O., & Chin, L. (2017). Wood fuel consumption, institutional quality, and forest degradation in sub-Saharan Africa: Evidence from a dynamic panel framework. *Ecological Indicators, 74*, 414–419. http://dx.doi.org/10.1016/j.ecolind.2016.11.045

Tee, N. T., Ancha, P. U., & Asue, J. (2009). Evaluation of fuelwood consumption and implications on the environment: Case study of Makurdi area in Benue state, Nigeria. *Journal of Applied Biosciences, 19*, 1041–1048.

UNEP, IISD (Ed.). (2004). *Exploring the links: Human well-being poverty and ecosystem services*. Nairobi and Winnipeg, Manitoba: Author.

United Nations Population Fund. (1991). *Population, resources and the environment: The critical challenges* (154 p.). New York, NY. ISBN 0-89714-101-6.

Watson, R. T., Noble, I. R., Bolin, B., Ravindranath, N. H., Verardo, D. J., & Dokken, D. J. (2000). *Land use, land-use change and forestry IPPC Report*. Retrieved from www.ipcc.ch/pdf/special-reports/spm/srl-en.pdf

White, L. J. T., & Oates, J. F. (1999). New data on the history of the plateau forest of Okomu, southern Nigeria: An insight into how human disturbance has shaped the African rain forest. RESEARCH LETTER. *Global Ecology and Biogeography, 8*, 355–361. http://dx.doi.org/10.1046/j.1365-2699.1999.00149.x

World resources. (1987). *An assessment of the resource base that supports the global economy*. New York, NY: International institute of environment and development and the world resources institute Basic Books, 369 pp. https://www.wri.org/our-work/project/world-resources-report/publications?page=2

Yeshaneh, E., Wagner, W., Exner-Kittridge, M., Legesse, D., & Blöschl, G. (2013). Identifying land use/cover dynamics in the Koga catchment, Ethiopia, from multi-scale data, and implications for environmental change. *ISPRS International Journal of Geo-Information, 2*, 302–323. http://dx.doi.org/10.3390/ijgi2020302

Zhu, X., Liu, D., & Chen, J. (2012). A new geostatistical approach for filling gaps in Landsat ETM+ SLC-off images. *Remote Sensing of Environment, 124*, 49–60. http://dx.doi.org/10.1016/j.rse.2012.04.019

Appendix 1

Trend of land cover changes (%) in the study area

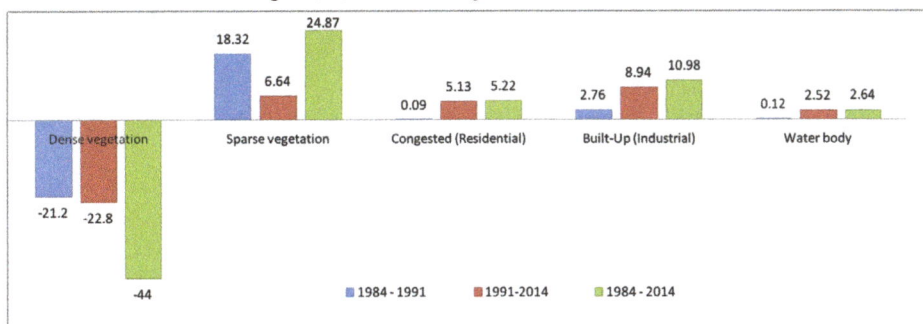

Appendix 2

Temporal pattern

Land use cover	1984	1991	2014
Vegetation resources	93.03	89.97	73.81
Urban encroachment	6.82	9.67	23.02
Water	0.24	0.36	2.88

2

Geological application of ASTER remote sensing within sparsely outcropping terrain, Central New South Wales, Australia

R. Hewson[1]*, D. Robson[2], A. Carlton[2] and P. Gilmore[2]

*Corresponding author: R. Hewson, Faculty of Geo-Information Science & Earth Observation (ITC), University of Twente, P.O. Box 217, 7500 AE, Enschede, The Netherlands
E-mail: r.d.hewson@utwente.nl
Reviewing editor: Louis-Noel Moresi, University of Melbourne, Australia

Abstract: One of the major problems faced by the application of geological remote sensing is its potential limitation in areas of a temperate climate with agricultural cultivation, limited outcrops and vegetation cover. This was the issue experienced when it was attempted to use the multi-spectral satellite Advanced Spaceborne Thermal Emission Reflectance Radiometer (ASTER) imagery to assist the updating of 1:100,000 geological mapping with the Ardlethan/Barmedman map sheets of central New South Wales (NSW), Australia. Most successful applications of geological remote sensing have been achieved in arid to semi-arid environments where vegetation and cultivation is minimal. Typically, day-time acquired ASTER visible to shortwave surface reflectance derived map products has extracted useful mineral related compositional information in such areas however in the studied areas of central NSW these techniques proved limited, particularly when using large mosaicked products such as the National Australia ASTER Geoscience Maps. Some improvement in geological discrimination was achieved using individual ASTER scenes, masked by high slope angle and processed into spectrally unmixed products. An alternative approach to extracting geoscience related products, utilised, night-time acquired ASTER thermal products. Their surface kinetic temperature products showed some potential for identifying the limited and sparse outcrops useful for field mapping geologists. Overall this study also showed the importance of the image spatial resolution in vegetated and cultivated areas with limited outcrop. Ideally

ABOUT THE AUTHOR

R. Hewson graduated with a MSc study in geophysics at Macquarie University followed by a PhD at the University of NSW within geological remote sensing. He worked from 1998 to 2010 within Australia's CSIRO Exploration and Mining Division, Perth, focusing on the geological case study development of airborne and satellite remote sensing, and its data integration with airborne geophysics. Since CSIRO, Hewson has undertaken remote sensing studies as a honorary research fellow for Professor Simon Jones at RMIT University, Melbourne and actively consulted with Australian mapping agencies. He currently works as an assistant professor in Geological Remote Sensing within ITC at University of Twente.

PUBLIC INTEREST STATEMENT

The present study is of interest for geoscientists attempting to apply remote sensing techniques for mapping in areas of limited outcrops within cultivated temperate terrain. Most published geological remote sensing applications have been typically undertaken within arid to semi-arid environments, however the described study investigates an application in a cultivated flat lying farm land environment with alluvial/soil cover. Attempts were undertaken to overcome these environmental factors by using DEM derived high slope defined masking to target spectrally defined outcrop anomalies from day-time acquisitions, and the application of night-time thermal infrared imagery for mapping limited surface and possible sub surface geological features associated with palaeochannels.

a finer spatial image product than available with ASTER's VNIR-SWIR combined products at 30 m is required.

Subjects: Geology - Earth Sciences; Applied & Economic Geology; Geophysics; Geomagnetics

Keywords: ASTER; geological mapping; night-time satellite; surface temperature; thermal inertia

1. Introduction

1.1. Previous examples of geological remote sensing using Advanced Spaceborne Thermal Emission Reflectance Radiometer (ASTER)

The results of fundamental research into mineral spectroscopy (Adams & Filice, 1967; Hunt & Ashley, 1979; Lyon & Burns, 1963; Vincent, Rowan, Gillespie, & Knapp, 1975; Vincent & Thomson, 1972) laid the basis for later geological remote sensing and prompted the development of such multi-spectral satellite sensors as ASTER, launched in December 1999 by Japan's METI and NASA. Since its operation in 2000, ASTER imagery has proved useful for discriminating and mapping several mineral groups (Rowan & Mars, 2003). However the majority of its applications for geological mapping studies have been in semi-arid to arid environments with minimal vegetation and cultivation. The issue of extending the application of such satellite multi-spectral imagery into less arid environments is important for the exploration of mineral resources in non-traditional areas. This particular study describes different approaches used to process and apply ASTER in such a non-arid environment, within central NSW, Australia. This study also follows on from previous studies by the author comparing ASTER applications in an arid environment with those from a temperate environment, albeit with standard processing methodologies (Hewson & Robson, 2014; Hewson, Robson et al., 2015).

This study also examines the application of seventeen compositional Australia wide map products, released by the Commonwealth Scientific Industrial Research Organisation (CSIRO) and Geoscience Australia (GA) from ASTER imagery (Caccetta, Collings, & Cudahy, 2013; Cudahy, 2012). These map products encompassed a wide variety of climatic environments and cultivation over the entire continental landmass of Australia and were utilised in the previous studies for the comparison between arid and temperate applications of ASTER (Hewson & Robson, 2014; Hewson, Robson et al., 2015). The processed digital values representing the seventeen map products assumed histogram stretched thresholds that at best qualitatively to semi-quantitatively represented mineral (group) composition across the entire Australian mosaicked ASTER data-set (Cudahy, 2012). The best results obtained from such ASTER derived geological mapping is undoubtedly within the arid to semi-arid exposed terrain (e.g. Mt Fitton, South Australia; Cobar, New South Wales–Hewson, Robson et al., 2015; Hewson & Robson, 2014). Hewson, Robson et al. (2015) also highlighted some of the environmental and geomorphological issues when studying temperate, cultivated and floodplain dominated terrain (Hewson, Robson et al., 2015). In particular, the studied temperate Wagga Wagga area of NSW exhibited moderate to limited rock outcrop exposure. The study showed that the generation of ASTER derived map products was improved by filtering or masking them for areas greater than 10% slope, availed by ENVI™ (http://www.harrisgeospatial.com/) processing of Shuttle Radar Topography Mission (SRTM) Digital Elevation Model (DEM) data (Hewson, Robson et al., 2015). This approach was based on a landform classification defined by topographic slope where there was an increased likelihood of erosional scarps or rocky outcrops for hilly slopes greater than 10% (Chen, 1997). Masking such ASTER products for vegetation fractional cover estimates, derived from Landsat (AusCover, http://data.auscover.org.au/), was also found to assist interpretation (Hewson, Robson et al., 2015). In particular, the Ferric oxide content, Fe oxide composition, Silica content and AlOH content products showed geologically associated anomalies either as trends within the alluvium floodplains or highlighted following the application of masks in outcropping areas (Hewson, Robson et al., 2015).

Several other intrinsic data issues can affect the interpretation of day-time ASTER imagery. As described in Hewson, Robson et al. (2015), ASTER's SWIR band spatial resolution of 30 m can be a problem in forested and woodland areas where few pixels would be free of canopy cover and its shadow components. Also, the detection of iron oxide mineralogy can be limited to where the ASTER VNIR spectral signatures are not dominated by those of vegetation (e.g. chlorophyll and pigment). ASTER's SWIR image products can also be affected by the crosstalk issue where stray light from band 4 "leaks" as an additive noise signal into bands 5 and 9 (Iwasaki, Fujisada, Akao, Shindou, & Akagi, 2005). An algorithm and software were developed to correct for this miscalibration issue (Iwasaki & Tonooka, 2005). However, in areas of low ground reflectance (e.g. shadow, thick vegetation, dark surfaces) residual crosstalk effect is apparent and can lead to "false anomalies" using spectral band indices (Hewson & Cudahy, 2011; Hewson, Cudahy, Mizuhiko, Ueda, & Mauger, 2005).

One important additional issue regarding the geological application of remote sensing is the spatial resolution of the imaging sensors. Studies by Kruse (2000), using various airborne hyperspectral VNIR-SWIR sensors, established the importance of image spatial resolution on the mapping of geology with scale dependent variations. In particular resampling airborne imagery of 2.4 m spatial resolution to 20 m showed there is a noticeable loss of discrete mapped occurrences of specific materials (Kruse, 2000). The spatial resolution of the ASTER SWIR sensor is 30 m and used as the overall resolution of many of the Australia ASTER Geoscience Maps. This is a potential limiting factor in its application in vegetated areas with limited geological outcrop.

1.2. Current ASTER mapping study
In this particular study, other additional approaches to aid geological interpretation of ASTER data in temperate areas, have been using on individual scenes, rather than large-scale processed map products, as attempted previously. Such scene specific processing avoids the issue of generalised and inappropriate product thresholds as well as offering greater enhancement of more subtle geological features in a non-arid environment. These approaches utilised a range of processing algorithms and tailored the histogram methods and thresholds for groundcover and geological terrain (Hewson, Cudahy, & Huntington, 2001; Hewson, Koch, Buchanan, & Sanders, 2002). Specifically, this study processed individual ASTER scenes within a cultivated and temperate region into map products, using:

- Colour composites images and band parameter ratios, similar to approaches applied by Cudahy (2012) and Hewson et al. (2005) but assuming different histogram stretches;
- Maximum Noise Fraction (MNF) analysis (Green, Berman, Switzer, & Craig, 1988).
- Surface reflectance and emissivity ASTER imagery with Mixture Tuned Matched Filter (MTMF) techniques (Kruse et al., 1993) to separate the spectral contributions within each image pixel signature into estimated proportions of geologically or vegetation related land cover (Hewson et al., 2002).

In addition to processing individual day-time ASTER imagery, night-time thermal ASTER imagery was also examined in the study area. Night-time thermal imagery, provided as surface kinetic temperature, is capable of (1) identifying areas of limited outcrop, useful for field geologists; and (2) potentially discriminate changes in the physical properties of soil, regolith and exposed outcrops such as density, porosity and thermal capacity when further processed into Apparent Thermal Inertia (ATI) (Kahle, 1987; Price, 1977). This can be useful in areas of extensive weathering and transported alluvial cover where there is limited surface geological outcrop.

Another issue in temperate and cultivated areas is often the limited and sparse rock outcrops for detailed identification and sampling. Access to farmland requires stricter protocols to be followed and targeting known or likely outcrops efficiently within the shortest field trip is a priority. The use of high spatial aerial digital imagery (e.g. 0.5 m resolution ADS40; Sandau et al., 2000) is typically used for identification of likely outcrops to field sample. However, the limited spectral resolution of aerial photography and the restriction of its bands to the visible wavelengths preclude a mineralogical based discrimination of such outcrops. Night-time ASTER TIR products were investigated for its ability to delineate such outcrops as an assistance for geological field survey mapping.

1.3. Study objectives & study outline

The principal aim of this study is to evaluate the potential of using multi-spectral ASTER imagery for geological mapping in a temperate and cultivated environment, with associated sparse outcrops. This required the applying individual scene specific histogram stretches, different processing methodologies and use of night-time ASTER thermal imagery.

A concurrent aim of this study was to provide additional geological information as part of the 1:1,00,000 (1:100 K) mapping update for the Ardlethan and Barmedman area of NSW, undertaken by the Geological Survey of New South Wales (GSNSW). It was hoped that such compositional information could be useful for field mapping activities. This study area was an extension to a previous study at Wagga Wagga (Hewson, Robson et al., 2015) albeit with less geological exposure. A detailed description of this study and its data is also available via the Geological of NSW Report, fulfilled as part of this study (Hewson, 2015), and also outlined in summary in Hewson, Carlton, Gilmore, Jones, and Robson (2015).

In this paper Section 2.0 describes the physiography, environment and geology of the study area of the Ardlethan and Barmedman areas of central NSW. Section 3.0 outlines the data-sets sourced in this project including the ASTER Australian Geoscience map products, and the ASTER data as used as for individual scene processing. The processing methodologies applied to the day- and night-time ASTER imagery are also described in Section 3.0. The results and conclusion to this study are explained in Sections 4.0 and 5.0, respectively.

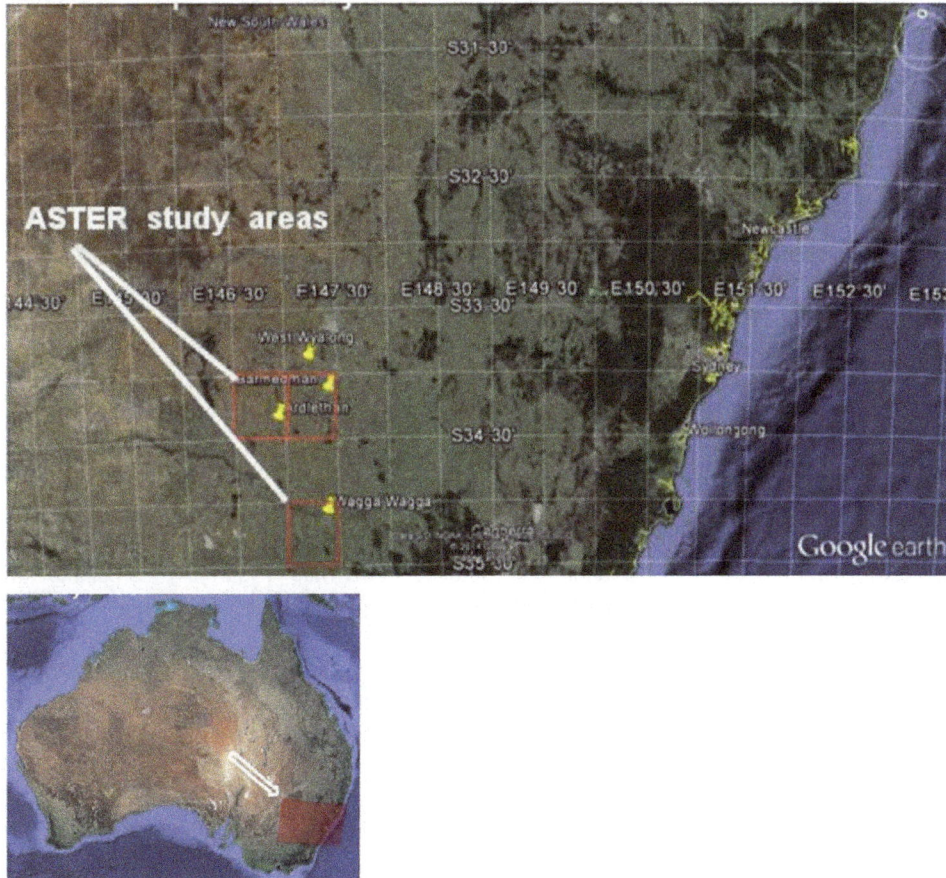

Figure 1. Location of the Ardlethan, Barmedman and Wagga Wagga 1:100,000 map sheet areas.

2. Study area

The study area used in this evaluation of remote sensing within a non-arid environment encompasses the Ardlethan and Barmedman 1:1,00,000 (1:100 K) map sheet areas of south central New South Wales (NSW), Australia (Figure 1). This area is of particular interest for intrusive sourced hydrothermal and alluvial hosted tin and gold deposits (Colquhoun, Meakin, & Cameron, 2005). However limited outcrop within a flat and heavily cultivated terrain has handicapped geological investigations and exploration.

2.1. Location and physiography

The Ardlethan (146.5°–147°E, 34°–34.5°S) and Barmedman (147°–147.5°E, 34°–34.5°S) 1:100 K map sheet areas of central southern NSW lie to the north and north-west of the Wagga Wagga 1:100 K map sheet area studied previously by (Hewson, Robson et al., 2015) (Figure 1). The area comprises extensive cultivated pasture and cropland within the floodplains and catchment of the Lachlan River. Limited hilltops and ridges occur within these areas and typically associated with open woodland or scrubby vegetation. The climate is typically Mediterranean with an average rainfall of 480 mm per year observed at Ardlethan (http://www.bom.gov.au/climate/data/). The False Colour imagery available by ASTER highlights these features (Figure 2). Areas of red hue highlight green photosynthetic vegetation (e.g. crops or trees). Dark red-brown areas within Figure 2(a) typically indicate native vegetated cover. Artificially illuminated SRTM DEM imagery highlights the low relief nature of the environment, varying from 145 to 430 m elevation within the two map sheets (Figure 2(b)). In particular slope analysis of the SRTM DEM reveals only a small proportion exhibited slopes greater than 10% (Figure 2(c)). Past published geological mapping for the Narrandera (Wynn, 1977) and Cootamundra (Warren, Gilligan, & Raphael, 1996) 1:2,50,000 map sheet areas, encompassing the Ardlethan and Barmedman 1:100 K sheets, indicates that outcrops and evidence for geological boundaries lay within mostly moderately topographic relief (Figure 2(c)). Significantly less topographic relief is present in this study area than apparent for the Wagga Wagga 1:100 K map sheet study area (Hewson, Robson et al., 2015).

The environment and native vegetation consists predominantly of eucalyptus/sheoak woodland and scrubland, mostly occupying the limited ridges and hilltops. The ASTER Green Vegetation product, generated as part of the National Australia ASTER Mapping, highlights these native vegetated areas, and to a lesser extent, the cultivated cropland (Figure 3). These woodland areas typically coincide with the areas of higher slope (Figure 2(c)), although they can also be associated with watercourses, tree plantations and crops (Figure 3).

2.2. Geology

The Ardlethan and Barmedman study areas contains a geological sequence of units within the central and western Lachlan Orogen, and consist of Ordovician to Devonian volcanic, sedimentary, meta-sedimentary and plutonic rocks. Tertiary to Quaternary deposits of the eastern Murray Basin are prevalent as cover over much of the Palaeozoic sequences (Clare, Fleming, & Glen, 1997; Colquhoun et al., 2005; Downes, McEvilly, & Raphael, 2004). Widespread unconsolidated Quaternary alluvial, colluvial and aeolian sediments blanket much of the terrain limiting the access to rock outcrops for direct geological mapping. The main units for the Ardlethan and Barmedman 1:100 K study areas are shown within the displayed geological/metallogenetic map (Figure 4) combining the Cootamundra (Fitzpatrick, 1979; Warren et al., 1996) and Narrandera (Heugh, 1979) 1:2,50,000 (250 K) published mapping.

The area has had an extensive history of tin/tungsten mining associated with various Silurian-Devonian intrusives (Colquhoun et al., 2005). In particular, the Ardlethan Tin Mine has been a significant producer (Figure 4). Its porphyry-style deposit has been described as having a magmatic and hydrothermal history associated with the Ardlethan Granite (Figure 4) (Ren, Walshe, Paterson, Both, & Andrew, 1995). Ren et al. (1995) described an alteration zonation of the Ardlethan tin deposit characterised by biotite, chlorite, sericite and tourmaline mineralogy. Scott and Rampe (1984) found mineralisation was a useful pathfinder indicator for the tin deposit than the geochemistry signature from

(a)

(b)

(c)

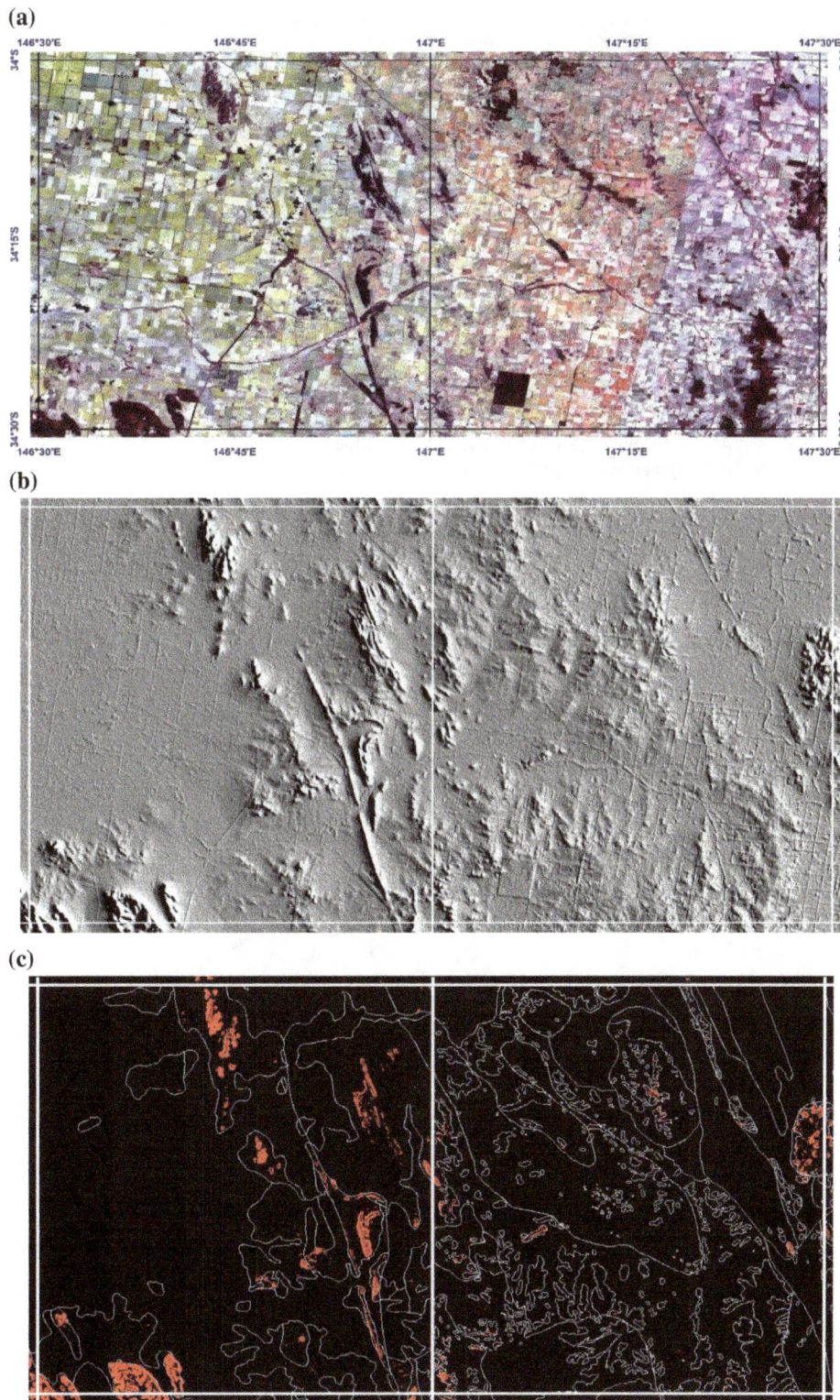

Figure 2. (a) Ardlethan and Barmedman 1:100 K map sheet areas–ASTER False Colour image; (b) Shaded SRTM DEM relief; (c) and areas of 10% slope (SRTM DEM) shown in red and previously published 1:250 K geological mapped boundaries (white boundaries).

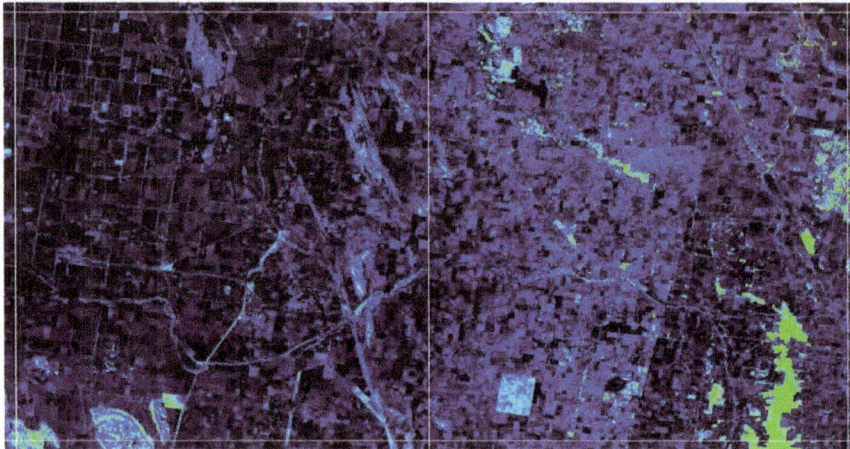

Figure 3. CSIRO-GA ASTER Green Vegetation product where blue is low, green is moderate and red is high green vegetation content. See Figure 2 for coordinate extents.

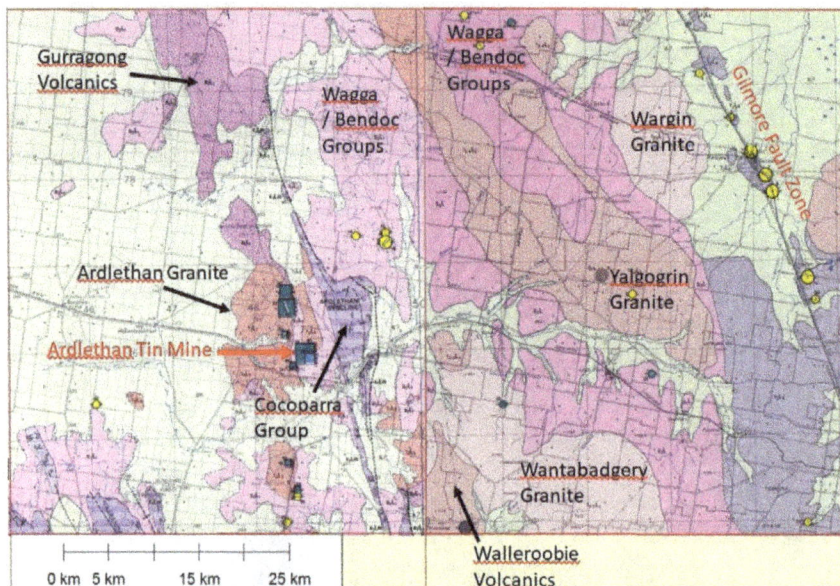

Figure 4. Past published geological mapping within the Ardlethan and Barmedman 1:100 K map sheet areas (Fitzpatrick, 1979; Heugh, 1979).

Notes: Round yellow symbols signify vein hosted gold deposits, square blue/green symbols signify disseminated tungsten/tin deposits. See Figure 2 for coordinate extents.

samples. Weathering was also found to range from up to 60 m but generally below 10 m depth and exhibited decreasing haematite/goethite and kaolinite development with depth (Scott & Rampe, 1984).

Sub-surface structures within the study area include palaeochannels within the Quaternary flood-plain deposits of the Lachlan River and its tributaries (Colquhoun et al., 2005). Considerable interest has been directed to such palaeochannels for their potential hosting of alluvial gold or tin deposits and also for dryland salinity and hydrological importance (Lawrie, Chan, Gibson, & de Souza Kovacs, 1999; Mackey et al., 2000). In particular, the Bland Creek Palaeochannel east of West Wyalong is an example of an alluviated valley hosting the Gibsonvale alluvial tin workings (Gibson & Chan, 1998). The present relief is generally low but it's suggested that the valley was originally 1 km wide and 35 m deep, dated to possibly the Oligocene (Gibson & Chan, 1998). Aero magnetics from airborne geophysical surveys has been found useful to delineate such palaeochannels, particularly where maghemite- and magnetite-bearing pisoliths and alluvium generates high spatial frequency magnetic anomalies (Gibson & Chan, 1998; Lawrie et al., 1999; Mackey et al., 2000). Magnetic model-ling of such palaeochannels has been attempted by Mackey et al. (2000). Alternatively, first vertical derivative (1VD) filtering of the measured Total Magnetic Intensity (TMI) has been demonstrated to be a useful qualitative technique to map near surface sourced magnetic anomalies associated with palaeochannels (Mackey et al., 2000).

3. Methodology

3.1. Data-sets

ASTER measures radiance from fourteen VNIR, SWIR and TIR bands at spatial pixel dimensions of 15 m (VNIR), 30 m (SWIR) and 90 m (TIR) (Table 1) (Fujisada, Sakuma, Ono, & Kudoh, 1998; Yamaguchi et al., 2001). In this study, both the processed ASTER Australian Geoscience products, as well as ASTER day-time (VNIR, SWIR and TIR bands) and ASTER night-time (TIR bands) image acquisitions were utilised. The ASTER Australian Geoscience map products were sourced for the past and current studies directly or indirectly from CSIRO and Geoscience Australia in either GeoTiff or ENVI™ (http://www.exelisvis.com/)/ER Mapper™ (http://www.hexagongeospatial.com/) software compatible for-mats. The Australian ASTER geoscience maps are available from: (i) the AuScope Discovery Portal (http://portal.auscope.org/portal/gmap.html); or (ii) via CSIRO (http://c3dmm.csiro.au/Australia_ASTER/index.html). The products were generated from traditional band (ratio) parameters (Crowley, Brickey, & Rowan, 1989), targeting mineral related spectral absorption features (albeit at multi-spectral ASTER resolution), followed by masking to exclude areas of cloud cover, deep shadow, wa-ter bodies and significant vegetation cover (Cudahy, 2012).

Individual ASTER image scenes were accessed at the time via Japan's METI Ground Data System (GDS) portal. NASA's Earth Explorer Reverb web portal now handles this ASTER data access (https://reverb.echo.nasa.gov/). Each ASTER image scene encompasses a 60 × 60 km area. Level 1b

Table 1. Spectral bandwidths and central band wavelengths of the ASTER sensor (Fujisada et al., 1998)

Module	VNIR	SWIR	TIR
Spectral bandwidth (μm) [Centre λ (μm)]	Band 1 0.52–0.60 [0.556]	Band 4 1.600–1.700 [1.656]	Band 10 8.125–8.475 [8.291]
	Band 2 0.63–0.69 [0.661]	Band 5 2.145–2.185 [2.167]	Band 11 8.475–8.825 [8.634]
	Band 3 N 0.78–0.86 [0.807]	Band 6 2.185–2.225 [2.209]	Band 12 8.925–9.275 [9.075]
	Band 3B 0.78–0.86 [0.804] (Backward looking)	Band 7 2.235–2.285 [2.262]	Band 13 10.25–10.95 [10.657]
		Band 8 2.295–2.395 [2.336]	Band 14 10.95–11.65 [11.318]
		Band 9 2.360–2.430 [2.400]	
Spatial resolution (m)	15	30	90

Note: 1 μm ≡ 1,000 nm.

Table 2. Acquisition dates of day- and night-time ASTER imagery used in study

Day-time surface temperature & VNIR-SWIR surface reflectance	21/3/2012	18/10/2013	
Night-time surface temperature	11/1/2012	29/11/2013	21/4/2008
Day-time VNIR-SWIR surface reflectance & TIR emissivity	13/1/2005		
GA-CSIRO ASTER inputs: Day-time VNIR-SWIR	16/11/2006 & 15/10/2006		

(radiance at sensor) and Level 2 (e.g. surface reflectance) products were used in this study. Table 2 lists the various day- and night-time ASTER acquisitions used in this particular study.

3.2. Day-time processing of ASTER imagery

Similar terrain issues were encountered within the Ardlethan and Barmedman 1:100 K map sheet areas, to those encountered in previous studies in the Wagga Wagga study, 100 km to the south east of (Hewson, Robson et al., 2015). It was also found that the Ardlethan and Barmedman study area had significantly less topographic relief and associated outcrop than the Wagga Wagga area. The nature of this terrain precluded the application of DEM derived slope masks to extract coherent geological information from the Australian ASTER products, as demonstrated in the Wagga Wagga area (Hewson, Robson et al., 2015). As a consequence, the following approaches were undertaken with the day-time ASTER acquisitions:

(1) Process specific individual subsets of ASTER surface reflectance (VNIR-SWIR) and emissivity (TIR) imagery of the Ardlethan–Barmedman area using band parameter processing, MNF style classification and spectral unmixing (Crowley et al., 1989; Green et al., 1988; Kruse et al., 1993).

(2) Field observations and sampling of rock outcrop and soils collected within the Ardlethan and Barmedman 1:100 K map sheet areas.

(3) Compare the spatial resolution of the ASTER products/imagery and the Airborne Digital Sensor 40 (ADS40) photogrammetry (Sandau et al., 2000) currently used for surface mapping by the GSNSW.

Processed subsets of day-time ASTER imagery within the Ardlethan–Barmedman area focused on key areas such as the Gurragongs Volcanics and the Wagga/Bendoc Group (Figure 4). Topographic slope masking was also incorporated with the MNF and spectral unmixing methodologies (Green et al., 1988; Kruse et al., 1993) applied to those areas within subsets, to target possible geological outcrops. ENVI™ software (http://www.harrisgeospatial.com/) was applied for this purpose Hewson, Robson, et al. (2015).

3.3. Night-time processing ASTER imagery

The basis of using night-time imagery, either processed to surface temperature, or as Thermal Inertia, as a geological mapping tool is explained in Sabins (1997) and Watson (1975). Thermal inertia (P) is a scalar volumetric physical property of a material, that cannot be measured directly but inferred from diurnal variations of temperature at the Earth's surface and related to the following other properties:

$$P = (K\rho c)^{1/2}$$

where K = thermal conductivity, ρ = density and c = specific heat.

Generally geological materials, particularly denser and non-porous, retain the heat acquired during the sunlight exposure more effectively (e.g. higher thermal inertia), and radiates its heat overnight. Unconsolidated and less dense materials however heat up and cool down more quickly (e.g. lower thermal inertia) in particular lower density regolith such as soils have lower thermal inertia than rocks and their constituent minerals (Watson, 1975). The addition of moisture to dry soil can also increase the thermal inertia due to both an effective increase in density as well as reduction in porosity affecting bulk thermal properties (Watson, 1975). Ideally therefore, night-time thermal imagery provides better results under dry summer conditions, or at least undertaken with a knowledge of local rainfall and moisture levels. Rainfall records from the Australian Bureau of Meteorology (http://www.bom.gov.au/climate/data/) were sourced for the nearby Tallimba weather station and precluded moisture as a significant influence on the ASTER acquisitions.

Night-time NOAA satellite thermal imagery has been previously used for geological mapping in Western Queensland (Russell & Lappi, 1988). Russell and Lappi (1988) noted that rock outcrops absorb and retain significantly more day-time thermal energy than the surrounding soil and alluvium, generating night-time temperature anomalies. They estimated that this solar heating effect could penetrate into unconsolidated material of 4 m, potentially indicating the presence of rock outcrops to this depth. Field observations in this study also identified areas with anomalously high night-time surface temperatures associated with dense forested woodland with no apparent geological outcrops. Although there is also the possibility of near-surface outcrop, the association of high temperature anomalies with dense vegetated areas acting as a thermal blanket, has been recognised previously by Fabris (2002).

Several different night-time ASTER acquisitions were obtained for this study to observe the possible effects of seasonal ground temperature and rainfall variations on the imagery (Table 2). Various histogram stretches were applied to the ASTER night-time surface temperature imagery expressed in Kelvin $\times$ 10 values to obtain the best qualitative information. The surface temperature ASTER products were based on the Temperature Emissivity Separation algorithm described by Gillespie et al. (1998) and validated by Hook, Vaughan, Tonooka, and Schladow (2007).

The derivation of Thermal Inertia is not trivial and an approximation is calculated as Apparent Thermal Inertia (ATI) (Kahle,1987; Price, 1977). The generation of ATI imagery requires both day-time T_1 (e.g. maximum) and night-time T_2 (e.g. minimum) time surface temperatures, and an estimate of albedo, ρ_{av} (e.g. average VNIR surface reflectance):

$$ATI \approx \frac{(1 - \rho_{av})}{\Delta T}$$

where $\Delta T = T_1 - T_2$

This approximation by Kahle (1987) and Price (1977) ignored for simplicity, topographic and atmospheric affects. The ATI approximation precludes the comparison of one ATI image values to another. Also, although the dates of ASTER day- and night-time acquisitions were not ideal for the Ardlethan–Barmedman areas, Kahle and Alley (1985) demonstrated that relative and apparent estimates of thermal inertia are still useful for the differentiation of outcrops and alluvium, even if several weeks' separate day- and night-time acquisitions. Another limitation in this particular application using ASTER TIR data is that the acquisition times are not likely to observe the maximum day-time and minimum night-time ground surface temperatures, at the 10:30 am and 10:30 pm acquisition times, rather than at approximately 1–3 pm and 3–5 am suggested by Price (1977).

The derivation of ATI products required sub-setting of all the input ASTER data into coincident overlapping imagery. Different spatial coverage of the ASTER data occurred from different acquisition dates, as well as from changes between their descending and ascending/day- and night-time orbits. Consequently the resulting image subset ATI product was less than 60 $\times$ 60 km coverage. The calculation derivation of the ATI and their image sub-setting was performed using ER Mapper™ software (http://www.hexagongeospatial.com/). An additional potential QC issue of these products was the 90 m spatial resolution of the ASTER TIR imagery, increasing the effective mixed observation of thermal properties between the vegetation, soil and outcrop radiant components.

4. Results

4.1. National Australia ASTER Geoscience maps

In this study both the ASTER Australian Geoscience products and independently processed ASTER day-time (VNIR, SWIR and TIR bands) and night-time (TIR bands) acquisitions were used in an attempt to extract geological mapping information. However, the ASTER False Colour (Figure 2(a)) and Green Vegetation (Figure 3) ASTER products for the Ardlethan and Barmedman 1:100 K areas highlight the issues of agricultural cropland and vegetation cover. Likewise, the ASTER Regolith product (Figure 5(a)) highlights the difficulty in its geological mapping application for this area and mostly highlights crop and paddock patterns and boundaries to variable vegetation and soil

(a)

(b)

(c)

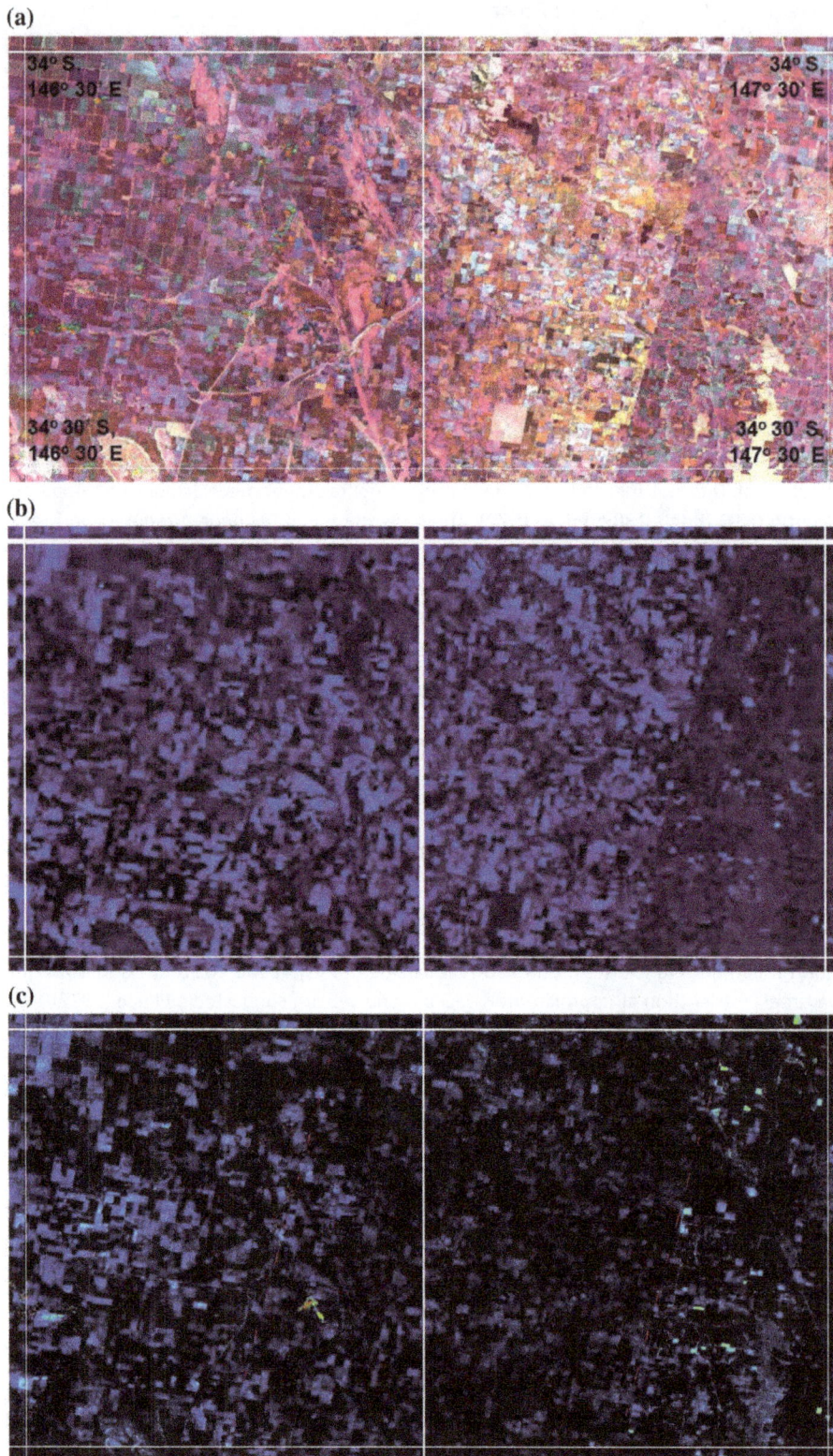

Figure 5. Ardlethan and Barmedman 1:100 K sheets: (a) National Australia ASTER Regolith product; (b) National Australia ASTER Silica Index; (c) National Australia ASTER AlOH. Derived map product values are represented by blue = low, green = moderate and red = high content. See Figure 2 for coordinate extents.

exposure. The ASTER Silica Index and AlOH Content products also show cropland patterns (Figure 5(b) and (c), respectively). Some variation in the silica/quartz content appears within the crop fields and possibly related to variable quartz sand content within the soils. Unlike previous studies within the nearby Wagga area, no lineaments or anomalies were apparent and suggestive of sub-surface geology from the ASTER ferric iron product within the soils and regolith within the cultivated areas (Hewson, Robson et al., 2015).

4.2. Spectral unmixing of individual ASTER surface reflectance and emissivity image imagery

The day-time ASTER VNIR-SWIR image acquisition obtained for the individual scene based analysis, straddled both the Ardlethan and Barmedman 1:100 K map sheets (Figure 6(a)). The chosen summer ASTER scene (13/1/2005) indicated the cropland showed high reflectance/albedo due to its dry and predominantly exposed soil/regolith (Figure 6(a)). MNF processing of the ASTER imager of the nine ASTER VNIR-SWIR bands for this summer acquisition generated several classified images combining landscape features of vegetation, soil and potentially outcrop. MNF band 5 generated from the whole scene appeared to highlight Quaternary alluvial cover although most MNF results were dominated by soil/regolith types and vegetated landforms. An image subset for the Wagga/Bendoc Group (blue box, Figure 6(b)) was also processed for its MNF results. A comparison of its MNF's 4, 5 and 6 composite imagery (Figure 6(c)) with an albedo image (greyscale background image of Figure 6(d)) highlighted subtle discrimination of the various beds and north westerly structural trends (493000 mE, 6220000 mN). Some of the apparent bedding features observed may also relate to topographic aspect illumination effects and associated vegetation variation and warrant further examination. MTMF spectral unmixing classification appears to highlight soil/regolith and/or crop vegetation anomalies rather than discriminate the actual outcropping Wagga/Bendoc beds (Figure 6(d)). However, the spectral signatures associated with each unmixed endmember predominantly mapped variety of vegetation and clay/soil types by their 0.8 µm and 2.2 µm spectral features, respectively.

A similar approach was applied at the Gurragong Volcanics subset (red box, Figure 6(b)) for both the day-time ASTER VNIR-SWIR data (Figure 7(a)–(d)). The ASTER vegetation map product (Figure 7(b)) and MNF processing of the subset (Figure 7(c)) again indicated the dominance of vegetation and cropping patterns. A more targeted approach to MNF processing was also attempted by further subsetting the imagery for elevated areas using a 5% slope mask, derived from the SRTM DEM, (Figure 7(d)). This decreased slope threshold, compared to the 10% used in the Wagga study, was assumed given the terrains more weathered and colluvial dominated outcropping slopes. A trend in the MNF spectral response suggests a change of composition and/or vegetation from the northerly Gurragong units (blue-green) to the yellow and red southerly occurrences (Figure 7(d)). However there is no current understanding to this apparent trend in processed ASTER spectral image response.

MNF and MTMF processed results of the ASTER TIR surface emissivity imagery proved noisy and again the results were dominated by vegetation and cropping patterns/soils. The variation of quartz sand within exposed or fallow soils would likely be a strong determinant in its response.

4.3. Effects of image spatial resolution on mapping

The effect of the spatial resolution of the ASTER sensor was examined in the study area within an area of a mixture of vegetation and exposed geology. A reduction in the surface feature spatial detectability was observed from the comparison of the 0.5 m resolution ADS40 photogrammetry (Sandau et al., 2000) and the 30 m ASTER products and imagery (VNIR-SWIR) within the study area (Figure 8(a) and (b)). Currently ADS40 is commonly used to assist surface mapping by the GSNSW, particularly for identifying the limited surface outcrops. Generally there are sparse and limited outcrops within the Ardlethan– Barmedman area, sometimes located within exposed cropland or on ridges and topographic highs with open woodland or scrubby vegetation. The differences between the ADS40 and ASTER imagery in their ability to highlight vegetation and outcrops is shown in Figure 8(a) and (b). The coarser 30 m ASTER False Colour imagery (Figure 8(a)) discriminates the predominantly woodland areas but not individual trees as discriminated by the ADS40 (Figure 8(b)).

(a)

(b)

(c)

(d)

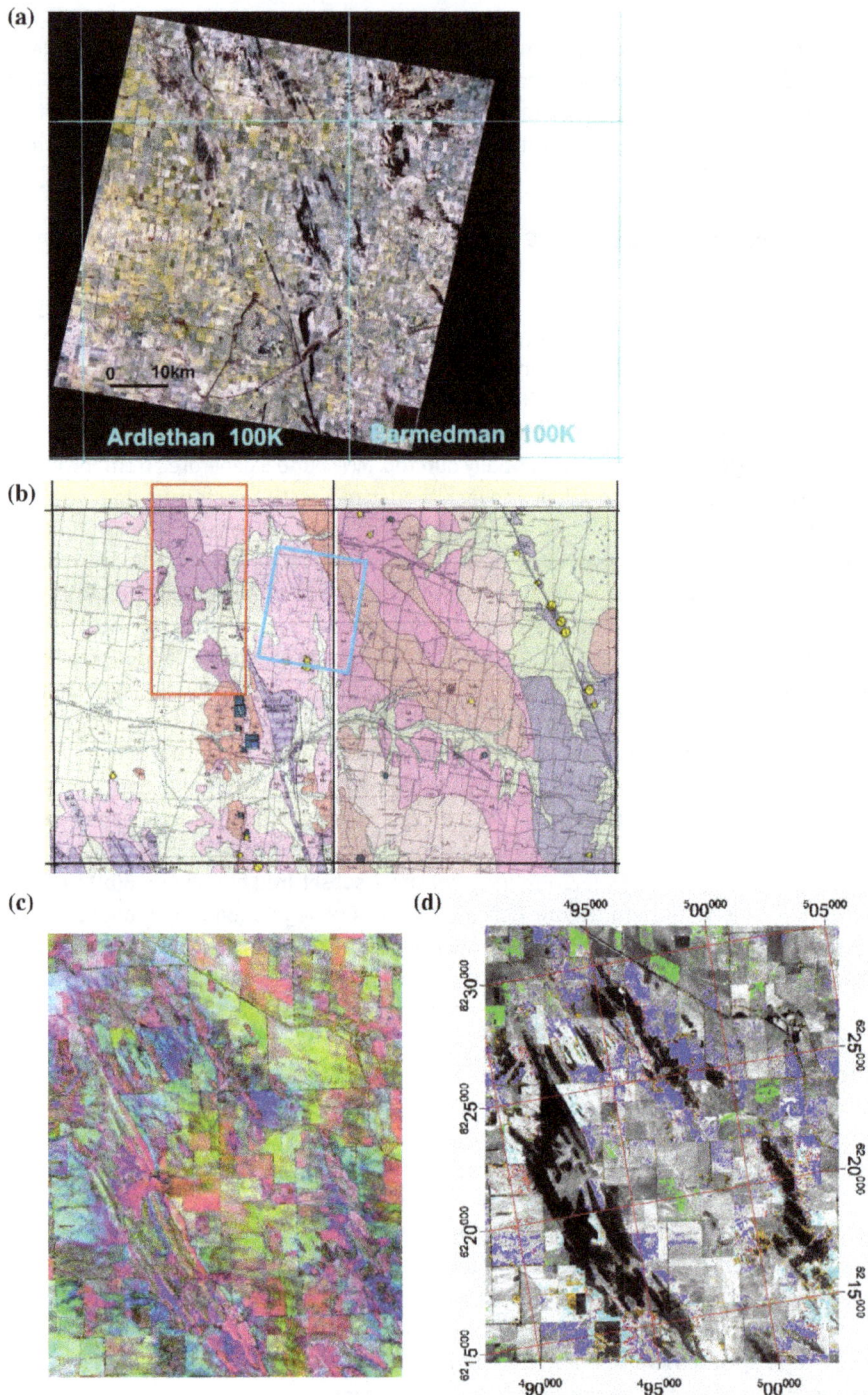

Figure 6. (a) ASTER False Colour imagery (13/1/2005) acquisition within the Ardlethan and Barmedman 1:100 K sheets; (b) Ardlethan–Barmedman 1:100 K sheet published geology showing the Gurragong Volcanics (red box) and Wagga/Bendoc Group (blue box) image subsets. See Figure 4 for definition of main geological units; (c) Wagga/Bendoc Group subset RGB composite of MNF bands 4, 5 and 6; (d) Wagga/ Bendoc Group subset MTMF spectral unmixed results, based on 8 spectral endmembers. Greyscale albedo image of ASTER band 4 overlain by examples of various subtle spectral endmembers dominated by clays in soil (blue, green, orange and cyan). See Figure 2 for coordinate extents.

Differences in the cropland soil colour and the presence of likely north-westerly trending Wagga/ Bendoc Group outcrops suggest that the ADS40 is more discriminatory for structural interpretation and outcrop delineation than the day-time ASTER products in this terrain (Figure 8(a) and (b)).

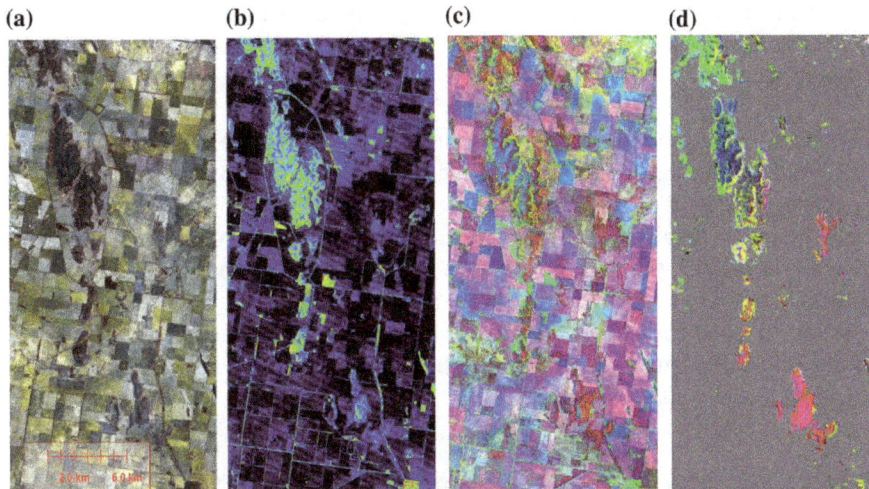

Figure 7. Gurragong Volcanics (a) subset of ASTER False Colour; (b) subset of ASTER Green Vegetation; (c) subset RGB composite of MNF bands 2, 5 and 6; (d) subset of RGB composite of MNF bands 5, 4, and 3 processed with DEM Slope greater than 5% mask. See Figure 6 for subset location.

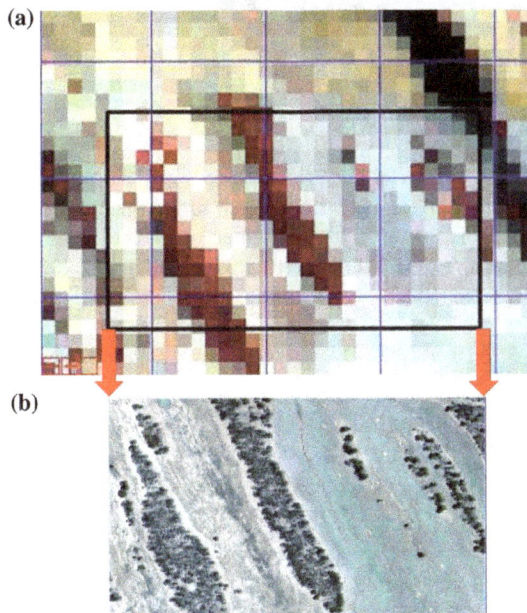

Figure 8. (a) Example of the 30 m ASTER False Colour product (~34°6′S, 146°57′) within the Wagga/Bendoc Group geological unit. The blue gridlines are 250 × 250 m; (b) ADS40 digital photogrammetry for the zoomed in area of (a) (black box).

4.4. ASTER night-time surface temperature imagery

The apparent limitation of using VNIR–SWIR and TIR day-time ASTER remote sensing for geological mapping for this study area's terrain suggests that alternative less surface affected techniques based more on physical properties rather than mineralogical be trialled. In particular this study examined the application of ASTER's surface temperature product, as supplied in values of Kelvin multiplied by 10. Such satellite surface temperature estimates are variable according to times, dates, winds, etc. so the displayed images, showing relative differences from low to high temperatures (e.g. blue to red), are useful more for their qualitative trends. A first pass examination of a night-time ASTER acquisition as surface temperature shows it typically has higher values on ridges and topographic rises between watercourses (Figure 9(a)). Past published 1:250 000 scale geological mapping also indicates many of the temperature anomalies are associated with the boundary of

Quaternary cover and contained within some of the mapped outcropping geological units (Figure 9(b)). The relationship between the night-time temperature anomalies and the topography is clearly illustrated in Figure 10.

The RGB composite imagery of thermal radiance bands 13, 12 and 10, for the night-time ASTER acquisition, also highlighted similar anomalous topographic relief and/or outcropping landforms (Figure 11). It appears that by using a combination of radiance bands, combining both the temperature and emissivity surface properties, the thermal and compositional nature was mapped (Figure 11). The red apron surrounding the anomalously high radiance also could be associated with the relatively higher quartz band 13 emissivity (e.g. red) in the sand colluvium/regolith/soil surrounding the outcrop associated topographic rises (Figure 11). This is consistent with the combined emissivity and temperature nature of radiance imagery where the spectral emissivity of surface materials is related to its compositional nature.

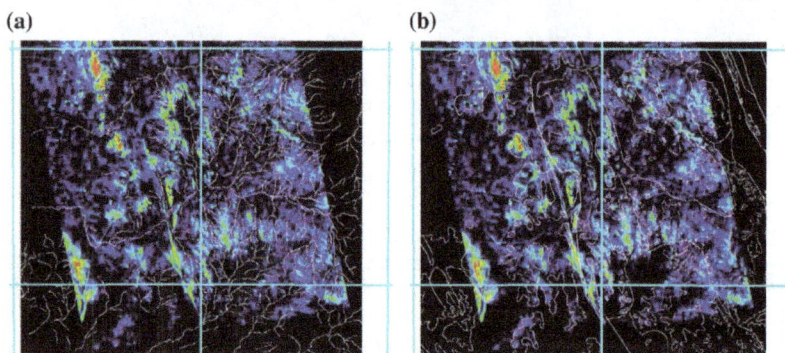

Figure 9. Ardlethan–Barmedman 1:100 K study area (cyan) (a) night-time ASTER surface temperature (99% linear stretch) in relation to 1:250 K watercourses (white), blue: low temperature, red: higher temperature; (b) night-time ASTER surface temperature (99.9% linear stretch) in relation to 1:250 K published geology (white), blue: low temperature, red: higher temperature.

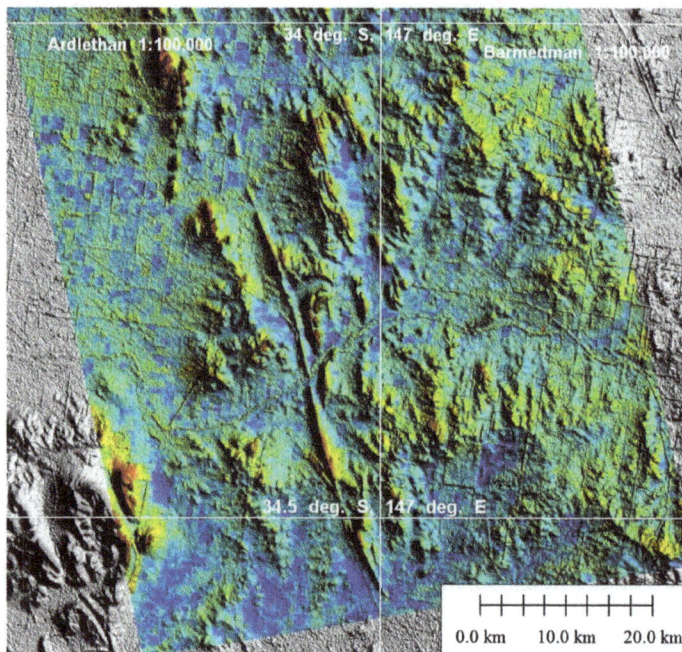

Figure 10. Shaded DEM relief overlain by night-time ASTER surface temperature within the Ardlethan–Barmedman 1:100 K study area, blue: low temperature, red: higher temperature.

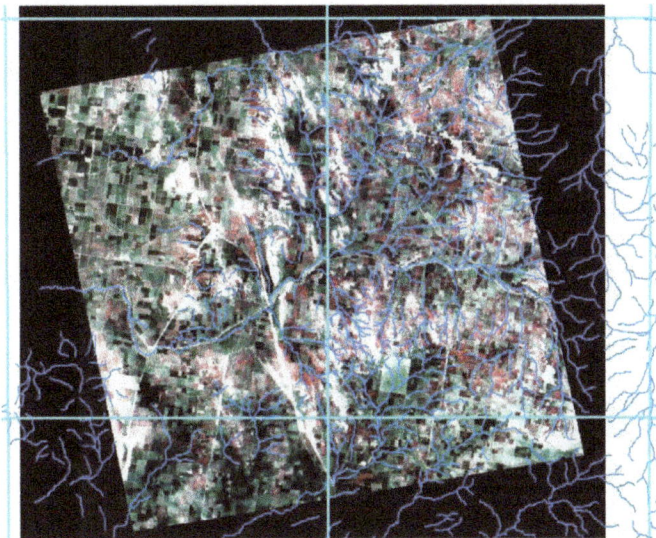

Figure 11. RGB composite image of night-time ASTER radiance bands 13, 12 and 10 with 1:250 K watercourses (blue) and Ardlethan–Barmedman 1:100 K sheet boundaries (cyan) as an overlay.

(a)

(b)

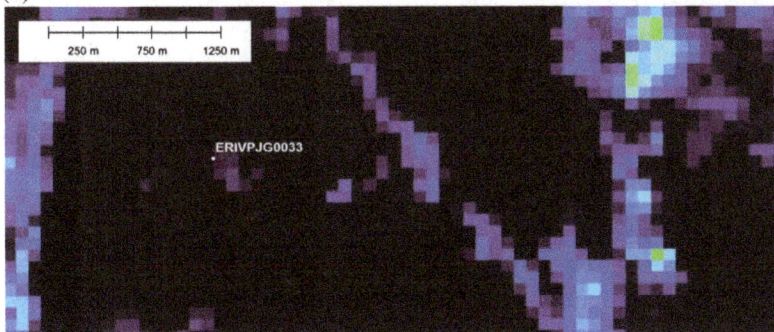

(c)

Figure 12. Site ERIVPJG0033: (a) ADS40 imagery of site; (b) Night-time ASTER Surface Temperature, blue: low temperature, red: higher temperature; (c) Panoramic photo of Ardlethan (?) Granite.

(a)

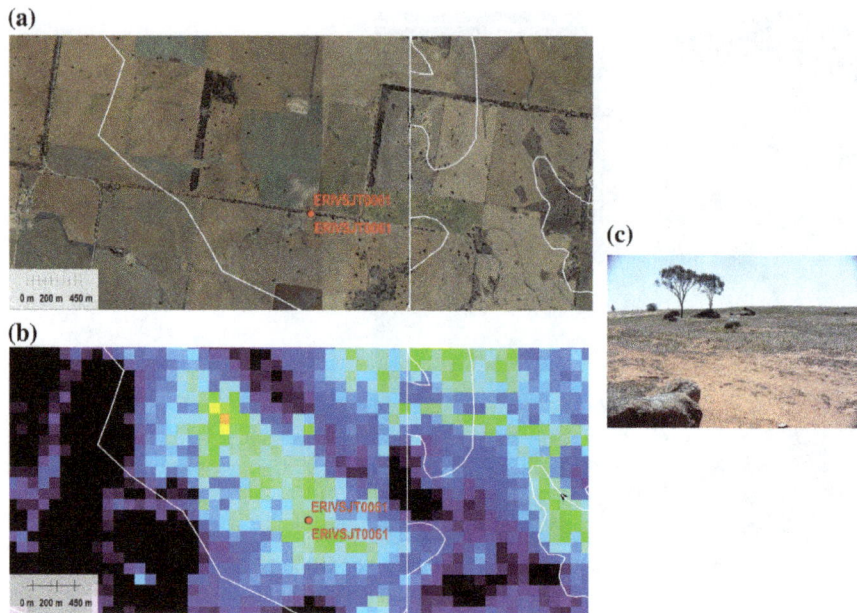

(b)

(c)

Figure 13. Site ERIVSJT0061: (a) ADS40 imagery of site; (b) Night-time ASTER Surface temperature with 1:250 K published geology (white boundary), blue: low temperature, red: higher temperature; (c) Photos of Ardlethan Granite outcrops.

Several field sites were examined for their ADS40 and night-time surface temperature ASTER imagery to investigate its ability to map outcrops. Site ERIVPJG0033 was located at an isolated possible outcrop of Ardlethan Granite. The ADS40 imagery (Figure 12(a)) did not appear to identify any granitic outcrop while the ASTER's surface temperature showed a subtle anomaly (Figure 12(b)). Panoramic photos of ERIVPJG0033 highlighted the isolated limited nature of this outcrop (Figure 12(c)). The night-time surface temperature imagery also appeared to highlight the nearby dams and a linear feature, possibly associated with a canal or road.

Site ERIVSJT0061 with some large outcrops were subtle within the ADSA40 imagery (Figure 13(a)) but part of a pronounced larger surface temperature anomaly striking to the northwest (Figure 13(b)). The outcrops on this hilltop are limited (Figure 13(c)) but were described as Ardlethan Granite within the previous 1:250 K geological mapping (Figure 13(b)).

4.5. ASTER derived apparent thermal inertia imagery

Two pairs of day- and night-time ASTER acquisitions were used to calculate Apparent Thermal Inertia (ATI) as described in Section 3.2. There was a limited choice of available pairs and day- and night-time ASTER acquisitions. However pairs of ASTER imagery were acquired on 18/10/2013 (Figure 14(a)) and on 21/3/2012 (Figure 14(b)) while night-time surface temperature imagery was obtained for 29/11/2013 and 11/1/2012. There was noticeably more vegetation present for the 2013 spring pair (e.g. increased red within the day-time false colour imagery, Figure 14(a)). The ATI was calculated for each pair using day- and night-time temperature products and estimate of the day-time albedo (Figure 14(c)).

Ambiguities due to areas associated with high vegetation cover, particularly woodland, were minimised by masking the ATI with an estimate of green vegetation cover derived by the Normalized Vegetation Difference Index (NDVI) from the ASTER VNIR imagery. The issue of increased green vegetation and spring soil moisture limited the usefulness of the 2013 ATI imagery which showed greater crop and paddock affected anomalies (Figure 14(c)).

The approach by Mackey et al. (2000) and Lawrie et al. (1999) for the detection of subsurface palaeochannels using aero magnetics anomalies was attempted to compare the results of surface

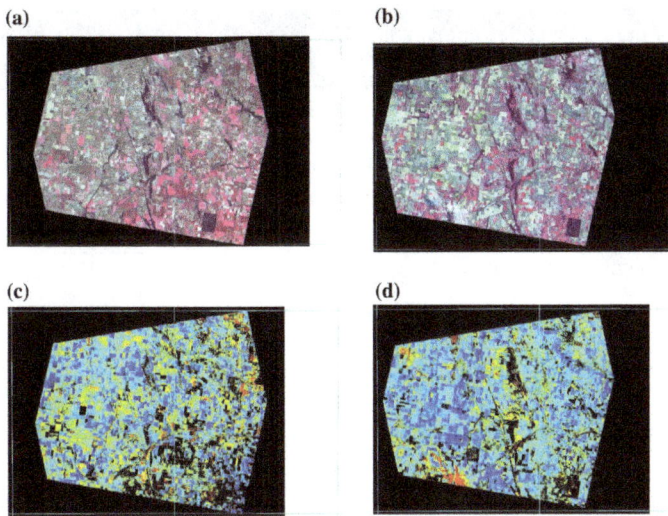

Figure 14. (a) False Colour imagery of day-time ASTER 18/10/2013 acquisition; (b) False Colour imagery of day-time ASTER 21/3/2012 acquisition; (c) derived ATI using ASTER night-time 29/11/2013 with NDVI mask; (d) derived ATI using ASTER night-time 11/1/2012 with NDVI mask. (c) and (d)–blue: low ATI, red: higher ATI.

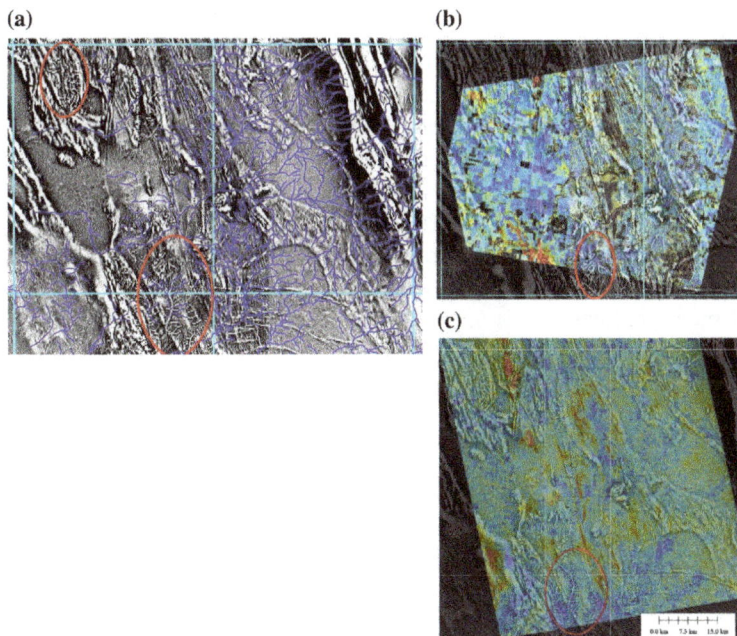

Figure 15. (a) 1VD TMI with 1:250 K watercourses within Ardlethan–Barmedman 1:100 K areas; (b) Combined ATI (ASTER night-time 29/11/2013) with NDVI mask overlain with 1VD TMI.

Notes: The dendritic 1VD TMI anomalies (red) in (a), and the corresponding ATI low thermal inertia anomaly (red); (c) 1VD TMI overlain by the night-time ASTER surface temperature.

temperature and ATI anomalies in this study. Several dendritic 1VD TMI anomalies were identified within the Ardlethan–Barmedman study area (red ellipses, Figure 15(a)). Comparison of these 1VD TMI anomalies with present-day watercourses showed a discrepancy and it is viable that this is a result of palaeochannel accumulations of maghemite (Figure 15(a)) as described elsewhere by Mackey et al. (2000) in the nearby area of West Wyalong. There was some limited correspondence to low ATI values in the area shown (blue, Figure 15(b)) however there was insufficient ATI coverage to fully confirm and completely map these features. A low surface temperature (Figure 15(c)) in the

Table 3. Correlation statistics of ASTER thermal products with topographic and vegetation information

Correlation matrix	SRTM slope%	ASTER (2008-04-21, pm) surf. temp. (Kelvin × 10)	NDVI (2012-03- 21, am)	ASTER (2012-03-21, am) surf. temp. Kelvin × 10	ASTER (2012-01-11, pm) surf. temp. Kelvin × 10	ATI (2012)
SRTM slope%	1.00	0.50	0.28	−0.29	0.18	0.36
ASTER (2008-04-21, pm) surf. temp. (Kelvin × 10)	0.50	1.00	0.24	−0.24	0.30	0.36
NDVI (2012-03-21, am)	0.28	0.24	1.00	−0.66	0.22	0.63
ASTER (2012-03-21, am) Surf. Temp. Kelvin × 10	−0.29	−0.24	−0.66	1.00	−0.26	−0.87
ASTER (2012-01-11, pm) surf. temp. Kelvin × 10	0.18	0.30	0.22	−0.26	1.00	0.60
ATI (2012)	0.36	0.36	0.63	−0.87	0.60	1.00

vicinity of these palaeochannels also suggested a cooler hydrological influence, although again, there is limited coverage in this study to be definitive. Although these preliminary results are quali-tatively suggestive of indicating sub surface structures, further studies over much larger areas is required to establish the benefits of a combined use of ATI and aero magnetics.

In order examine the likely effects of high relief associated outcrops (e.g. high slope %) and veg-etation (NDVI) on the calculated ASTER ATI products, correlation statistics were calculated for spa-tially coincident data-sets (Table 3). The slope % showed a slightly higher correlation of R = 0.5 to the 2008 night-time surface temperature compared to the ATI (R = 0.36). The ATI seemed more affected by the vegetation (R = 0.63) although coincident 2008 NDVI imagery wasn't available. An interesting result was the cooling effect of the vegetation on day-time surface temperature (R = −0.66). Further work is required and recommended to more clearly distinguish the effects of vegetation and rocky outcrops on such thermal image products.

5. Conclusions
This study tested the ability of ASTER remote sensing imagery to assist the geological mapping of a vegetated and cultivated temperate environment with limited outcrop exposure. Limited informa-tion was extracted in the study area using the large-scale regional ASTER compositional maps that used band parameter style processing and uniform histogram methods and thresholds. There ap-peared improved qualitative results for discriminating subtle geological/regolith variations using spectral processing and the application of masks, generated from higher topographic slopes, on in-dividual day-time ASTER acquired scenes. Overall however, it appeared that the coarse spatial reso-lution of satellite ASTER (e.g. 30 m) compared to airborne ADS40 imagery (e.g. 0.5 m) significantly limited the delineation of surface feature boundaries and the separation from vegetation cover. Night-time ASTER thermal imagery appeared useful for locating sparse rock outcrops in areas of low or high relief although anomalous "warm" forested areas require the use of day-time imagery to mask their effects. Further studies into the ability of ASTER or other night-time thermal imagery and its integration with geophysics (e.g. 1VD TMI) would be useful to assess its potential for the delinea-tion of sub-surface geological structures in such areas.

In summary, the study's evaluation of the application of ASTER for using day- and night-time ac-quisitions, showed limited to modest success for routine geological mapping within an area of low relief, sparse outcrops and cultivated terrain. Although a freely available satellite image source im-agery, ASTER's spatial resolution limit its application in this such an example of a non-arid environment. Also ASTER's limited day- and night-time paired acquisitions handicap its potential for the reliable interpretation of thermal inertial properties. The availability of the recently launched higher spatial resolution WorldView-3 sensor (Kruse, Baugh, & Perry, 2015) and future NASA HySpiri mission thermal capability (https://hyspiri.jpl.nasa.gov/) offers a possible partial solution to the geological remote sensing technical issues in such a challenging terrain.

Acknowledgements

This study has been supported by the Geological Survey of New South Wales, in part, for the purpose of contributing to the knowledge of the Ardlethan and Barmedman 100 000 map sheet areas geology for mineral exploration. ASTER GeoTiffs map products were also gratefully obtained from Geoscience Australia and CSIRO-Earth Science & Resource Engineering, whilst the original ASTER imagery was made available by NASA and METI of Japan. GIS shapefiles for the Narranadera and Cootamundra 1:250 000 map sheet areas were also gratefully obtained from Geoscience Australia. Assistance from Simon Jones and Laurie Buxton of RMIT University was also forthcoming and appreciated during this study.

Funding

This study was initially instigated and funded by the Geological Survey of New South Wales, NSW Trade & Investment, Division of Resources & Energy.

Author details

R. Hewson[1]

E-mail: r.d.hewson@utwente.nl

D. Robson[2]

E-mail: Robodavidf@gmail.com

A. Carlton[2]

E-mail: Astrid.carlton@trade.nsw.gov.au

P. Gilmore[2]

E-mail: Phil.gilmore@trade.nsw.gov.au

[1] Faculty of Geo-Information Science & Earth Observation (ITC), University of Twente, P.O. Box 217, 7500 AE, Enschede, The Netherlands.

[2] Geological Survey of NSW, NSW Trade & Investment, Division of Resources & Energy, P.O. Box 344, Hunter Region Mail Centre, 2310, Sydney, NSW, Australia.

References

Adams, J. B., & Filice, A. L. (1967). Spectral reflectance 0.4 to 2.0 microns of silicate rock powders. *Journal of Geophysical Research, 72*, 5705–5715. https://doi.org/10.1029/JZ072i022p05705

Caccetta, M., Collings, S., & Cudahy, T. (2013). A calibration methodology for continental scale mapping using ASTER imagery. *Remote Sensing of Environment, 139*, 306–317. https://doi.org/10.1016/j.rse.2013.08.011

Chen, X. Y. (1997). Quaternary sedimentation, parna, landforms, and soil landscapes of the Wagga Wagga 1:1,00,000 map sheet area, south-eastern Australia. *Australian Journal of Soil Research, 35*, 643–668. https://doi.org/10.1071/S96071

Clare, A. P., Fleming, G. D., & Glen, R. A. (1997, May). *Geology of the Cargelligo and Narrandera map sheet areas: Removing the cover using Discovery 2000 Geophysics* (No 104, 22 pp.). Quarterly Notes–Geological Survey of New South Wales.

Colquhoun, G. P., Meakin, N. S., & Cameron, R. G. (2005). *Explanatory notes Cargelligo 1:250 000 geological sheet* (3rd ed., SI/55-6). Geological Survey of NSW.

Crowley, J. K., Brickey, D. W., & Rowan, L. C. (1989). Airborne imaging spectrometer data of the Ruby Mountains, Montana: Mineral discrimination using relative absorption band-depth images. *Remote Sensing of Environment, 29*, 121–134. https://doi.org/10.1016/0034-4257(89)90021-7

Cudahy, T. J. (2012). Australian ASTER Geoscience Product Notes, (Version 1), 7th August, 2012–CSIRO, ePublish No. EP-30-07-12-44.

Downes, P. M., McEvilly, R., & Raphael, N. M. (2004, February). Mineral deposits and models, Cootamundra 1:250,000 map sheet area. *Quarterly Notes, Geological Survey of New South Wales, No, 116*, 37.

Fabris, A. (2002). Thermal satellite imagery–An aid to heavy mineral sand discoveries. *MESA Journal, 24*, 24–26.

Fitzpatrick, K. R. (1979). *Cootamundra 1:2,50,000 metallogenic Map explanatory notes*. Sydney: Geological Survey of New South Wales.

Fujisada, H., Sakuma, F., Ono, A., & Kudoh, M. (1998). Design and preflight performance of ASTER instrument protoflight model. *IEEE Transactions on Geoscience and Remote Sensing, 36*, 1152–1160. https://doi.org/10.1109/36.701022

Gibson, D. L., & Chan, R. A. (1998). Aspects of palaeodrainage in the North Lachlan Fold Belt Region, cooperative research centre for landscape environments and mineral exploration, In G. Taylor, & C. Pain (Eds), *Regolith 98 Proceedings volume, Program and Abstracts volume* (pp. 23–37). Kalgoorlie: Field Trip Guide.

Gillespie, A., Rokugawa, S., Matsunaga, T., Cothern, J. S., Hook, S., & Kahle, A. B. (1998). A temperature and emissivity separation algorithm for Advanced Spaceborne Thermal Emission and Reflection Radiometer (ASTER) images. *IEEE Transactions on Geoscience and Remote Sensing, 36*, 1113–1126. https://doi.org/10.1109/36.700995

Green, A. A., Berman, M., Switzer, P., & Craig, M. D. (1988). A transformation for ordering multispectral data in terms of image quality with implications for noise removal. *IEEE Transactions on Geoscience and Remote Sensing, 26*, 65–74. https://doi.org/10.1109/36.3001

Heugh, J. P. (1979). *Cargelligo-Narrandera 1:250 000 Metallogenic Map* (1st ed.). Sydney: Geological Survey of New South Wales.

Hewson, R. D. (2015). *Using remote sensing, spectral and geophysical information to assist exploration within the Ardlethan and Barmedman 1:100,000 map sheet areas* (49 pp.). New South Wales Geological Survey (Report GS2015/0185). Retrieved from http://digsopen.minerals. nsw.gov.au/

Hewson, R., Carlton, A., Gilmore, P., Jones, S., & Robson, D. (2015, July 7–10). *Geological mapping within NSW using remotely sensed and proximal spectral data*. In proceedings near surface geophysics conference, Society of Exploration Geophysicists, Waikoloa, Hawaii, 4 pp.

Hewson, R. D., & Cudahy, T. J. (2011). Issues affecting geological mapping with ASTER data: A case study of the Mt Fitton Area, South Australia. In B. Ramachandran (Ed.), *Land remote sensing and global environmental change: NASA's earth observing system and the science of ASTER and MODIS: Applications in ASTER*. New York, NY: Springer-Verlag. ISBN: 978-1-4419-6748-0.

Hewson, R. D., Cudahy, T. J., Huntington, J. F. (2001, July 9–13). *Geologic and alteration mapping at Mt Fitton, South*

Australia, using ASTER satellite-borne data. In IEEE 2001 international geoscience and remote sensing symposium (IGARSS), p. 3.

Hewson, R. D., Cudahy, T., Mizuhiko, S., Ueda, K., & Mauger, A. J. (2005). Seamless geological map generation using ASTER in the Broken Hill-Curnamona province of Australia. *Remote Sensing of Environment, 99*, 159–172. https://doi.org/10.1016/j.rse.2005.04.025

Hewson, R., Koch, C., Buchanan, A., & Sanders, A. (2002). *Detailed geological and regolith mapping in the Bangemall Basin, WA, using ASTER multi-spectral satellite-borne data.*. Eleventh Australasian Remote Sensing and Photogrammetric Conference, Brisbane, 2–5th September, pp. 110–125.

Hewson, R., & Robson, D. (2014). *Applying ASTER and geophysical mapping for mineral exploration in the Wagga Wagga and Cobar areas.* Quarterly Notes, Geological Survey of New South Wales, No 140. Retrieved from http://www.resourcesandenergy.nsw.gov.au/about-us/quarterly-notes

Hewson, R., Robson, D., Mauger, A., Cudahy, T., Thomas, M., & Jones, S. (2015). Using the Geoscience Australia-CSIRO ASTER maps and airborne geophysics to explore Australian geoscience. *Journal of Spatial Science.* doi:10.1080/14498596.2015.979891

Hook, S. J., Vaughan, R. G., Tonooka, H., & Schladow, S. G. (2007). Absolute radiometric in-flight validation of mid infrared and thermal infrared data from ASTER and MODIS on the terra spacecraft using the Lake Tahoe, CA/NV, USA, automated validation site. *IEEE Transactions on Geoscience and Remote Sensing, 45*, 1798–1807. https://doi.org/10.1109/TGRS.2007.894564

Hunt, G. R., & Ashley, R. P. (1979). Spectra of altered rocks in the visible and near infrared'. *Economic Geology, 74*, 1613–1629. https://doi.org/10.2113/gsecongeo.74.7.1613

Iwasaki, A., & Tonooka, H. (2005). Validation of a crosstalk correction algorithm for ASTER/SWIR. *IEEE Transactions on Geoscience and Remote Sensing, 43*, 2747–2751. https://doi.org/10.1109/TGRS.2005.855066

Iwasaki, A., Fujisada, H., Akao, H., Shindou, O., & Akagi, S. (2005). Enhancement of spectral separation performance for ASTER/SWIR. *Proceedings of SPIE–The International Society for Optical Engineering, 4486*, 42–50.

Kahle, A. B. (1987). Surface emittance, temperature, and thermal inertia derived from thermal infrared multispectral scannner (TIMS) for death valley. *Geophysics, 52*, 858–874. https://doi.org/10.1190/1.1442357

Kahle, A. B., & Alley, R. E. (1985). Calculation of thermal interia from day-night measurements separated by day or weeks. *Photogrammetric Engineering and Remote Sensing, 51*, 73–75.

Kruse, F. A. (2000). *The effects of spatial resolution, spectral resolution, and signal-to-noise on geologic mapping using hyperspectral data, Northern Grapevine Mountains, Nevada.* Proceedings 9th JPL Airborne Earth Science workshop (pp. 261–269). JPL Publications.

Kruse, F. A., Baugh, W. M., & Perry, S. L. (2015). *Validation of DigitalGlobe WorldView-3 Earth imaging satellite shortwave infrared bands for mineral mapping,* (Vol. 9, pp. 17).

Kruse, F. A., Lefkoff, A. B., Boardman, J. B., Heidebrecht, K. B., Shapiro, A. T., & Barloon, P. J. (1993). The spectral image processing system SIPS–Interactive visualization and analysis of imaging spectrometer data. *Remote Sensing of Environment, 44*, 145–163. https://doi.org/10.1016/0034-4257(93)90013-N

Lawrie, K. C., Chan, R. A., Gibson, D. L., & de Souza Kovacs, N. (1999). *Alluvial gold potential in buried palaeochannels in the Wyalong district, Lachlan Fold Belt,* (Vol. 30, 5 pp.). New South Wales: AGSO Research Newsletter.

Lyon, R. J. P., & Burns, E. A. (1963). Analysis of rocks and minerals by reflected infrared radiation. *Economic Geology, 58*, 274–284. https://doi.org/10.2113/gsecongeo.58.2.274

Mackey, T., Lawrie, K., Wilkes, P., Munday, T., Kovacs, N., Chan, R., ... Evans, R. (2000). Palaeochannels near West Wyalong, New South Wales: A case study in delineation and modelling using aero magnetics. *Exploration Geophysics, 31*, 1–7. https://doi.org/10.1071/EG00001

Price, J. C. (1977). Thermal inertia mapping: A new view of the Earth. *Journal of Geophysical Research, 82*, 2582–2590. https://doi.org/10.1029/JC082i018p02582

Ren, S. K., Walshe, J. L., Paterson, R. G., Both, R. A., & Andrew, A. (1995). Magmatic and hydrothermal history of the porphyry-style deposits of the Ardlethan tin field, New South Wales, Australia. *Economic Geology, 90*, 1620–1645. https://doi.org/10.2113/gsecongeo.90.6.1620

Rowan, L. C., & Mars, J. C. (2003). Lithologic mapping in the Mountain Pass, California area using advanced spaceborne thermal emission and reflection radiometer (ASTER) data. *Remote Sensing of Environment, 84*, 350–366. https://doi.org/10.1016/S0034-4257(02)00127-X

Russell, R., & Lappi, D. W. (1988, February 14–21). *NOAA satellite thermal interpretation of ATP 354P, western Queensland,* Proceedings ASEG/SEG conference, Adelaide, pp. 141–147.

Sabins, F. F. (1997). *Remote sensing–Principles and interpretation* (3rd ed., p. 432). Long Grove, IL: Waveland Press.

Sandau, R., Braunecker, B., Driescher, H., Eckardt, A., Hilbert, S., Hutton, J., Kirchhofer, W., ... Wicki, S. (2000). Design principles of the LH systems ADS40 airborne digital sensor. *International Archives of Photogrammetry and Remote Sensing, XXXIII*, Part B1, 258–265.

Scott, K. M., & Rampe, M. (1984). Integrated mineralogical and geochemical exploration for tin in the bygoo region of the ardlethan tin field, Southern N.S.W., Australia. *Journal of Geochemical Exploration, 20*, 337–354. https://doi.org/10.1016/0375-6742(84)90075-X

Vincent, R. K., Rowan, L. C., Gillespie, R. E., & Knapp, C. (1975). Thermal-infrared spectra and chemical analyses of twenty-six igneous rock samples. *Remote Sensing of Environment, 4*, 199–209. https://doi.org/10.1016/0034-4257(75)90016-4

Vincent, R. K., & Thomson, F. (1972). Spectral compositional imaging of silicate rocks. *Journal of Geophysical Research, 77*, 2465–2472. https://doi.org/10.1029/JB077i014p02465

Warren, A. Y. F., Gilligan, L. B., & Raphael, N. M. (1996). *Cootamundra 1:2,50,000 Geological Sheet SI/55-11* (2nd ed.). Sydney: Geological Survey of New South Wales.

Watson, K. (1975). Geologic applications of thermal infrared images. *Proceedings of the IEEE, 63*, 128–137. https://doi.org/10.1109/PROC.1975.9712

Wynn, D. W. (1977). *Narrandera 1:2,50,000 Geological Sheet SI/55-10* (2nd ed.). Sydney: Geological Survey of New South Wales.

Yamaguchi, Y., Fujisada, H., Kahle, A., Tsu, H., Kato, M., Watanabe, H., ... Kudoh, M., (2001, July 9–13), *ASTER instrument performance, operation status, and application to earth sciences.* Proceedings IEEE 2001 International geoscience and remote sensing symposium.

3

Geospatial approach to study the spatial distribution of major soil nutrients in the Northern region of Ghana

Mary Antwi[1]*, Alfred Allan Duker[2], Mathias Fosu[3] and Robert Clement Abaidoo[4,5]

*Corresponding author: Mary Antwi, Department of Crop and Soil Sciences, Kwame Nkrumah University of Science and Technology, Kumasi, Ghana
E-mails: martwi2007@yahoo.com, maryamoah1982@gmail.com
Reviewing editor: Saied Pirasteh, University of Waterloo, Canada

Abstract: Spatial distribution of soil nutrients is not normally considered for smallholder farms in Ghana resulting in blanket fertilizer application which leads to low efficiencies of some applied nutrients. This study focuses on applying geospatial analyses to map 120 maize farms in 16 districts of the Northern region of Ghana to identify nutrient distribution. Soil samples were taken from these 120 locations and analysed for contents of nitrogen (N), phosphorus (P) and potassium (K). Spatial models of the contents were generated through geostatistical analysis to map the status of N, P and K nutrients across the locations. Study results indicated that proportion of area deficient in N is 97%, P is 72% and K is 12%. Distribution pattern for N and K nutrients were clusters of low or high contents at specific locations; and that of P was random. Outcome of this study could enhance site-specific nutrient recommendation in Ghana.

Subjects: Agriculture; Soil Sciences; Statistics; Technology

Keywords: smallholder farmers; geospatial analysis; spatial distribution; major soil nutrients; soil fertility

1. Introduction
Agriculture is the main economic activity in the Northern region of Ghana. According to statistics provided by FAO (AQUASTAT, 2005), majority of the population in the region are smallholder farmers

ABOUT THE AUTHORS

The author's key research activities are focused on the use of Geographical Information Systems to study the spatial pattern within natural resources. These activities include modelling and mapping underlying spatial features and factors that contribute to the continuous degradation of these natural resources in order to make decisions that could control and monitor the resources. The author's research interest in the presented paper, therefore, falls in line with these activities and presents the models and maps of major soil nutrients that will aid appropriate fertilizer application at different locations according to their initial contents. The paper relates to the broader call on research to improve soil fertility and control the continuous soil degradation in Ghana and Sub-Saharan Africa due to the inappropriate fertilizer application by smallholder farmers resulting in low crop yields which also contributes to food insecurity in the region.

PUBLIC INTEREST STATEMENT

The article describes the pattern of distribution of soil nitrogen (N), phosphorus (P) and potassium (K) nutrients contents in 16 districts of the Northern region of Ghana. Although the region is noted for producing most of the major food crops in Ghana, this study showed that the soils in most of the districts have low levels of N, P and K nutrients. The article, therefore, provides maps of distribution of these major soil nutrients contents at different locations after assessing the extent of their levels in the study districts. The generated maps showed locations where the nutrients levels are very low, low or adequate to support the production of maize crops in the study area. This study is to aid decision-making during application of N, P and K fertilizers to the soil so that high efficiencies of applied fertilizers would be attained to give desired crop yields.

and about 80% of the land area is used for cropping. The soils, however, in these land areas of Sub-Saharan Africa are poor, especially in nutrient levels (Wairegi, 2011). Improper fertilizer rates application (Martey et al., 2014) and poor land management (Nkonya, 2004) contribute to the low contents of nutrients in these soils. This had resulted to persistently low crop yields. To increase the productivity of these soils for enhanced crop yields and increase the income of smallholder farmers, spatial distribution of the major soil nutrients needs to be mapped and based on the results of the distribution, appropriate fertilizer needs could be recommended for localised intervention. Since soil is not a renewable resource (Haghdar, Malakouti, Bybordi, & Ali, 2012), the concept of soil nutrient content assessment becomes very important for higher agricultural productivity and the economic development of every country.

In addition, adoption and implementation of soil fertility management concepts may vary, it is, therefore, necessary to assess the soil nutrient levels within locations where smallholder farmers are cropping. The mapped results of the soil nutrient assessment could then be used for effective monitoring of changes that might occur between cropping systems' and seasons' overtime. Monitoring of the nutrient levels will enable stakeholders to assess soil fertility improvement or otherwise in such localities. The tedious and costly conventional methods needed to obtain soil nutrient information will also be reduced when nutrient levels are mapped because those conventional methods are no more affordable (Behrens & Scholten, 2006). Accordingly, mapping of the nutrient levels will provide spatial soil nutrients information that can be used as a decision support tool. Hence, developing spatial distribution maps of soil nutrients is important in the "breadbasket" regions of which Northern region of Ghana is one (Adesina, 2009), since it will help refine agricultural management practices, improve sustainable resource use as well as provide a base against which future soil nutrients can be recommended at site-specific locations (Fairhurst, 2012; Reetz & Rund, 2004). Mapping of soil nutrient levels, especially nitrogen (N), phosphorus (P) and potassium (K) would also facilitate proper monitoring and review of recommended farming technologies at locations from time to time. This might also help in the evaluation of the impact of a particular technology at a particular time (e.g. every 10 years) in a particular location depending on assessment of the soil quality (Wang & Gong, 1998). In addition, due to the growing knowledge in precision agriculture (Buick, 1997; Ping, Wang, & Jin, 2009), researchers and decision-makers in soil science would be in a better position to implement location-based technologies, if approximate levels of soil nutrients in specific locations in the region are mapped. This approach will improve soil fertility management results and increase the interest of smallholder farmers to invest more in the agricultural sector.

The aims of this study were therefore (i) to generate appropriate nutrient models and parameters in order to produce a spatial distribution map of major soil nutrient contents across the Northern region of Ghana and (ii) to evaluate their pattern of distribution through spatial modelling of the N, P and K contents. The results of the study would, therefore, reveal the spatial variation and pattern of distribution of N, P and K nutrient contents across the study area and their evaluation would help to make appropriate site-specific nutrient analysis.

2. Materials and methods

2.1. Study area
The study was carried out in 16 out of 22 districts in the Northern region of Ghana (Figure 1), which is one of the regions classified as the "breadbasket" area of Ghana (Adesina, 2009). This is because most of the major food crops in the country are cultivated in this region and they include maize, rice, cowpea and yam. The region covers an area of about 70,384 km² and is the largest of the 10 regions in Ghana. The study districts, however, covered an area of approximately 40,000 km². It lies in a geographical location of latitudes N9° 30′ and N10° 00′ and longitudes W0° 51′ and W1° 00′ with a mean elevation of 149 m above sea level (Getamap, 2006). The mean annual rainfall of the area ranges from 750 to 1,050 mm and the mean temperature is 28°C which can fall as low as 14°C in the night of December/January and rise as high as 40°C during the day in February/March. The region is located in the Guinea savanna agro-ecological zone. Some of the major soils found in the region

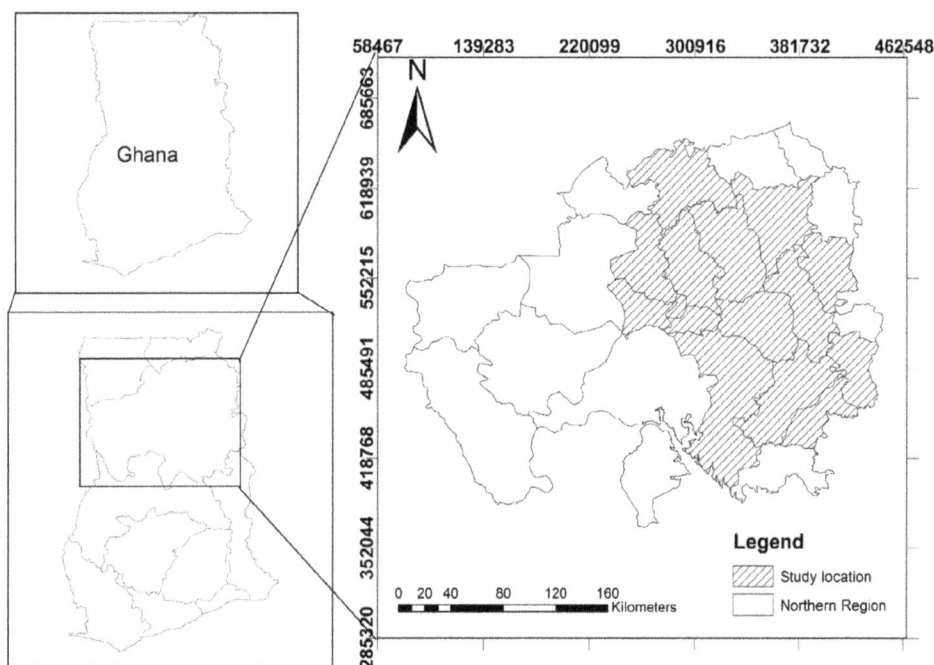

Figure 1. Map of study area.

include Lixisols, Luvisols, Acrisols and Gleysols (Dedzoe, Senayah, & Asiamah, 2001). The study, how-ever, considered maize-cultivated fields since maize crop is one of the most cultivated cereals in the region and is regarded as one of the fundamental crops in the "food security equation" (Sauer, Hardwick, & Wobst, 2006).

2.2. Methods

The fields of eight maize smallholder farmers were chosen from each district that has data on par-ticulars of farmers regarding their farming activities that have been recorded in the database pro-vided by Savanna Agricultural Research Institute (SARI), as well as farmers who do not have their particulars included in the database. Each field was selected from different communities within a district to ensure uniformity of dispersion in the distribution of selected fields. Locations of selected maize fields, based on the objective of this study and for mapping purposes, were obtained with the Garmin GPS. Fifteen districts were found to have complete smallholder farmers' data on maize pro-duction from 2012 to 2014 cropping seasons in the database provided by SARI and theses districts were selected to be used in this study. A total of 120 locations were therefore obtained. A map show-ing the locations of the farms within the districts was produced (Figure 2).

2.2.1. Soil sampling and analysis

Soil samples were taken from each of these 120 locations previously cropped to maize, between May 2013 and March 2014 with the soil auger. Twenty cores of 0–20 cm depth of soil were taken and hand-mixed thoroughly in a bucket to homogenise the sample. A composite sample was then taken from the bulk to represent that location as shown in Figure 2. The soils were analysed for total N, available P and exchangeable K contents; total N was determined by Kjedahl's method, available P by Bray I method and exchangeable K by ammonium acetate extraction (Matula, 2009).

2.2.2. Statistical and geostatistical analysis of nutrient contents

The soil nutrient contents obtained through the laboratory analysis were subjected to Genstat (twelfth edition) statistical descriptive analysis. Some statistical parameters that were observed for the purpose of this study were the mean, standard deviation (SD), coefficient of variation (CV), skew-ness and kurtosis of the data distribution (Table 1).

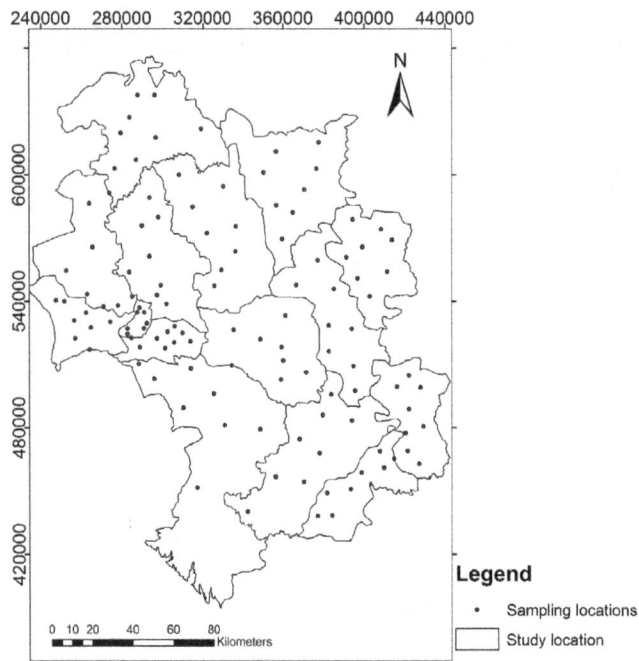

Figure 2. Map of soil sampling locations.

Geostatistical analysis that employs the use of simple point kriging and simulations (ESRI, 2010) was used to model the total N, available P and exchangeable K contents to produce the spatial distribution maps. The simple kriging model is expressed as follows:

$$Z(s) = \mu + \varepsilon(s) \tag{1}$$

where $Z(s)$ = the predicted value at the prediction location

μ = a known constant

$\varepsilon(s)$ = estimated error

Simple kriging assumes normality within the data before modelling. However, in this study, data exploration of the soil nutrient contents revealed that the levels were not normally distributed as shown by the elements of test of normality, skewness and kurtosis values generated by the data statistics (Table 1).

Table 1. Statistical parameters of major soil nutrient contents (n = 120)			
Variable	**Total N (%)**	**P (mg kg⁻¹)**	**K (cmol_c kg⁻¹)**
S.D.	0.02	3.61	0.11
C.V. (%)	29.85	70.56	67.30
Mean	0.07	5.12	0.16
Minimum	0.03	1.24	0.06
Maximum	0.13	15.57	0.73
Skewness	0.11	0.99	2.53
Kurtosis	2.43	2.96	7.71

A data transformation was therefore applied to the original nutrient levels to render the contents normalised before modelling. A logarithmic transformation was applied to the total N and exchangeable K contents and a normal score transformation (ESRI, 2010; Harter, 1961; Royston, 1982) was applied to the available P contents.

The normalised soil nutrient contents were then analysed geostatistically by fitting different semi-variogram models iteratively, to measure the spatial variation within the soil nutrient contents (Liu et al., 2006; Matheron, 1963). In addition, the semi-variogram provided the necessary input parameters for spatial interpolation of kriging (Krige, 1951). ESRI (2010) ArcGIS defines the semi-variogram as follows:

$$\gamma(s_i, s_j) = 1/2 \, var(Z(s_i) - Z(s_j))$$ (2)

where var is the variance, s_i and s_j are two different locations and; Z is the difference in their values.

The semi-variogram model that fitted the soil nutrients phenomena more accurately and provided the least root mean square error (RMSE) compared to others was selected for each of the nutrient levels. The RMSE was evaluated using the formula by Chai and Draxler (2014) as follows:

$$RMSE = \sqrt{\frac{1}{N} \sum_{i=1}^{N} (x_i - \overline{x}_i)^2}$$ (3)

where N is the sample size, x_i is the observed value and $\overline{x}_i$ is the mean value for the observed sample values.

The parameters that were obtained from the semi-variogram were the sill, *nugget* and *range*. The sill represented the amount of variation defined by the spatial correlation structure and it is the value of the semi-variogram at which the model first levels out (given as partial sill plus the nugget). The range is the lag distance from where the model levels off and the nugget is the variability error (measurement error) obtained at shorter distances than the typical sampling interval (Bohling, 2005). The nugget-to-sill ratio was then used to classify the spatial dependence within the nutrient contents.

The prediction model for the soil nutrients was finally validated to measure the accuracy of the prediction map generated showing the distribution of the nutrients. The prediction model that gave the minimum average standard error (i.e. which fits the distribution more accurately) and the least RMSE was considered to be simulated as a measure of uncertainty within the predictions.

The simple kriged maps of the major soil nutrients across the study area were then simulated (generated from 10 realisations from different statistical parameters, i.e. mean, median, SD, upper value, lower value, first and second quartiles, minimum and maximum values and the percentile) to generate stochastic models of the surfaces since simple kriged maps produced smooth surfaces.

Spatial autocorrelation, using the Moran's index (Moran, 1948) was then calculated to assess the significance of the pattern of distribution within the nutrient contents. The calculation was done as follows:

$$I_{Nu} = \frac{n_{(loc)}}{S_0} \frac{\sum_{i=1}^{n} \sum_{j=1}^{n} (Nu_{(content)i} - \overline{Nu}_{(content)}) \times loc_{ij} \times (Nu_{(content)j} - \overline{Nu}_{(content)})}{\sum_{i=1}^{n} (Nu_{(content)i} - \overline{Nu}_{(content)})^2}$$ (4)

where $n_{(loc)}$ is the number of farm locations where soil samples were taken, loc_{ij} is the element in the spatial weights matrix corresponding to the pairs of locations i, j and $Nu_{(content)i}$ and $Nu_{(content)j}$ are nutrient contents in location i, j, respectively; and $\overline{Nu}_{(content)}$ are the mean nutrient content values.

The spatial weight matrix was generated for each of the nutrient contents and denoted by

$$S_0 = \sum_{i=1}^{n} \sum_{j=1}^{n} loc_{ij} \tag{5}$$

where loc_{ij} collectively defined the neighbourhood structure over the entire study location.

The z-score value for each of the nutrient contents and the Moran's index, I obtained from the spatial autocorrelation analysis were then used to specify the pattern of distribution that existed within the soil N, P and K contents. The probability value obtained was then used to assess the significance of the distribution, whether dispersed, clustered or of a random nature.

3. Results and discussion

3.1. Statistical description of major soil nutrient contents in the study area

Description of the untransformed N, P and K contents in the study area showed that their statistical distributions were all positively skewed (with skewness values of N = 0.11, P = 0.99 and K = 2.53; Figure 3). K contents had higher peak (Kurtosis = 7.71) with N and P contents showing near normal peaks (Kurtosis for N = 2.43; P = 2.96). For a standard normal distribution, skewness should be equal to zero and kurtosis equals to three as reported by Jondeau and Rockinger (2003). After data transformations, the N, P and K contents followed a near normal distribution which rendered the data values appropriate for modelling.

N and K contents followed log-normal distribution with positive skewness (see Table 1), an indication that large proportions of the study location have low to moderate concentrations (found within the range of 0.05–0.1% for N and 0.1 and 0.25 $cmol_c$ kg^{-1} for K; Table 1) whether clustered or random.

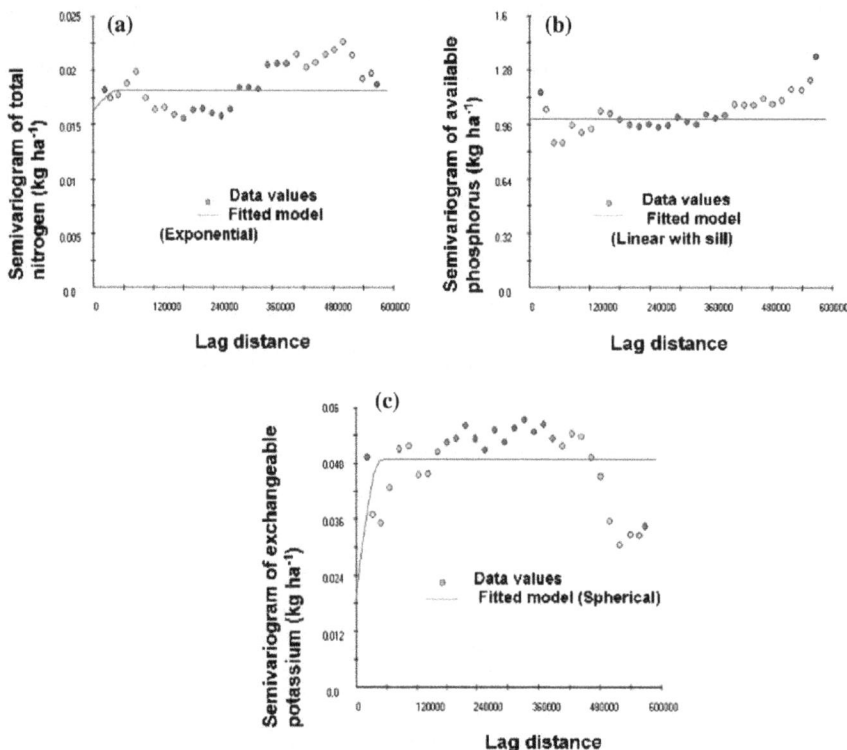

Figure 3. Fitted semi-variograms illustrating the strength of statistical correlation between major soil nutrients; (a) Nitrogen, (b) Phosphorus, and (c) Potassium in selected districts of the Northern region of Ghana.

Few locations recorded relatively high contents (i.e. N being more than 0.1% and K more than 0.25 cmol$_c$ kg^{-1}) as reported by Wopereis, Defoer, Idinoba, Diack, and Dugué (2009) (see Table 2).

The CV determined for the nutrient concentrations showed that N contents had high variations within its distribution (CV value 29.85%; see Table 1), which signifies relatively low dispersion across the districts; those for P and K concentrations were of very high values (CV values of 70.56 and 67.30%, respectively) indicating more dispersion in their distributions. High CV value implies that the data distribution is more variable (dispersed) and, hence, less stable and less uniform (nCalculators, 2013). However, the contents normalised after applying the data transformations.

The near normal distribution of K after log-transformation might be due to the fact that the difference between the minimum and maximum contents were not so large (0.06 and 0.73 cmol$_c$ kg^{-1}, respectively); as compared to that of P contents (Table 1). The fact that some locations recorded P contents as low as 1.24 mg kg^{-1} and others, as high as 15.57 mg kg^{-1} might account for the reason why P contents followed a normal score distribution.

The locations that recorded relatively high p-values were those locations where the smallholder farmers were incorporating about 3 t ha^{-1} of only cattle manure or including NPK (15:15:15) fertilizers plus sulphate of ammonia to the soil (according to the data provided by SARI and interviews with farmers). According to Zhang, Johnson, and Fram (2002), available P builds up in soil when animal manure is applied at a high rate to meet nitrogen requirements and that could have contributed to the high P contents in those locations. The differences in the spatial distribution of the soil nutrient concentrations across the region may thus be attributed to differences in nutrient management practices (Tsirulev, 2010), differences in soil forming processes, inherent heterogeneity in parent material at the different locations, as well as land use pattern and amount of fertilizer used (Liu et al., 2006) by the smallholder farmers. The distribution of the major nutrients confirms the assertion that spatial variation of soil nutrients exist even in neighbouring fields as has been previously reported by Goovaerts (1998), van der Zaag (2010) and Voortman, Brouwer, and Albersen (2002).

When the nutrient contents in the study area were compared with the soil fertility status table (Table 4) produced by Wopereis et al. (2009), only 3% of the locations had N contents within average range for maize production leaving about 97% of the locations having N contents in the soil below recommended average. 28% of the locations studied had levels of P within average contents, whilst 72% had P contents below average. Twelve per cent of the located areas had good K concentrations, 61% were within average, whilst 27% had K contents below average. The low to moderate contents might have resulted from continuous cropping of maize on the same piece of farmland (as confirmed by the farmers to be their practice), low rates of nutrient fertilizers applied at such locations (Mwangi, 1996), soil nutrient loss through soil erosion (Barrows & Kilmer, 1963) and export of nutrients through harvested produce (including straw and stover collection) from the farm (Doran, Wilhelm, & Power, 1984).

3.1. Models of the spatial dependence of major soil nutrients

The semi-variogram models that were used to derive parameters needed to explain the spatial dependence of the soil nutrient contents are presented in Figure 3. These models were obtained based on the semi-variogram model that presented the least RMSE as a measure of uncertainty as shown in Table 3.

Table 2. Nutrient concentration levels indicating soil fertility status for maize production at a depth of 0–20 cm (Wopereis et al., 2009)			
Nutrient level	**N (%)**	**P (mg kg^{-1})**	**K (cmol$_c$ kg^{-1})**
Good	>0.1	>25	>0.25
Adequate	–	6–25	0.10–0.25
Low	0.05–0.1	3–6	0.05–0.10
Very low	<0.05	<3	<0.05

In general, variables of nutrient contents that have a nugget-to-sill ratio less than 0.25 are regarded to have strong spatial dependence within them (i.e. spatial relationship that exists in variable pattern) (Cambardella et al., 1994; Liu et al., 2006). The spatial dependence is considered moderate if the ratio is between 0.25 and 0.75 and weak if it more than 0.75. The nugget-to-sill ratio obtained from the semi-variogram model (Figure 3) for N, P and K contents in the study were 0.64, 0.39 and 0.62, respectively, an indication that their spatial dependencies were moderate. The moderate spatial dependencies within the nutrient contents imply that the degree of association between the variables at different locations may increase as the distances become close to each. This suggests that there could be a possible continuity of the N, P and K variables exhibiting similarities in their values at shorter distances (less than 50 km as shown by the range distance in this study; Table 3). Smallholder farmers located within shorter distances are likely to adopt similar fertilizer management strategies regardless of their soil nutrient variations, which might affect N, P and K contents in the soils in a similar pattern (Adler, Raff, & Lauenroth, 2001). As distances increase, fertilizer management strategies may differ and the dependencies could become weaker or stronger depending on impact of the management (Jonsson & Moen, 1998). Therefore, as previously reported by Luo, Ding, Mi, Yu, and Wu (2009), Pringle, Doak, Brody, Jocqué, and Palmer (2010) soil fertility management should be consistent within patterns of spatial distribution of nutrient contents in the soil in order to manage the considerable variation of the nutrient contents in the study area.

Nitrogen and K nutrient contents recorded in the study area showed a positive low nugget, an indication that sampling error, random and other inherent variations that existed in the variables (Bohling, 2005; Clark, 2010; Liu et al., 2006) were minimal. Phosphorus contents, however, showed a high nugget effect indicating random and inherent variations within the variables. The considerable range of variations within the N, P and K contents might be caused by effects of variable farm level soil fertility management (Tittonell, Vanlauwe, Leffelaar, Rowe, & Giller, 2005; Trangmar, Yost, & Uehara, 1985) across considerable distances from the locations.

The prediction uncertainty generated by the cross-validation of the model were 0.02 kg ha^{-1} for N, 0.98 kg ha^{-1} for P and 0.11 kg ha^{-1} for K (Table 4). These values were less than 1 and hence considered appropriate for the model. The obtained root mean square standardised were also close to 1 for the N, P and K content variables suggesting that none of the variables were under-estimating or over-estimating the predictions as reported by (Hawkins & Sutton, 2011).

Table 3. Parameters for variogram model for major soil nutrients (NPK)

Major soil nutrient	Model	RMSE*	Nugget	Partial sill	Range (m)
Total N (kg ha^{-1})	Spherical	0.00213	–[a]	–	–
	Exponential	0.00211	0.0159	0.009	50000
	Gaussian	0.00212	–	–	–
	Linear with sill	0.00212	–	–	–
Available P (kg ha^{-1})	Spherical	0.1120	–	–	–
	Exponential	0.1186	–	–	–
	Gaussian	0.1179	–	–	–
	Linear with sill	0.1120	0.3666	0.582	50000
Exchangeable K (kg ha^{-1})	Spherical	0.0083	0.0316	0.019	50000
	Exponential	0.0084	–	–	–
	Gaussian	0.0084	–	–	–
	Linear with sill	0.0084	–	–	–

*RMSE (root mean square error).

[a]Values that were not considered in the model.

Table 4. Measure of uncertainties in the prediction estimates of N, P and K contents in 15 districts of the Northern region of Ghana		
Transformed major soil nutrient	Average standard error	Root mean square standardised
Total nitrogen (kg ha⁻¹)	0.02	0.97
Available phosphorus (kg ha⁻¹)	0.98	0.97
Exchangeable potassium (kg ha⁻¹)	0.11	0.99

3.2. Spatial distribution and autocorrelation of major soil nutrients

The simulated maps from the mean values of the nutrient contents (generated from 10 realisations from different statistical parameters) are presented in Figures 4a–4c. The means were presented because according to simulation concepts by ESRI (2010), the means do not change over the spatial domain of the data. In addition, the mean has a Gaussian distribution around the true value, as stated by the central limit theorem (Engblom, Ferm, Hellander, & Lötstedt, 2009) and will therefore provide a better representation of the distribution.

The nutrients contents ranged from very low to adequate levels for maize cultivation in the study area. The differences in the variation within the distribution might be attributed to factors such the differences in elevation topography of the study area (McKenzie, 2013), soil pH that might influence nutrient levels (Wang, Bai, Huang, Deng, & Xiao, 2011) as well as different fertilizer application strategies (Bationo, Waswa, Okeyo, Maina, & Kihara, 2011; Zingore, Murwira, Delve, & Giller, 2007) as practiced by smallholder farmers in the different districts.

The spatial autocorrelation test that was done to test the significance of the distribution of the soil major nutrient contents are presented in Table 5. The hypothesis for the pattern analysis was that the nutrients levels across the study area were randomly distributed. In the theory of random patterns described by ESRI (2010), when p-value is very small (in this study $p < 0.05$) and z-value is either

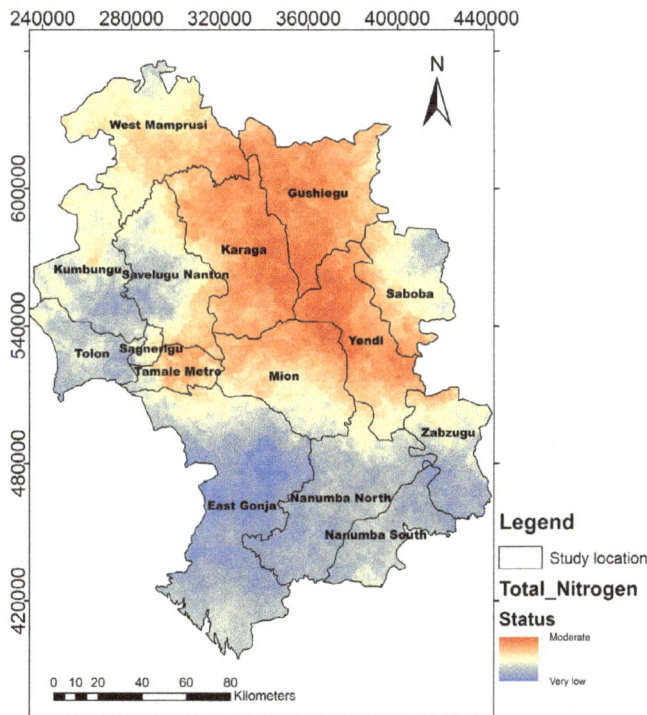

Figure 4a. Spatial distribution of total soil nitrogen contents in 16 districts within the Northern region of Ghana.

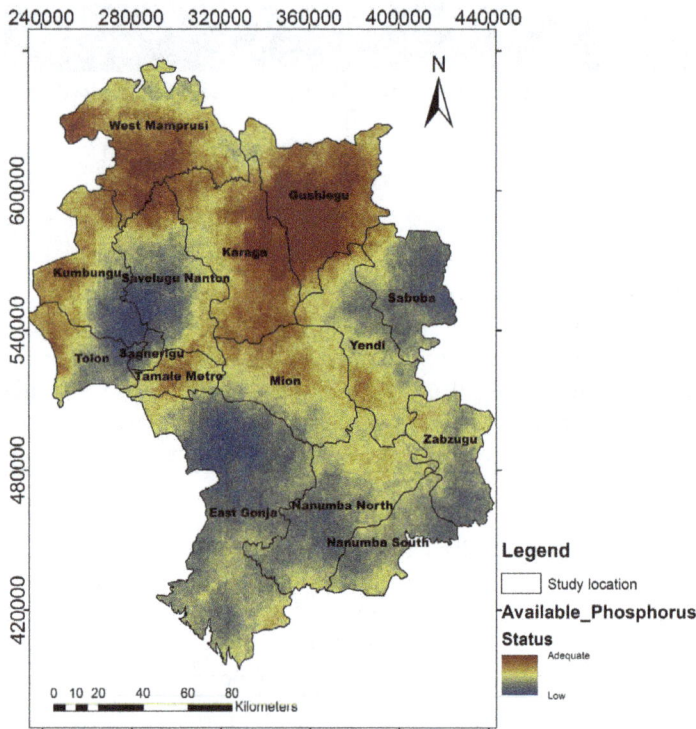

Figure 4b. Spatial distribution of available soil phosphorus contents in 16 districts within the Northern region of Ghana.

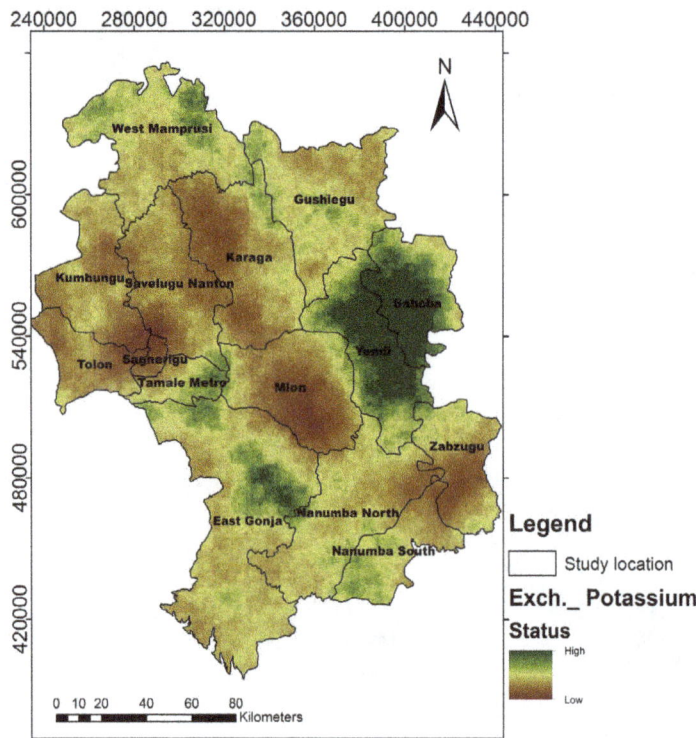

Figure 4c. Spatial distribution of exchangeable soil potassium contents in 16 districts within the Northern region of Ghana.

Table 5. Test of significance of pattern analysis for soil nutrient concentration; ($p < 0.05$) and ($1.96 < z < -1.96$)			
	Nitrogen (N)	**Phosphorus (P)**	**Potassium (K)**
Moran's Index	0.28	0.04	0.27
Expected Index	−0.01	−0.01	−0.01
Variance	0.002	0.002	0.002
z-score	6.73	1.18	6.70
p-value	0.0003	0.24	0.0001

very high or very low ($1.96 < z < -1.96$), the spatial pattern is not likely to reflect a random form of distribution. In addition, a negative Moran's I index value indicates that the data are dispersed and a positive value indicates a tendency of clustering (clusters of high values only or low values only) at particular locations (Anselin, 1996). Test of significance for values returned by the analysis of the major soil nutrients indicated that N and K have clustered distributions in the study area (Table 5); with low levels clustered at one location and high levels at the other. On the other hand, the pattern of distribution of P did not appear to be significantly different from a random distribution.

Management strategies towards soil N, P and K nutrients enhancement could be implemented in the districts using the spatial distribution maps (Figures 4a–4c) as guide. Soil spatial distribution maps, therefore, provide a quick reference and reliable means by which variability within soil nutrients can be assessed to make decisions on fertilizer allocation at specific locations (Schnug, Panten, & Haneklaus, 1998).

4. Conclusion

Geospatial analysis of soil nutrient contents in the study area has proved to be essential in identifying locations in the Northern region of Ghana, where N, P and K levels are relatively low, moderate and high, respectively. Large proportions of the area recorded nutrient levels below average (N = 97%, P = 72% and K = 12%) which indicated that the study area has low nutrient levels. Models of the distribution maps suggest that N and K nutrients levels were clustered spatially and the distribution pattern of P in the study area was a random one. The soil nutrient information on low levels of N, P and K contents in the study area could be improved using the spatial distribution maps as a guide for fertilizer allocation and management taking into account the pattern within the distribution. The spatial distribution maps generated through this study therefore provided foreknowledge of the N, P and K nutrients status in the districts which could be used by research scientists as bases for fertilizer recommendations. When these considerations are made, proper site-specific nutrient recommendations could be promoted in order to increase soil nutrient fertility in the region.

Acknowledgement
Support: Secondary data were obtained from the Savanna Agricultural research institute (SARI).

Funding
The work was supported by the Alliance for Green Revolution in Africa (AGRA) project.

Author details
Mary Antwi[1]
E-mails: martwi2007@yahoo.com, maryamoah1982@gmail.com
ORCID ID: http://orcid.org/0000-0002-6226-5464
Alfred Allan Duker[2]
E-mails: duker@itc.nl, duker@alumni.itc.nl

Mathias Fosu[3]
E-mail: mathiasfosu@yahoo.co.uk
Robert Clement Abaidoo[4,5]
E-mail: abaidoorc@yahoo.com

[1] Department of Crop and Soil Sciences, Kwame Nkrumah University of Science and Technology, Kumasi, Ghana.

[2] Department of Geomatic Engineering, Kwame Nkrumah University of Science and Technology, Kumasi, Ghana.

[3] Council for Scientific and Industrial Research, Savanna Agriculture Research Institute, Tamale, Ghana.

[4] College of Agriculture and Natural Resources, Kwame Nkrumah University of Science and Technology, Kumasi, Ghana.

[5] International Institute of Tropical Agriculture, Ibadan, Nigeria.

References

Adesina, A. (2009). *Taking advantage of science and partnerships to unlock growth in Africa's breadbaskets*. (A. Adesina, Vice President, Speaker). Wageningen, The Netherlands: Alliance for a Green Revolution in Africa (AGRA) at the Science Forum. Retrieved November 10, 2015, from http://www.scienceforum2009.nl/Portals/11/Adesina-pres.pdf

Adler, P., Raff, D., & Lauenroth, W. (2001). The effect of grazing on the spatial heterogeneity of vegetation. *Oecologia, 128*, 465–479.

Anselin, L. (1996). The Moran scatterplot as an ESDA tool to assess local instability in spatial association. *Spatial Analytical Perspectives on GIS, 111*, 111–125.

AQUASTAT. (2005). *Food and agriculture organization's information system on water and agriculture* Retrieved June 7, 2013, from http://www.fao.org/nr/water/aquastat/countries_regions/GHA/index.stm

Barrows, H. L., & Kilmer, V. J. (1963). Plant nutrient losses from soils by water erosion. *Advanced Agronomy, 15*, 303–316. http://dx.doi.org/10.1016/S0065-2113(08)60401-0

Bationo, A., Waswa, B., Okeyo, J. M., Maina, F., & Kihara, J. M. (2011). *Innovations as key to the green revolution in Africa: Exploring the scientific facts*. Berlin: Springer Science & Business Media. http://dx.doi.org/10.1007/978-90-481-2543-2

Behrens, T., & Scholten, T. (2006). Digital soil mapping in Germany—A review. *Journal of Plant Nutrition and Soil Science, 169*, 434–443. http://dx.doi.org/10.1002/(ISSN)1522-2624

Bohling, G. (2005). Introduction to geostatistics and variogram analysis. *Kansas Geological Survey*, 1–20. Retrieved from http://people.ku.edu/~gbohling/cpe940

Buick, R. D. (1997). *Precision Agriculture: An integration of information technologies with farming*. Paper presented at the 50th N.Z. Plant Protection Conference, New Zealand.

Cambardella, C. A., Moorman, T. B., Novak, J. M., Parkin, T. B., Karlen, D. L., Turco, R. F., ... Konopka, A. E. (1994). Field-scale variability of soil properties in Central Iowa soils. *Soil Science Society of America Journal, 58*, 1501–1511. http://dx.doi.org/10.2136/sssaj1994.03615995005800050033x

Chai, T., & Draxler, R. R. (2014). Root mean square error (RMSE) or mean absolute error (MAE)?–Arguments against avoiding RMSE in the literature. *Geoscientific Model Development, 7*, 1247–1250. http://dx.doi.org/10.5194/gmd-7-1247-2014

Clark, I. (2010). Statistics or geostatistics? Sampling error or nugget effect? *The Journal of the South African Institute of Mining and Metallurgy, 110*, 307–312.

Dedzoe, C., Senayah, J., & Asiamah, R. (2001). Suitable agro-ecologies for cashew (Anacardium occidetale L) production in Ghana. *West African Journal of Applied Ecology, 2*, 103–115.

Doran, J. W., Wilhelm, W. W., & Power, J. F. (1984). Crop residue removal and soil productivity with no-till corn, sorghum, and soybean1. *Soil Science Society of America Journal, 48*, 640–645. doi:10.2136/sssaj1984.03615995004800030034x

Engblom, S., Ferm, L., Hellander, A., & Lötstedt, P. (2009). Simulation of stochastic reaction-diffusion processes on unstructured meshes. *SIAM Journal on Scientific Computing, 31*, 1774–1797. http://dx.doi.org/10.1137/080721388

ESRI. (2010). *The Principles of Geostatistical Analysis (3)*. Retrieved January 20, 2015, from *http://maps.unomaha.edu/Peterson/gisII/ESRImanuals/Ch3_Principles.pdf*

Fairhurst, T. (2012). *Handbook for integrated soil fertility management*. Nairobi: Africa Soil Health Consortium.

Getamap. (2006). *Northern Region/Ghana*. Retrieved October 22, 2013, from www.getamap.net

Goovaerts, P. (1998). Geostatistical tools for characterizing the spatial variability of microbiological and physico-chemical soil properties. *Biology and Fertility of Soils, 27*, 315–334. doi:10.1007/s003740050439

Haghdar, A., Malakouti, J. Mohammad, Bybordi, A., & Ali, K. (2012). Study on spatial variation of some chemical characteristics of dominant soil series using geostatistics in Iran: Case study of Heris region. *Journal of Food, Agriculture & Environment, 10*, 977–982.

Harter, H. L. (1961). Expected values of normal order statistics. *Biometrika, 48*, 151–165. doi:10.2307/2333139

Hawkins, E., & Sutton, R. (2011). The potential to narrow uncertainty in projections of regional precipitation change. *Climate Dynamics, 37*, 407–418. doi:10.1007/s00382-010-0810-6

Jondeau, E., & Rockinger, M. (2003). Conditional volatility, skewness, and kurtosis: Existence, persistence, and comovements. *Journal of Economic Dynamics and Control, 27*, 1699–1737. doi:10.1016/S0165-1889(02)00079-9

Jonsson, B. G., & Moen, J. (1998). Patterns in species associations in plant communities: The importance of scale. *Journal of Vegetation Science*, 327–332. http://dx.doi.org/10.2307/3237097

Krige, D. G. (1951). A statistical approach to some basic mine valuation problems on the Witwatersrand. *Journal of Chemical and Metallurgical Mining Society of South Africa, 52*, 119–139.

Liu, D., Wang, Z., Zhang, B., Song, K., Li, X., Li, J., ... Duan, H. (2006). Spatial distribution of soil organic carbon and analysis of related factors in croplands of the black soil region, Northeast China. *Agriculture, Ecosystems and Environment, 113*, 73–81. http://dx.doi.org/10.1016/j.agee.2005.09.006

Luo, Z., Ding, B., Mi, X., Yu, J., & Wu, Y. (2009). Distribution patterns of tree species in an evergreen broadleaved forest in eastern China. *Frontiers of Biology in China, 4*, 531–538. http://dx.doi.org/10.1007/s11515-009-0043-4

Martey, E., Wiredu, A. N., Etwire, P. M., Fosu, M., Buah, S. S. J., Bidzakin, J., ... Kusi, F. (2014). Fertilizer adoption and use intensity among smallholder farmers in Northern Ghana: A case study of the AGRA soil health project. *Sustainable Agriculture Research, 3*, 24–36. doi:10.5539/sar.v3n1p24.

Matheron, G. (1963). Principles of geostatistics. *Economics Geology, 58*, 1246–1266.

Matula, J. (2009). A relationship between multi-nutrient soil tests (Mehlich 3, ammonium acetate, and water extraction) and bioavailability of nutrients from soils for barley. *Plant Soil Environment, 55*, 173–180.

McKenzie, R. H. (2013). Understanding soil variability to utilize variable rate fertilizer technology. *Government of Alberta, Agriculture and Rural Development*. Retrieved from http://www.farmingsmarter.com/understanding-soil-variability/

Moran, P. A. (1948). The interpretation of statistical maps. *Journal of the Royal Statistical Society. Series B (Methodological), 10*, 243–251.

Mwangi, W. M. (1996). Low use of fertilizers and low productivity in sub-Saharan Africa. *Nutrient Cycling in Agroecosystems, 47*, 135–147. http://dx.doi.org/10.1007/BF01991545

nCalculators. (2013). *Coefficient of variation*. Retrieved Novembr 07, 2013, from http://ncalculators.com/math-worksheets/coefficient-variation-example.htm

Nkonya, E. (2004). *Strategies for sustainable land management and poverty reduction in Uganda* (Vol. 133, pp. 1–3). Washington, DC: International Food Policy Research Institute.

Ping, H. E., Wang, H., & Jin, J. (2009, November 5–7). *GIS based soil fertility mapping for SSNM at village level in China*. This presentation was made at the IPI-OUAT-IPNI International Symposium, OUAT, Bhubaneswar. *The role and benefits of potassium in improving nutrient management for food production, quality and reduced environmental damage*. Beijing: International Plant Nutrition Institute.

Pringle, R. M., Doak, D. F., Brody, A. K., Jocqué, R., & Palmer, T. M. (2010). Spatial pattern enhances ecosystem functioning in an African Savanna. *PLoS Biology, 8*, e1000377. doi:10.1371/journal.pbio.1000377

Reetz, H. F., & Rund, Q. B. (2004). *GIS in nutrient management–A 21st century paradigm shift.* Paper presented at the Proceedings of ESRI International User Conference, 24th, San Diego, CA 1328.

Royston, J. P. (1982). Algorithm AS 177: Expected normal order statistics (exact and approximate). *Journal of the Royal Statistical Society. Series C (Applied Statistics), 31*, 161–165. doi:10.2307/2347982

Sauer, J., Hardwick, T., & Wobst, P. (2006). Alternate soil fertility management options in Malawi: An economic analysis. *Journal of Sustainable Agriculture, 29*, 29–53.

Schnug, E., Panten, K., & Haneklaus, S. (1998). Sampling and nutrient recommendations-the future. *Communications in Soil Science & Plant Analysis, 29*, 1455–1462.

Tittonell, P., Vanlauwe, B., Leffelaar, P., Rowe, E. C., & Giller, K. E. (2005). Exploring diversity in soil fertility management of smallholder farms in western Kenya: I. Heterogeneity at region and farm scale. *Agriculture, Ecosystems & Environment, 110*, 149–165.

Trangmar, B. B., Yost, R. S., & Uehara, G. (1985). Application of geostatistics to spatial studies of soil properties. *Advances in agronomy, 38*, 45–94.

Tsirulev, A. (2010). Spatial variability of soil fertility parameters and efficiency of variable rate fertilizer application in the Trans-Volga Samara region. *Better Crops, 94*, 26–28.

van der Zaag, P. (2010). Viewpoint—Water variability, soil nutrient heterogeneity and market volatility—Why Sub-Saharan Africa's green revolution will be location-specific and knowledge-intensive. *Water Alternatives, 3*, 154–160.

Voortman, R. L., Brouwer, J., & Albersen, P. J. (2002). *Characterization of spatial soil variability and its effect on Millet Yield on Sudano-Sahelian coversands in SW Niger* (S. W. P. Centre for World Food Studies). *Stichting Onderzoek Wereldvoedselvoorziening van de Vrije Universiteit.*

Wairegi, L. (2011, May 25–26). *Framework for decision support tools for integrated soil fertility management in sub-saharan Africa (Draft).* Paper presented at the African Soil Health Consortium inaugral workshop.

Wang, Q., Bai, J., Huang, L., Deng, W., & Xiao, R. (2011). Soil nutrient distribution in two typical paddy terrace wetlands along an elevation gradient during the fallow period. *Journal of Mountain Science, 8*, 476–483. http://dx.doi.org/10.1007/s11629-011-1122-y

Wang, X., & Gong, Z. (1998). Assessment and analysis of soil quality changes after eleven years of reclamation in subtropical China. *Geoderma, 81*, 339–355. doi:10.1016/S0016-7061(97)00109-2

Wopereis, M. C. S., Defoer, T., Idinoba, P., Diack, S., & Dugué, M. J. (2009). Integrated soil fertility management. *Participatory learning and action research (PLAR) for integrated rice management (IRM) in Inland Valleys of Sub-Saharan Africa:Technical manual.* Cotonou, Benin: Africa Rice Centre.

Zhang, H., Johnson, G. V., & Fram, M. (2002). *Managing phosphorus from animal manure.* Stillwater: Division of Agricultural Sciences and Natural Resources, Oklahoma State University.

Zingore, S., Murwira, H. K., Delve, R. J., & Giller, K. E. (2007). Influence of nutrient management strategies on variability of soil fertility, crop yields and nutrient balances on smallholder farms in Zimbabwe. *Agriculture, Ecosystems & Environment, 119*, 112–126. doi:10.1016/j.agee.2006.06.019

4

Contour tracing for geographical digital data

Tatsuya Ishige[1]*

*Corresponding author: Tatsuya Ishige, Department of Industrial and System Engineering, Hosei University, Tokyo, Japan
E-mail: tatsuya.ishige0809@gmail.com
Reviewing editor: Louis-Noel Moresi, University of Melbourne, Australia

Abstract: Our purpose is to trace a contour in the form of a polygon. In this research, we use a bicubic spline function for interpolation of the elevation data on a grid covering the area of concern. We construct the polygon as a data consisting of ordered contour points on sides of the grid. The contour enters a cell at an entry point and goes out at an exit point on its sides. The polygon is formed connecting these points. A problem occurs as to which two points should be connected when a cell of the grid has more than three contour points on its sides. As for existing methods of the differential geometry such as discretization using tangential increments, it is difficult to predetermine a suitable step size to arrive at a next contour point correctly if several contour components wind closely to each other within a cell. As a solution, we take an algebraic approach exploiting a simple fact that a bicubic function is viewed as a univariate cubic function with a parameter. From this perspective, we identify the exit point examining the behavior of the real roots of the cubic equation for the contour in terms of the numerical order. Our method enables us to faithfully trace the contour of bicubic spline functions which provide smoother and better fitting curves than bilinear spline functions used by the other authors. Computation time is exhibited in the numerical experiment for an island in Japan.

Subjects: Computer Mathematics; CAD CAE CAM - Computing & Information Technology; Computational Numerical Analysis; Simulation & Modeling; Computer Graphics & Visualization; Computer Science (General); Engineering Mathematics

Keywords: contour trace; polygon; cubic equation; geographical digital data; bicubic spline; Sturm's theorem

ABOUT THE AUTHOR

Tatsuya Ishige is a graduate student and a member of Management Science Laboratory in Department of Industrial and System Engineering, Faculty of Science and Engineering, Hosei University. In addition to this research, the author studies the geographical properties of the Voronoi territories for the convenience stores in Japan and lectured in APORS 2015.

PUBLIC INTEREST STATEMENT

This research paper is intended for researchers and practitioners who may be interested in making a contour map from geographical digital data. The data-set consists of altitudes at vertices of a rectangular grid. We interpolate these data by a bicubic spline function. We then calculate the locations of contour points on the sides of the grid. These points are connected to form a polygon which represents a contour of the altitude.

1. Introduction

We develop a method to trace a contour for the surface interpolated by a bicubic spline function in the form of a polygon. The principal features of this method are:

- We use the elevation data given on a rectangular grid.
- We interpolate the elevation data by a bicubic spline function which provides fitting and smooth approximation with moderate computational load.
- We detect all the contour points on the sides of the grid and connect them to trace a contour.
- In connecting contour points on the sides of a cell, contour points can be linked correctly even if there are more than three points.
- We can obtain the same polygon no matter which point we choose on the contour as a starting point, different from iterative methods using discretization.

We store contours in the form of polygon data which is easy to handle when applying to various engineering needs such as calculation of the volume of a water reservoir.

The algorithm of tracing and drawing contour is sometimes referred to as "contour tracing" and "contouring." The pioneering work is Cottafava and Moli (1969). Their method is to compute contour points on a rectangular grid and connect them together to form a contour. Since they use the intermediate value theorem to detect a contour point, the number of contour points on a side of a cell is always presumed to be no more than one. McLain (1971) uses the weighted least square method to approximate the elevation function. This method cannot express the contour winding intricately since the direction of increment vector is presumed to change at an angle of zero or 45 degrees to the previous one. Lopes and Brodlie (1998) use bilinear splines to interpolate the elevation function. By incorporating one or two contour points in the interior of a cell, they enhance the preciseness of contour tracing for the bilinear hyperbola. Maple (2003) uses the method called marching squares which approximates the contour in a rectangular cell by 16 possible patterns.

Recently, in geographical modeling and analysis, the importance of application using discrete altitude data such as geographic information system is widely recognized. In particular, the Triangular Irregular Network (TIN) expressing a terrain surface in a set of triangles is frequently used. The TIN allows the arbitrarily scattered points. For terrain representation, they use only the key points which are less than those aligned regularly in Digital Elevation Model (DEM). Gold, Charters, and Ramsden (1977) propose the method of contour trace using TIN, and Dobkin, Wilks, Levy, and Thurston (1990) extend it to mappings between higher dimensional spaces. Saux, Thibaud, Li, and Kim (2004) extract topographic properties on scattered elevation data model by TIN. Goto and Shimakawa (2015) propose the iterative method of discretization using gradient vectors of an elevation function which is constructed by interpolation with a bicubic spline function.

The preceding works mentioned above do not faithfully represent contours winding intricately in a rectangular cell. In this case, McCullagh (1981) subdivides a grid cell into a series of subcells enabling contours to be laced smoothly through the cell. In our research, however, we presume a fixed rectangular grid and still maintain the accuracy. The outline of our process is illustrated as follows. The altitude data on a rectangular grid are given in Figure 1. To reproduce the surface, we interpolate the altitude data on the grid by bicubic spline in Figure 2. We then calculate the contour points which are intersections between the rectangular grid and the contour. We finally produce a polygon of the

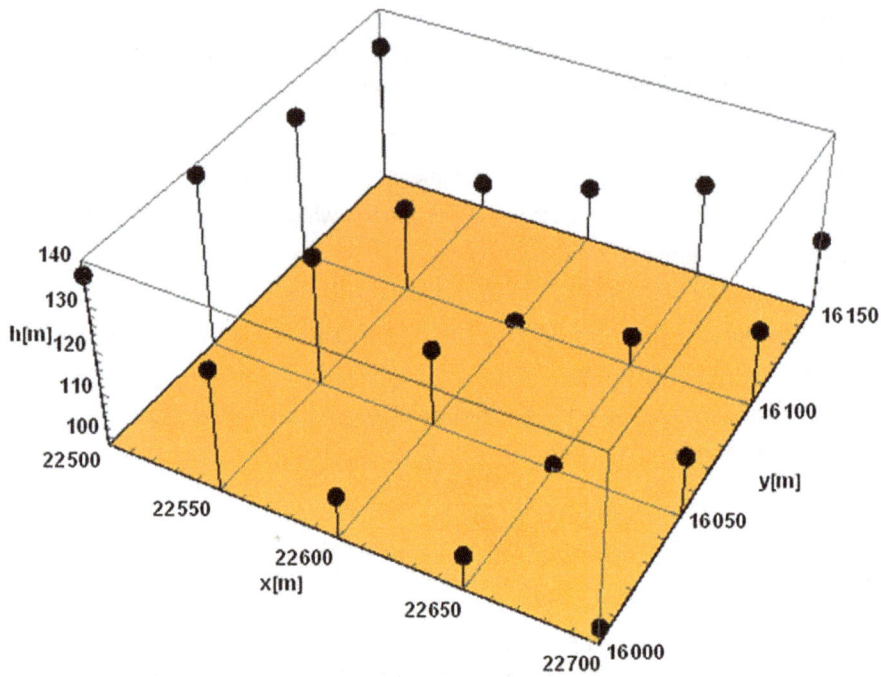

Figure 1. Geographical digital data of altitudes.

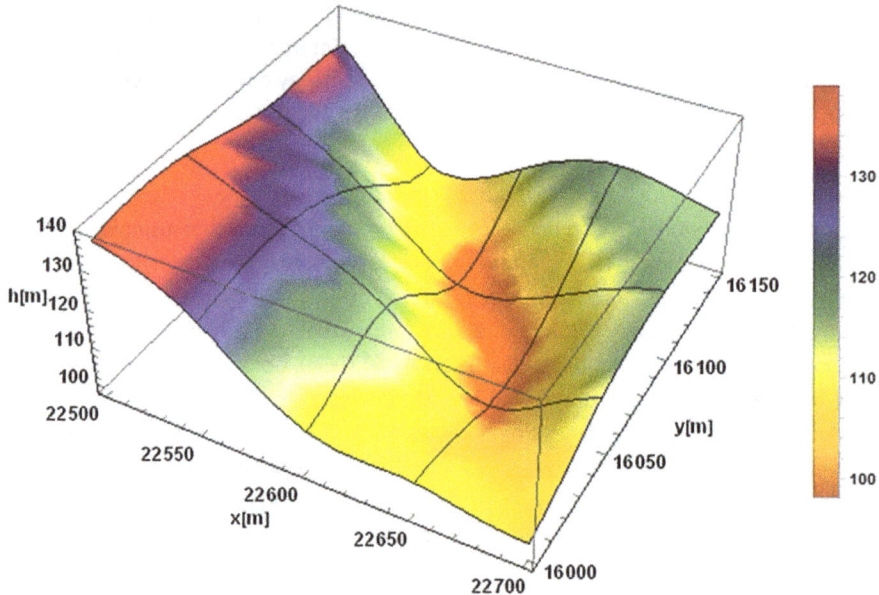

Figure 2. Elevation function interpolated.

contour for an altitude as in Figure 3. Although the process is same as that of Lopes and Brodlie (1998), we adopt bicubic spline functions for interpolating elevation data. Taking full advantage of their accuracy, we achieve representation of complicated contours such as having multiple points on a side of a cell shown in Figure 4 connecting A to B, C to D, and E to F. Our key idea which makes it possible is to connect the calculated contour points using the order of real roots for cubic equations.

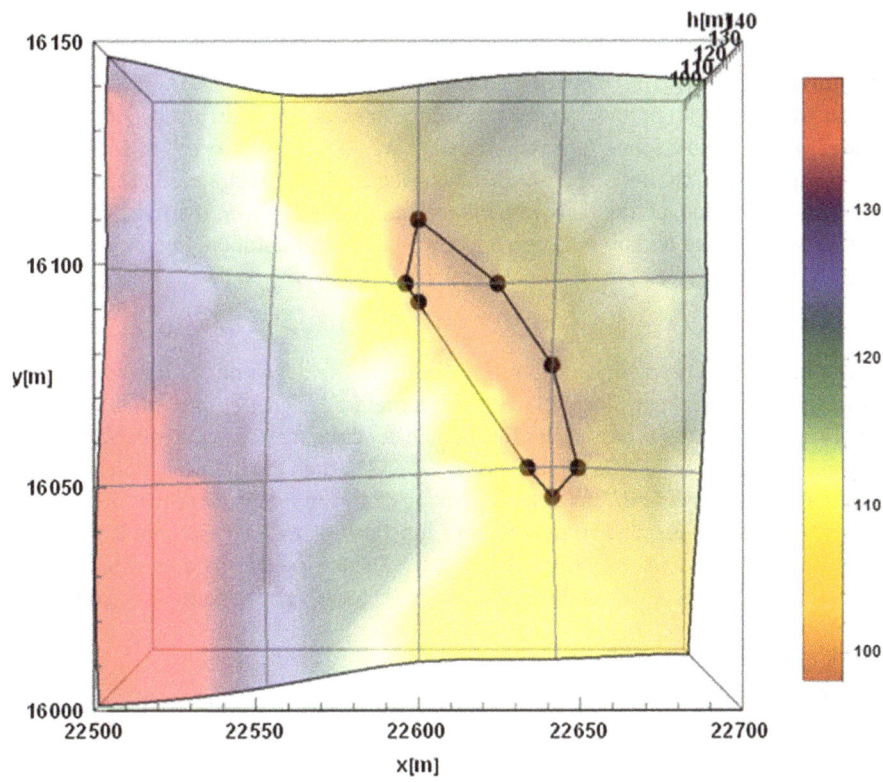

Figure 3. Polygon of the contour.

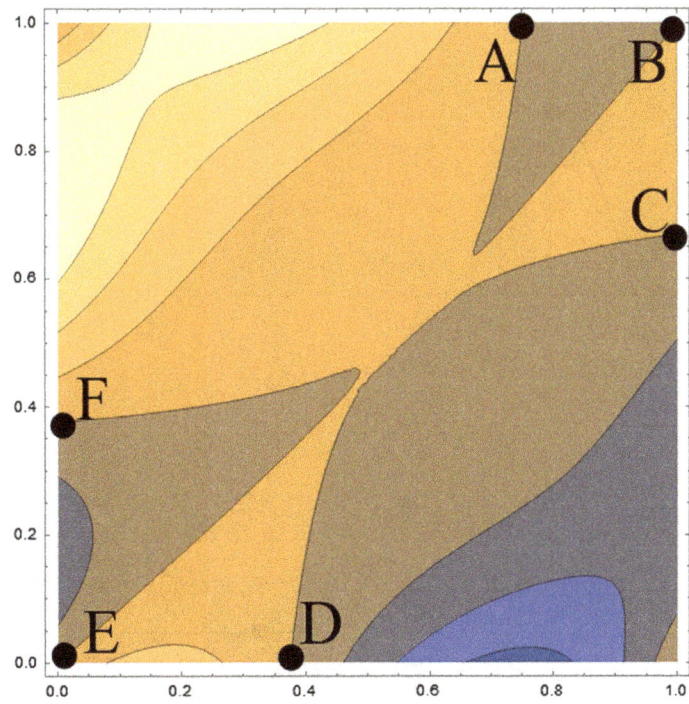

Figure 4. Complicated contours in a cell.

2. Preliminaries

2.1. Elevation function

We construct an elevation function over an $m \times n$ rectangular grid with interval δ by interpolating data of altitudes with a bicubic spline function. Although we suppose that a cell of the grid is a square $\delta \times \delta$ for simplicity, the discussion below can be applied easily to a rectangle $\delta_x \times \delta_y$. There are various types of spline polynomial functions with different degrees. However, the most commonly used are cubic splines because of their precision and efficiency in computation. In particular, cubic B-splines are preferred. A base function of one-dimensional B-spline is given by

$$M_{k,i}(x) = M_k(x; \xi_i, \xi_{i+1}, \cdots, \xi_{i+k}) \quad (i = 1, \cdots, m). \tag{1}$$

For bicubic spline functions, k is set to 4 in expression (1). As for the definition of right-hand side of (1), readers should refer to Piegl and Tiller (2012). The parameters $\xi_i, \xi_{i+1}, ..., \xi_{i+k}$ are called knots and used for controlling the smoothness of the base function. In our case, they coincide with consecutive nodes of the grid. If there are no identical knots in expression (1), then the base function is C^2-class at each knot. We generate base functions of bicubic spline by taking a tensor product of two base functions of one-dimensional B-spline. Thus, the base functions of two-dimensional B-spline are $M_{4,i}(x) M_{4,j}(y)$ $(i = 1, ...,m; j = 1, ..., n)$. Put $(x_i, y_j) = ((i - 1)\delta, (j - 1)\delta)$ for the coordinates of (i, j)-th grid point, set $\xi_{i+l} = x_{i+l}$ $(l = 0, ..., k)$ for $M_{4,i}(x)$, and $\xi_{j+l} = y_{j+l}$ $(l = 0, ..., k)$ for $M_{4,j}(y)$. The form of $M_{4,i}(x)$ is given by

$$M_{4,i}(x) = \sum_{l=-2}^{2} \frac{[|x_{i+l} - x| + (x_{i+l} - x)]^3}{w_i'(x_{i+l})}, \tag{2}$$

where $w_i(x) = \prod_{l=-2}^{2} (x - x_{i+l})$. The function $M_{4,j}(y)$ is given by replacing x with y. The graph of $M_{4,i}(x)$ is illustrated in Figure 5. Taking a suitable linear sum, we obtain a bicubic spline function:

$$h(x,y) = \sum_{i=1}^{m} \sum_{j=1}^{n} c_{ij} M_{4,i}(x) M_{4,j}(y). \tag{3}$$

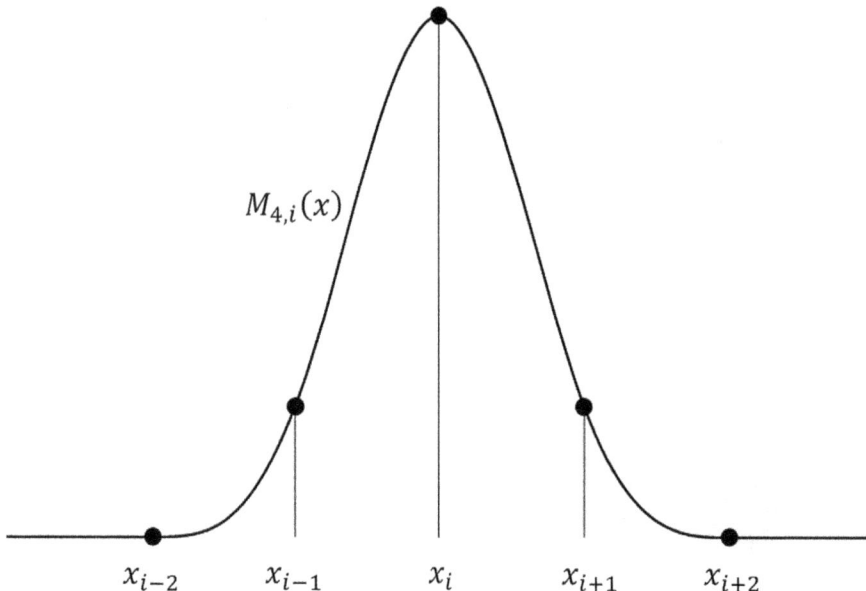

Figure 5. One-dimensional B-spline base function.

2.2. Roots of the cubic equation

The contour of an altitude h_0 is given by the equation $h(x, y) = h_0$ which is a cubic equation in one variable while fixing the other. We use some aspects of cubic equations to draw contours. Thus in this section, we consider a general cubic equation

$$ax^3 + bx^2 + cx + d = 0. \tag{4}$$

Putting

$$x = x' - b/3a, \tag{5}$$

we obtain

$$x'^3 + px' + q = 0, \tag{6}$$

Where

$$p = -1/3(b/a)^2 + c/a,$$
$$q = -2/27(b/a)^3 - bc/3a^2 + d. \tag{7}$$

Let the cubic and quartic roots of 1 with the positive imaginary part be denoted by ω and i, respectively. According to the Cardano's method, we put

$$D = -q^2/4 - p^3/27, \tag{8}$$

$$U = \begin{cases} -q/2 + \sqrt{D}i & (D \geq 0) \\ -q/2 + \sqrt{|D|} & (D < 0) \end{cases},$$

$$V = \begin{cases} \bar{U} & (D \geq 0) \\ -q/2 - \sqrt{|D|} & (D < 0) \end{cases}, \tag{9}$$

$$u = \begin{cases} |U|^{1/3} \exp(1/3(\text{Arg}U)i) & (D \geq 0) \\ U^{1/3} & (D < 0) \end{cases},$$

$$v = \begin{cases} \bar{u} & (D \geq 0) \\ V^{1/3} & (D < 0) \end{cases}. \tag{10}$$

The roots of (6) are given as follows.

$$x_1' = \omega u + \omega^2 v,$$
$$x_2' = \omega^2 u + \omega v, \tag{11}$$
$$x_3' = u + v,$$

and $x_i = x_i' - b/3a$ (i = 1, 2, 3) for Equation (4).

In the case of $D < 0$, the equation has one real root and two imaginary roots.

In the case of $D > 0$, we have $\text{Im}(U) > 0$ and $0 < \text{Arg}(U) < \pi$. Then the ranges of the arguments for u, ωu, and $\omega^2 u$ are $0 < \text{Arg}(u) < \pi/3$, $2\pi/3 < \text{Arg}(\omega u) < \pi$, and $4\pi/3 < \text{Arg}(\omega^2 u) < 5\pi/3$, respectively, as shown in Figure 6. Since they are separated, we obtain three distinct real roots $x_1 < x_2 < x_3$. In this case, we call the suffix i of x_i the order of the root.

In the case of $D = 0$, we have $U = V = -q/2$ from expression (9). If $q < 0$, x_3 is a simple root and x_1, x_2 are a multiple root as illustrated in Figure 7. If $q > 0$, x_1 is a simple root and x_2, x_3 are a multiple root as illustrated in Figure 8.

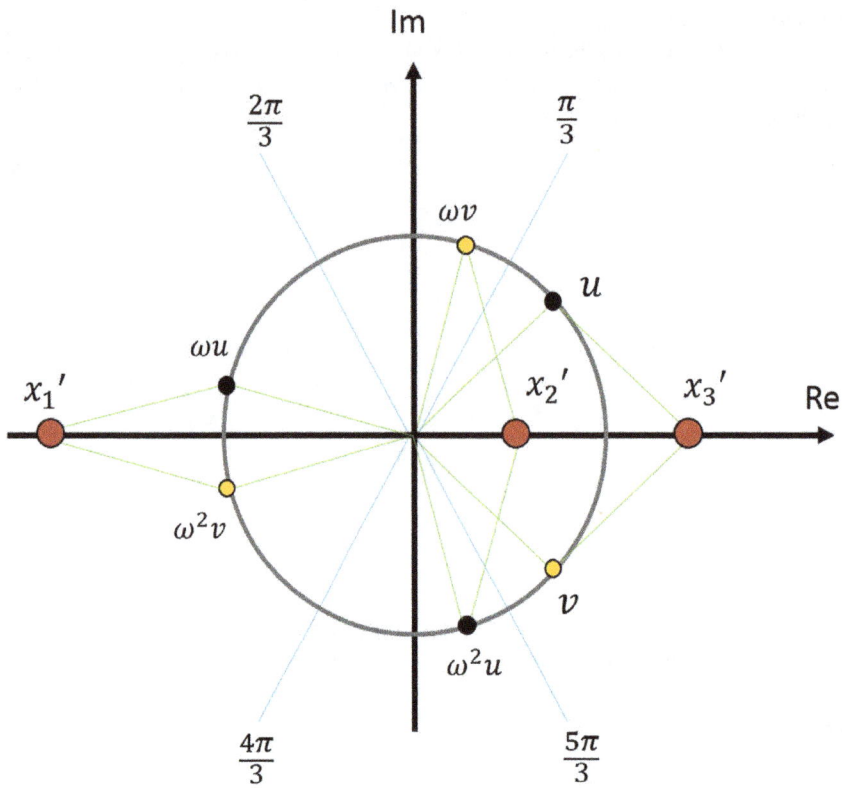

Figure 6. Three real roots (*D* > 0).

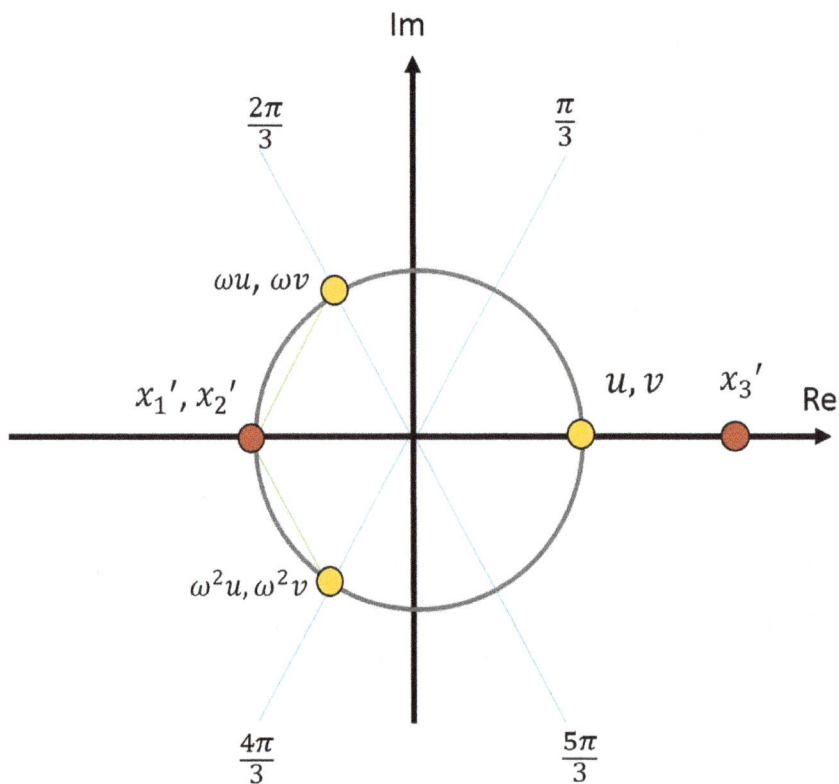

Figure 7. One simple root and one multiple root (*D* = 0, *q* < 0).

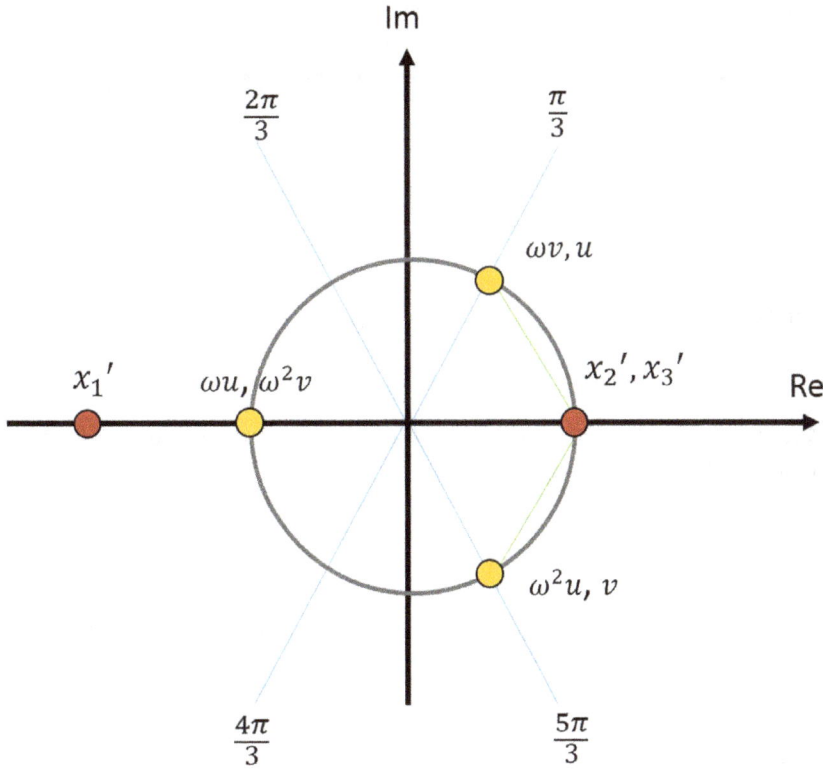

Figure 8. One simple root and one multiple root ($D = 0$, $q > 0$).

2.3. Behavior of the roots

We here generally consider a parameterized cubic equation

$$f(s, t) = a(t)s^3 + b(t)s^2 + c(t)s + d(t) = 0. \tag{12}$$

Each coefficient is a continuous function of t. Solving Equation (12) for s, we obtain the roots $s_1(t)$, $s_2(t)$, $s_3(t)$ corresponding to x_1, x_2, x_3 for (4), respectively. The counterparts of q and D are parameterized as $q(t)$ and $D(t)$. The behavior of the roots is classified as follows.

(i) In the case of $D(t) < 0$, one root is real and the other roots are imaginary.

(ii) In the case of $D(t) > 0$, we obtain three distinct real roots $s_1(t) < s_2(t) < s_3(t)$.

(iii) In the case of $D(t) = 0$ for $t = t_d$, the Equation (12) have a multiple root. We then have $U(t_d) = V(t_d) = - q(t_d)/2$. If $q(t_d) < 0$, then $s_3(t_d)$ is a simple root and $s_1(t_d)$, $s_2(t_d)$ are confluent as a multiple root. If $q(t_d) > 0$, then $s_1(t_d)$ is a simple root and $s_2(t_d)$, $s_3(t_d)$ are confluent as a multiple root.

(iv) In the case of $a(t) = 0$ for $t = t_a$, the Equation (12) degenerates to a quadratic equation. Substituting $1/r$ into s of (12), we obtain an equation

$$d(t)r^3 + c(t)r^2 + b(t)r + a(t) = 0. \tag{13}$$

Since $a(t_a) = 0$, one of the roots for (13) denoted by $\varepsilon(t)$ approaches to zero as t approaches to t_a. For the roots $\alpha(t)$, $\beta(t)$, $\varepsilon(t)$ of (13) we have relations

$$\alpha(t) + \beta(t) + \varepsilon(t) = -c(t)/d(t), \tag{14}$$

$$\alpha(t)\beta(t) + \beta(t)\varepsilon(t) + \alpha(t)\varepsilon(t) = b(t)/d(t), \tag{15}$$

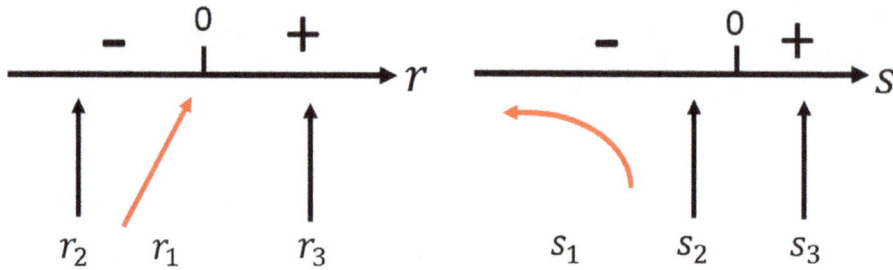

Figure 9. Behavior of the real roots around t_a.

$$\alpha(t)\beta(t)\varepsilon(t) = -a(t)/d(t). \tag{16}$$

From (15) and (16), as $0 < |\varepsilon(t)| \ll 1$, we obtain $\alpha(t)\beta(t) \simeq b(t)/d(t)$, $\varepsilon(t) \simeq -a(t)/b(t)$. If $a(t)b(t) > 0$, then $\varepsilon(t) < 0$ and the smallest of the roots of (12) satisfies $s_1(t) = 1/\varepsilon(t) \to -\infty$ $(t \to t_a)$ as shown in Figure 9. On the other hand, if $a(t)b(t) < 0$, then $\varepsilon(t) > 0$ and the largest of the roots of (12) satisfies $s_3(t) = 1/\varepsilon(t) \to +\infty$ $(t \to t_a)$.

3. Proposed algorithm

We trace the contour points on the sides of a grid cell based on the behavior of the real roots of the cubic equations. To make a polygon, we propose an algorithm composed of the following four phases:

(i) Interpolation with bicubic spline functions.

(ii) Detection of the grid cells with relevant contour points on their sides.

(iii) Computation of the coordinates for the contour points.

(iv) Selection of the next contour point on the sides of a grid cell.

More precisely, (i) we interpolate the altitudes for the grid points with a bicubic spline function. (ii) we count the number of contour points on each side of a grid cell by Sturm's method (Bourdon, 1853) and detect the grid cells which the contour traverses. (iii) we compute the coordinate values of con-tour points on the sides of the grid cells detected above using a numerical method. (iv) we determine the exit point corresponding to the entry point and append it to the polygon. In the case that the cell has more than three contour points on the sides, we select the exit point according to the behavior of the real roots of the cubic equations. We elaborate on the first and fourth phases below.

3.1. Interpolation

We describe below an algorithm for determining coefficients c_{ij} in expression (3). We give matrix representations $\mathbf{A}$ and $\mathbf{B}$ of the system of base functions for one-dimensional B-spline in x and y, respectively, as follows.

$$\mathbf{A} = \begin{bmatrix} M_{4,1}(x_1) & M_{4,2}(x_1) & \cdots & M_{4,m}(x_1) \\ \vdots & \ddots & \ddots & \vdots \\ M_{4,1}(x_m) & M_{4,2}(x_m) & \cdots & M_{4,m}(x_m) \end{bmatrix}, \tag{17}$$

$$\mathbf{B} = \begin{bmatrix} M_{4,1}(y_1) & M_{4,2}(y_1) & \cdots & M_{4,n}(y_1) \\ \vdots & \ddots & \ddots & \vdots \\ M_{4,1}(y_n) & M_{4,2}(y_n) & \cdots & M_{4,n}(y_n) \end{bmatrix}, \tag{18}$$

where (x_i, y_j) $(i = 1, ..., m; j = 1, ..., n)$ represent locations of grid points. Let the altitude at a grid point (x_i, y_j) be denoted by h_{ij}. Since the values of the spline function coincide with the altitudes at the grid points, we have

$$h(x_i, y_j) = h_{ij}. \tag{19}$$

We may put $c = (c_{11}, c_{12}, ..., c_{mn})^T$, $h = (h_{11}, h_{12}, ..., h_{mn})^T$ and rewrite (19) as

$$(\mathbf{A} \otimes \mathbf{B})\mathbf{c} = \mathbf{h}. \tag{20}$$

Moreover, instead of using the tensor product in the left-hand side of (20), we obtain an equivalent equation

$$\mathbf{ACB}^T = \mathbf{H}, \tag{21}$$

where

$$\mathbf{C} = \begin{bmatrix} c_{11} & \cdots & c_{1n} \\ c_{21} & \cdots & c_{2n} \\ \vdots & \ddots & \vdots \\ c_{m1} & \cdots & c_{mn} \end{bmatrix}, \mathbf{H} = \begin{bmatrix} h_{11} & \cdots & h_{1n} \\ h_{21} & \cdots & h_{2n} \\ \vdots & \ddots & \vdots \\ h_{m1} & \cdots & h_{mn} \end{bmatrix}. \tag{22}$$

The equivalence above is clear when expressed in summations:

$$\sum_{p=1}^{m} \sum_{q=1}^{n} a_{ip} b_{jq} c_{pq} = h_{ij}. \tag{23}$$

Putting $W = CB^T$, the equation of (21) becomes $AW = H$. The jth vertical vector of W is the solution of

$$\mathbf{A} \begin{pmatrix} w_{1j} \\ \vdots \\ w_{mj} \end{pmatrix} = \begin{pmatrix} h_{1j} \\ \vdots \\ h_{mj} \end{pmatrix}. \tag{24}$$

We may employ LU-decomposition to solve (24). The computation is made efficient taking account of the fact L and U are band matrices in the form of

$$\mathbf{L} = \begin{bmatrix} * & 0 & \cdots & \cdots & 0 \\ * & * & \ddots & & 0 & \vdots \\ 0 & * & * & \ddots & \vdots \\ \vdots & \ddots & \ddots & \ddots & 0 \\ 0 & \cdots & 0 & * & * \end{bmatrix}, \mathbf{U} = \begin{bmatrix} 1 & * & 0 & \cdots & 0 \\ 0 & 1 & * & 0 & \vdots \\ 0 & 0 & 1 & \ddots & 0 \\ \vdots & \ddots & \ddots & \ddots & * \\ 0 & \cdots & 0 & 0 & 1 \end{bmatrix}. \tag{25}$$

Likewise, we compute C by solving $BC^T = W^T$ with LU-decomposition. Thus, we obtain the spline function $h(x, y)$ in (3).

3.2. Selection of the next contour point

We focus on the fourth phase which is the mainstay of the proposed algorithm. In the course of construction, the polygon of a contour enters a cell at a contour point on a side which we call an entry point. It goes out of the cell at one of the other contour points on the sides which we call an exit point for the entry point. Thus, if the number of the contour points on the sides is two, the exit point is automatically determined. Otherwise, we select the exit point using the following algorithm.

3.2.1. Coordinate transformation

To facilitate the algorithm, each time we enter a new (i, j)-cell, we transform the coordinates from (x, y) to (s, t) as specified in Table 1. Note that the range of the (i, j)-cell is $\left\{ (x, y) \in \mathbb{R}^2 | (i-1)\delta \leq x \leq i\delta, (j-1)\delta \leq y \leq j\delta \right\}$. As a result, the cell is rescaled as a unit square and the entry point is relocated on the bottom side. By substituting (s, t) into (x, y) in expression (3), the elevation function is formulated by

$$h_{ij}(s, t) = a_{ij}(t)s^3 + b_{ij}(t)s^2 + c_{ij}(t)s + d_{ij}(t). \tag{26}$$

All the coefficients in expression (26) are polynomials in t of degree at most 3. The range of the (i, j)-cell is rescaled as $\left\{ (s, t) \in \mathbb{R}^2 | 0 \leq s \leq 1, 0 \leq t \leq 1 \right\}$. We put $f_{ij}(s, t) = h_{ij}(s, t) - h_0$. The contour of altitude h_0 within the (i, j)-cell is equal to the solution curve determined by

$$f_{ij}(s, t) = 0. \tag{27}$$

Thus, we are prepared to proceed to the selection of the exit point. Various cases are described below along with illustration in figures where an entry point is indicated by a triangle and the selected exit point is encircled by a dashed circle.

3.2.2. The case $D(t) < 0$ for $0 \leq t \leq 1$

For the exit point, we select a point with the smallest t among the remaining contour points on the sides (Figure 10).

Table 1. Transformation of the coordinates (x, y) to (s, t)		
Side containing the entry point	**s**	**T**
South	$x/\delta - i + 1$	$y/\delta - j + 1$
West	$j - y/\delta$	$x/\delta - i + 1$
North	$i - x/\delta$	$j - y/\delta$
East	$y/\delta - j + 1$	$i - x/\delta$

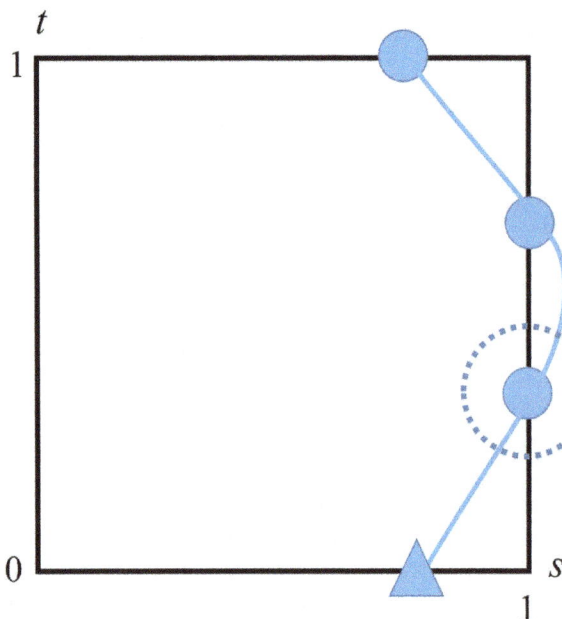

Figure 10. The case $D(t) < 0$ for $(0 \leq t \leq 1)$.

3.2.3. The case D(t) > 0 for 0 ≤ t ≤ 1

We select the exit point using the numerical order of the three real roots of the cubic equations. The exit point is determined as one with the smallest t among the contour points on the sides which have the same order as the entry point. To obtain the order of a root, no information is required on the other roots. We just use the following fact:

For a monic cubic function

$$f(x) = (x - \alpha)(x - \beta)(x - \gamma), \ (\alpha < \beta < \gamma),$$

we have

$$f'(x) = (x - \beta)(x - \gamma) + (x - \alpha)(x - \gamma) + (x - \alpha)(x - \beta), \text{ and}$$

$$f''(x) = 2(3x - \alpha - \beta - \gamma).$$

Thus, we determine the order according to Table 2.

In the case of Figure 11, if the contour line enters the right-hand cell at the point A which order is 2, then it leaves the cell at the point D which order is also 2. On the other hand, if the contour line enters the right cell at the point B which order is 3, then it goes out of the cell at the point C which order is also 3. Note that the order of a root is determined over the range $-\infty < s < +\infty$.

Table 2. Sign of the derivatives at the three real roots

Derivative	α	β	γ
$f'(x)$	+	−	+
$f''(x)$	−	N/A	+

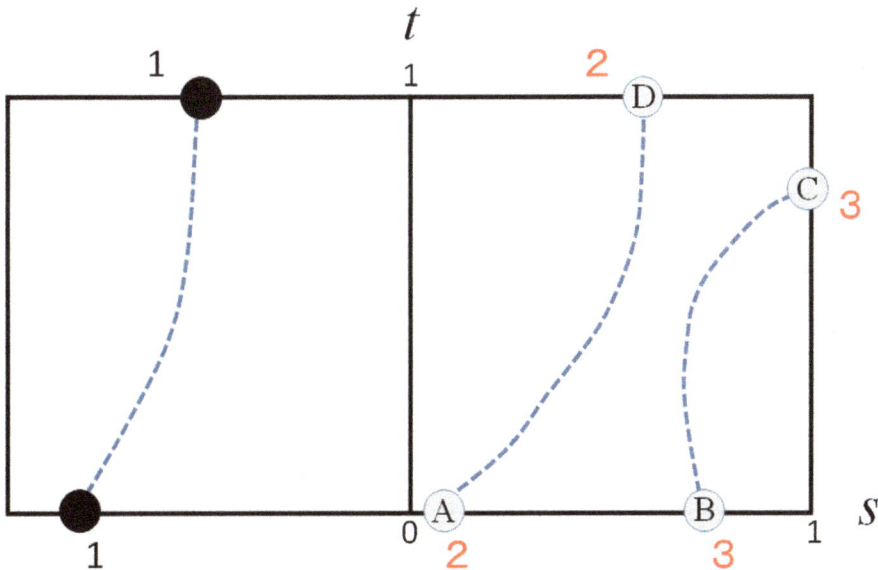

Figure 11. The case D(t) > 0 for (0 ≤ t ≤ 1).

3.2.4. The case $D(t_d) = 0$ for some t_d with $0 < t_d < 1$

The process of selecting the exit point varies according to the sign of $D'(t_d)$ and $q(t_d)$:

(i) The case $D'(t_d) < 0$ and $q(t_d) > 0$

If the entry point is of order 1, among the other contour points, we select a point of order 1 with the smallest $t < t_d$ if any (Figure 12). If there is no such point, we select a point with the smallest $t > t_d$ (Figure 13).

If the entry point is of order 2(3), we select a point of order 2(3) with the smallest $t < t_d$ if any (Figure 14). If there is no such point, we select a point of order 3(2) with the largest $t < t_d$ (Figure 15).

(ii) The case $D'(t_d) < 0$ and $q(t_d) < 0$

If the entry point is of order 3, among the other contour points, we select a point of order 3 with the smallest $t < t_d$ if any (Figure 16). If there is no such point, we select a point with the smallest $t > t_d$ (Figure 17).

If the entry point is of order 1(2), we select a point of order 1(2) with the smallest $t < t_d$ if any (Figure 18). If there is no such point, we select a point of order 2(1) with the largest $t < t_d$ (Figure 19).

(iii) The case $D'(t_d) > 0$ and $q(t_d) > 0$

Among the contour points other than the entry point, we select a point with the smallest $t < t_d$ if any (Figure 20). If there is no such point, we select a point of order 1 with the smallest $t > t_d$ (Figure 21).

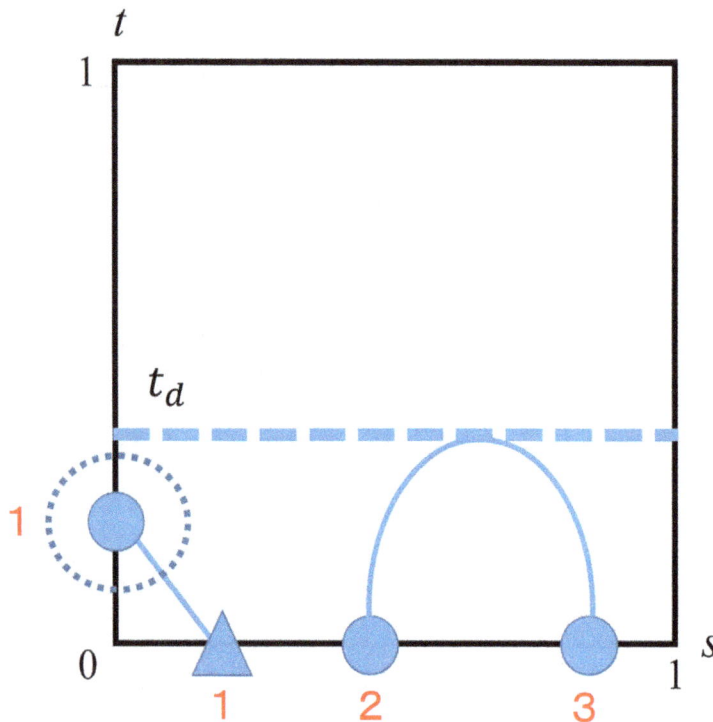

Figure 12. The case $D'(t_d) < 0$ and $q(t_d) > 0$, enter at order 1, exit at $(t < t_d)$.

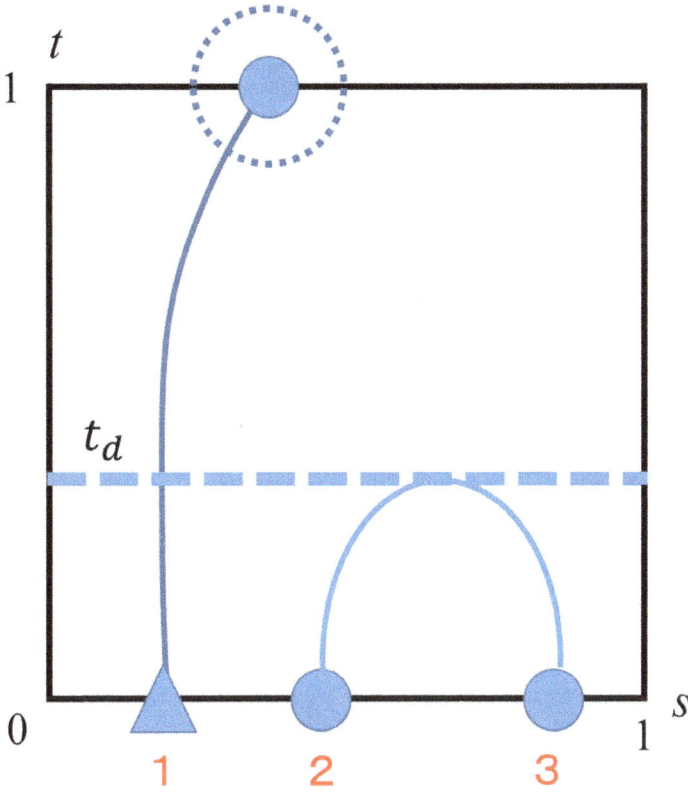

Figure 13. The case $D'(t_d) < 0$ and $q(t_d) > 0$, enter at order 1, exit at $(t > t_d)$.

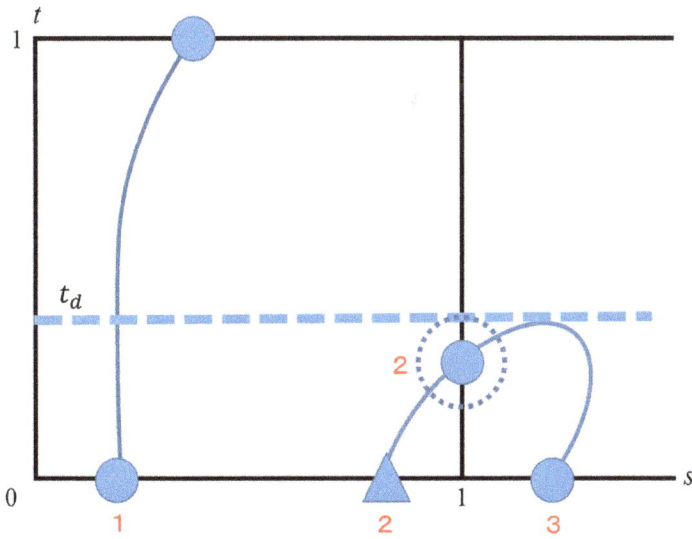

Figure 14. The case $D'(t_d) < 0$ and $q(t_d) > 0$, enter at order 2(3), exit at order 2(3) $(t < t_d)$.

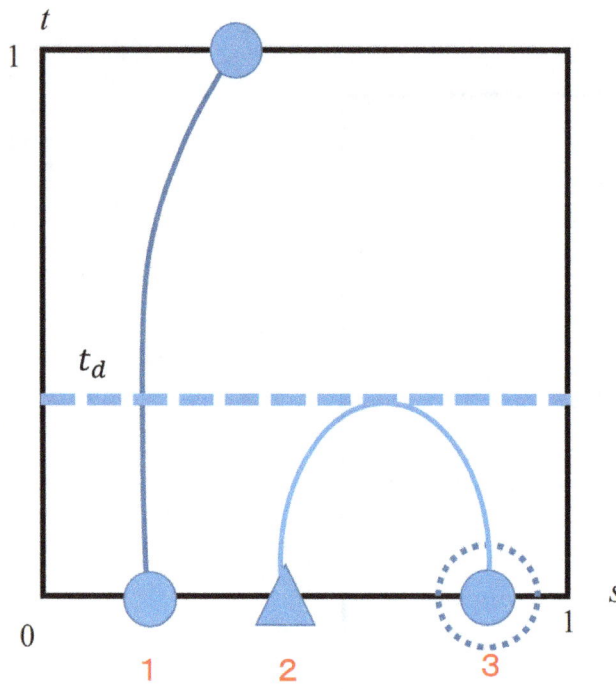

Figure 15. The case $D'(t_d) < 0$ and $q(t_d) > 0$, enter at order 2(3), exit at order 3(2) $(t < t_d)$.

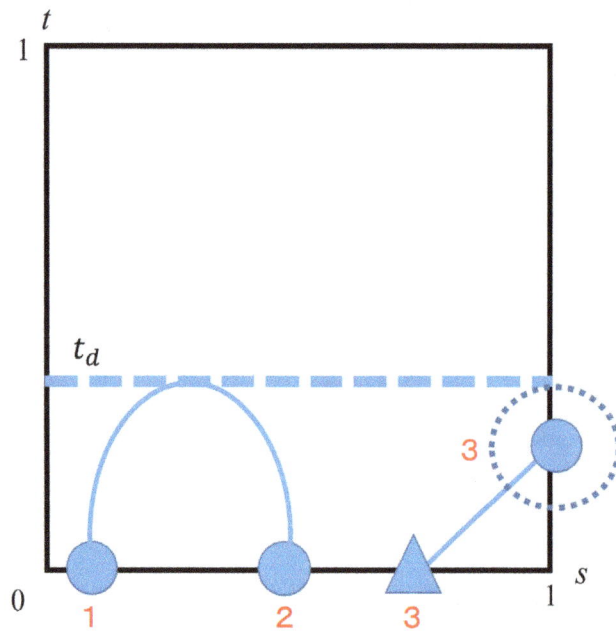

Figure 16. The case $D'(t_d) < 0$ and $q(t_d) < 0$, enter at order 3, exit at order 3 $(t < t_d)$.

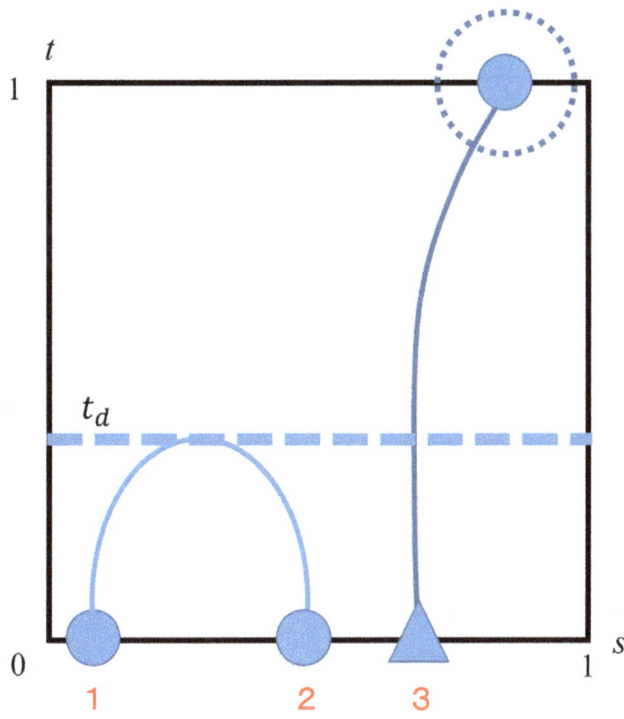

Figure 17. The case $D'(t_d) < 0$ and $q(t_d) < 0$, enter at order 3, exit at $(t > t_d)$.

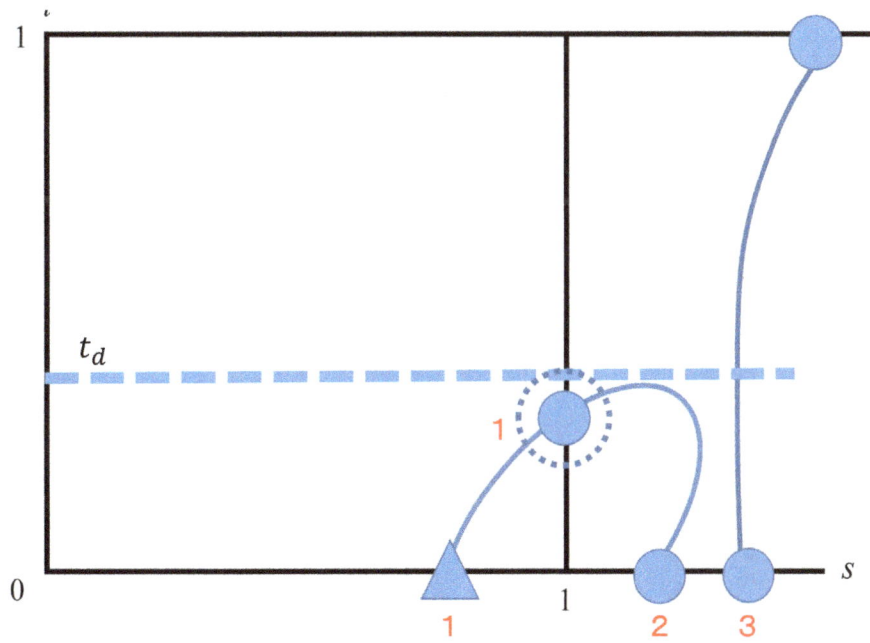

Figure 18. The case $D'(t_d) < 0$ and $q(t_d) < 0$, enter at order 1(2), exit at order 1(2) $(t < t_d)$.

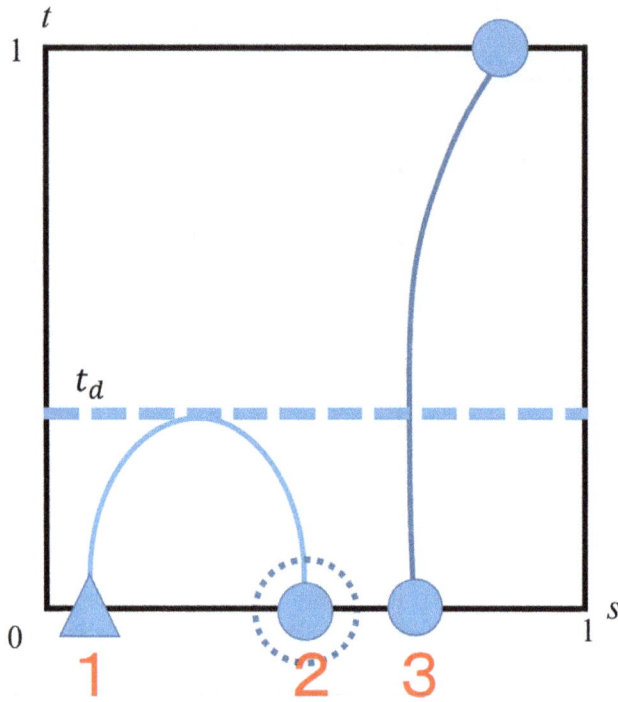

Figure 19. The case $D'(t_d) < 0$ and $q(t_d) < 0$, enter at order 1(2), exit at order 2(1) ($t < t_d$).

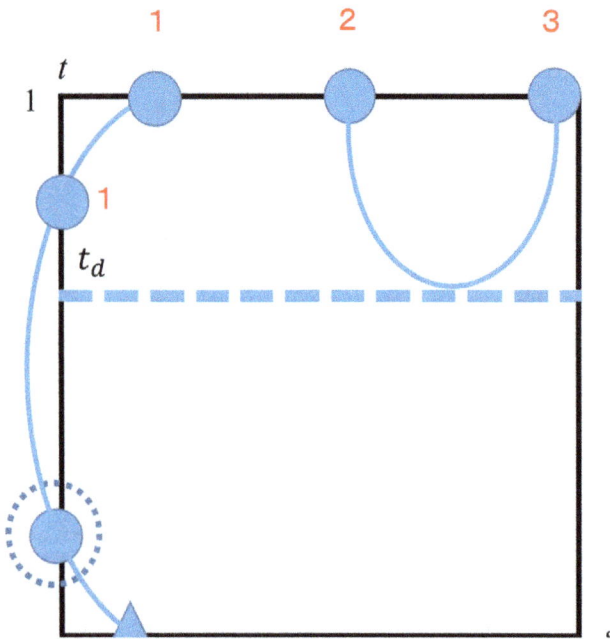

Figure 20. The case $D'(t_d) > 0$ and $q(t_d) > 0$, enter as a unique real root ($t < t_d$), exit at ($t < t_d$).

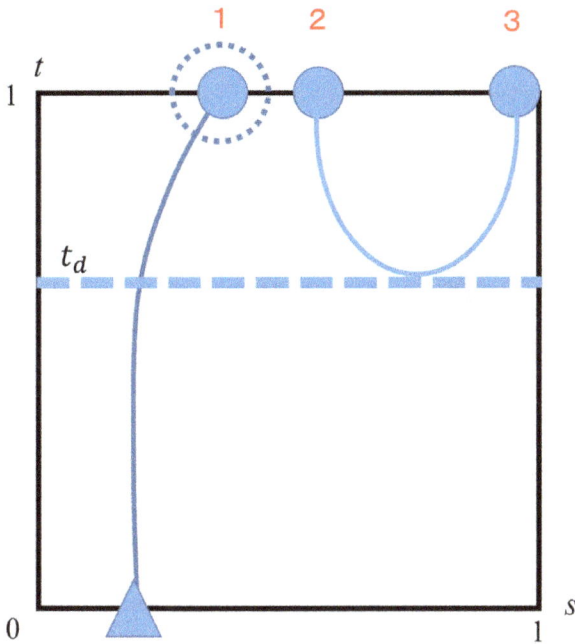

Figure 21. The case $D'(t_d) > 0$ and $q(t_d) > 0$, enter as a unique real root ($t < t_d$), exit at order 1 ($t > t_d$).

(iv) The case $D'(t_d) > 0$ and $q(t_d) < 0$

Among the contour points other than the entry point, we select a point with the smallest $t < t_d$ if any (Figure 22). If there is no such point, we select a point of order 3 with the smallest $t > t_d$ (Figure 23).

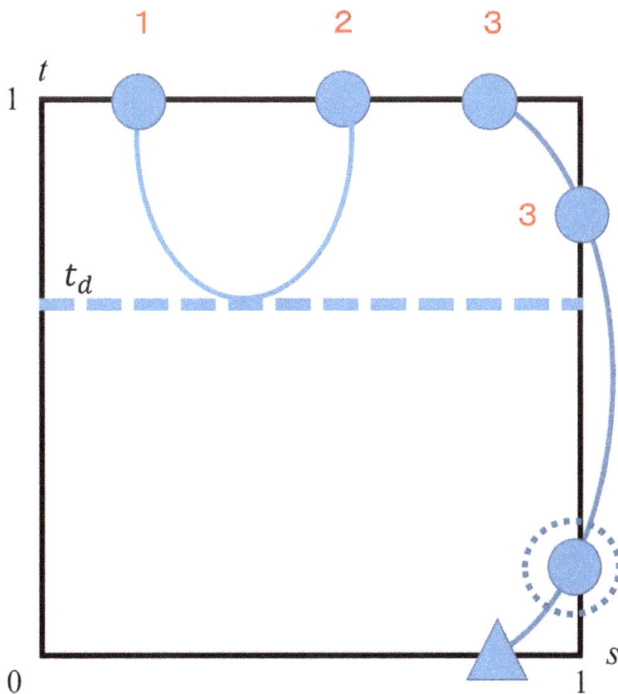

Figure 22. The case $D'(t_d) > 0$ and $q(t_d) < 0$, enter as a unique real root ($t < t_d$), exit at ($t < t_d$).

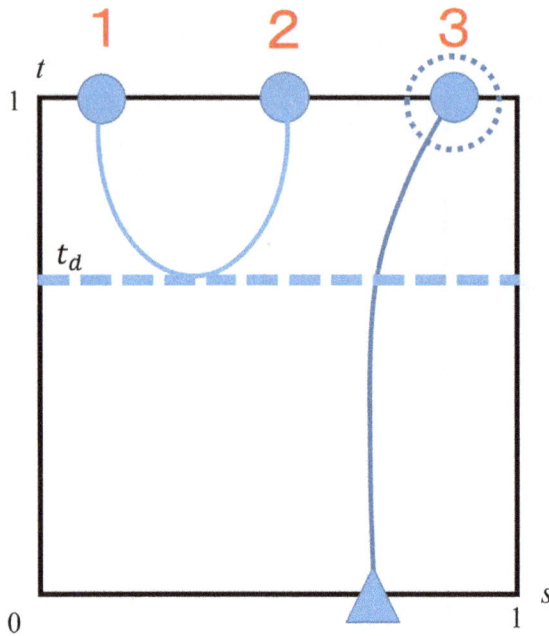

Figure 23. The case $D'(t_d) > 0$ and $q(t_d) < 0$, enter as a unique real root ($t < t_d$), exit at order 3 ($t > t_d$).

3.2.5. The case $a(t_a) = 0$ for some t_a with $0 < t_a < 1$
The process of selecting the exit point varies according to the sign of $a'(t_a)b(t_a)$ and $D(t)$:

(i) The case $D(t) < 0$ for $0 < t < 1$

Among the contour points other than the entry point, we select a point with the smallest $t < t_a$ (Figure 24).

(ii) The case $a'(t_a)b(t_a) > 0$ and $D(t) > 0$ for $0 < t < 1$

If the entry point is of order 1(2), among the other contour points, we select a point of order 1(2) with the smallest $t < t_a$ if any (Figure 25). If there is no such point, we select a point of order 2(3) with the smallest $t > t_a$ (Figure 26).

If the entry point is of order 3, we select a point of order 3 with the smallest $t < t_a$ (Figure 27).

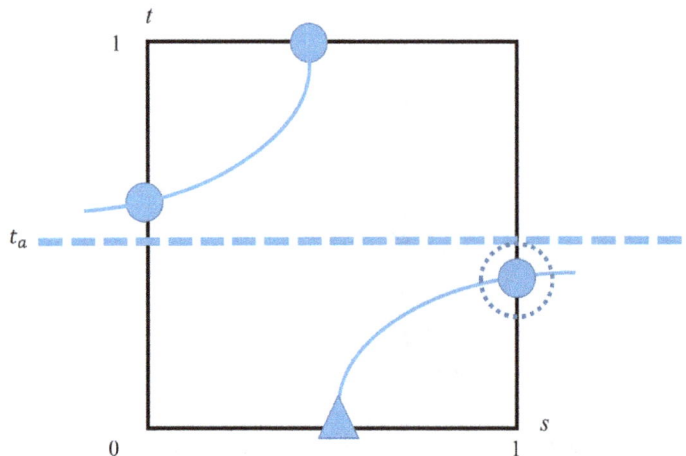

Figure 24. The case $D(t) < 0$, enter as a unique real root ($t < t_a$), exit at ($t < t_a$).

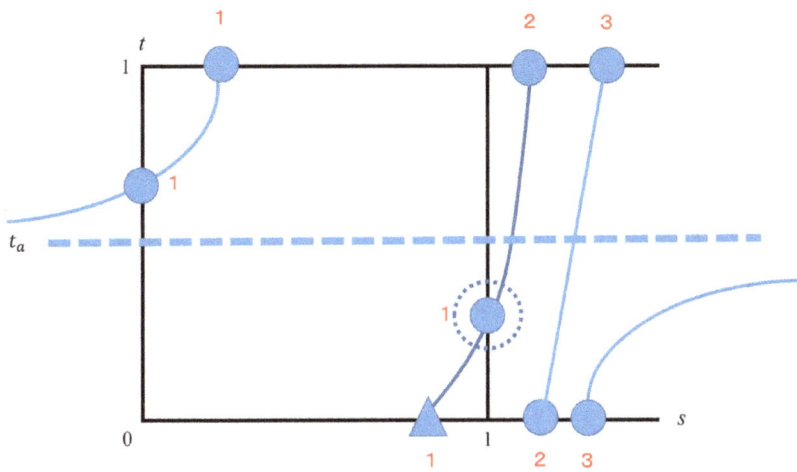

Figure 25. The case $a'(t_a)b(t_a) > 0$ and $D(t) > 0$, enter at order 1(2), exit at order 1(2) ($t < t_a$).

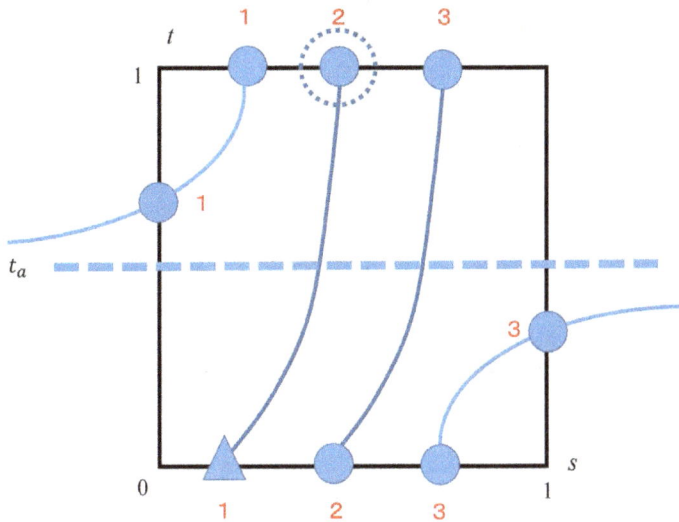

Figure 26. The case $a'(t_a)b(t_a) > 0$ and $D(t) > 0$, enter at order 1(2), exit at order 2(3) ($t > t_a$).

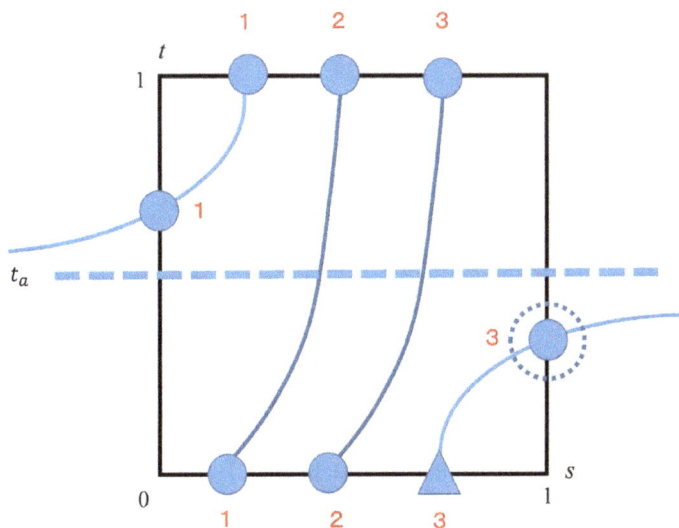

Figure 27. The case $a'(t_a)b(t_a) > 0$ and $D(t) > 0$, enter at order 3, exit at order 3 ($t < t_a$).

(iii) The case $a'(t_a)b(t_a) < 0$ and $D(t) > 0$ for $0 < t < 1$

If the entry point is of order 2(3), among the other contour points, we select a point of order 2(3) with the smallest $t < t_a$ if any (Figure 28). If there is no such point, we select a point of order 1(2) with the smallest $t > t_a$ (Figure 29).

If the entry point is of order 1, we select a point of order 1 with the smallest $t < t_a$ (Figure 30).

3.2.6. The case $D(t_d) = 0$ and $a(t_a) = 0$ for some t_d, t_a with $0 < t_d$, $t_a < 1$
We perform the procedures of 3.2.4 and 3.2.5 according to the numerical order of t_d, t_a. In the case of more t_d and t_a, we iterate the procedures accordingly.

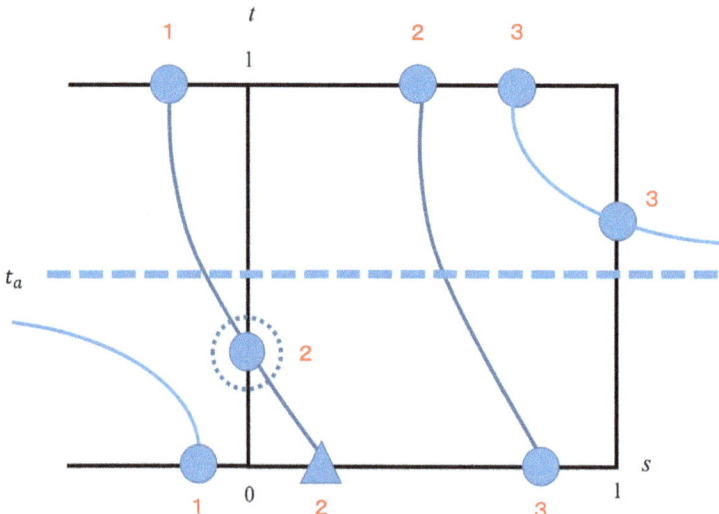

Figure 28. The case $a'(t_a)b(t_a) < 0$ and $D(t) > 0$, enter at order 2(3), exit at order 2(3) ($t < t_a$).

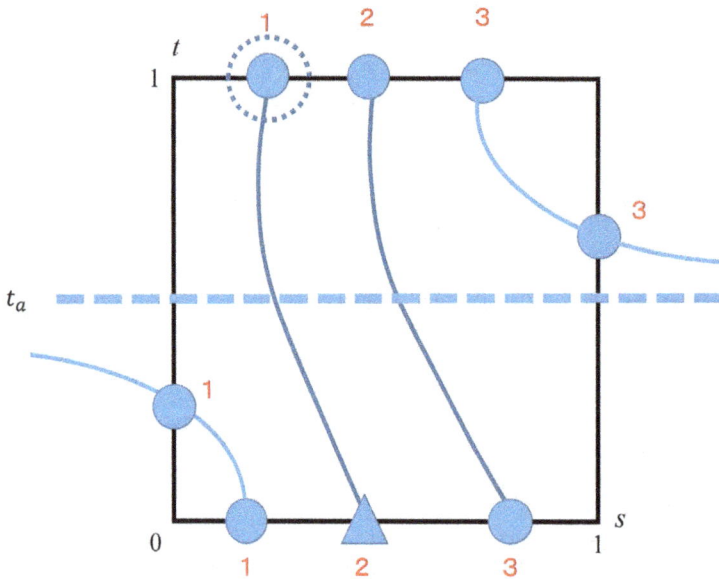

Figure 29. The case $a'(t_a)b(t_a) < 0$ and $D(t) > 0$, enter at order 2(3), exit at order 1(2) ($t > t_a$).

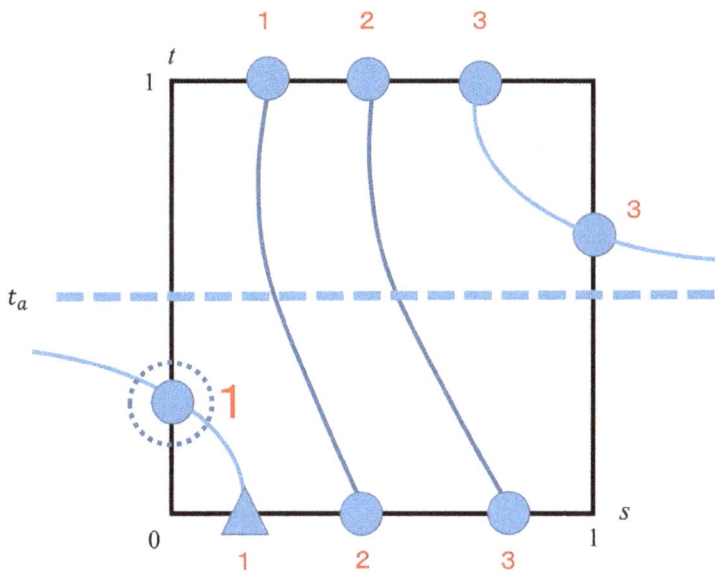

Figure 30. The case $a'(t_a)b(t_a) < 0$ **and** $D(t) > 0$**, enter at order 1, exit at order 1** $(t < t_a)$**.**

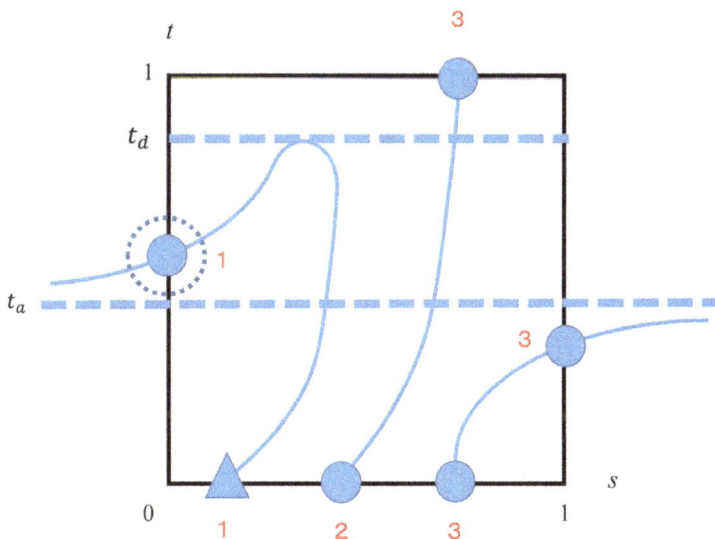

Figure 31. The composite procedure of *3.2.5-(ii)* **and** *3.2.4-(ii)*.

To illustrate the process, we consider a cell of the grid with solutions t_a, t_d for $a(t_a) = 0$ and $D(t_d) = 0$, respectively, assuming $a'(t_a)b(t_a) > 0$ and $D(t) > 0$ $(0 < t < t_d)$, $D'(t_d) < 0$ and $q(t_d) < 0$. In this case, we perform a procedure composed of 3.2.5-(ii) and 3.2.4-(ii). We start from an entry point of numerical order 1 shown in Figure 31. Firstly, we perform the process of 3.2.5-(ii) since $t_a < t_d$. No contour points of numerical order 1 are within the range of $0 < t < t_a$. Thus, the root curve containing the entry point is connected to the root curve of order 2 within the range of $t_a < t < t_d$. Secondly, we perform the process of 3.2.4-(ii) considering the intersection between the root curve and line $t = t_a$ as an entry point of order 2. No contour points of numerical order 2 are within the range of $t_a < t < t_d$. Thus, we select the contour point of numerical order 1 within the range of $t_a < t < t_d$ as the exit point.

3.2.7. Selection of the next cell
In generating a polygon for a connected component of contour, we start from the entry point in the initial cell, and link the entry point of the current cell to the correct exit point and add it to the polygon. This process is repeated until returning to the initial point. In this process, we need to select the new cell for the entry point which is also the exit point in the current cell. In other words, we transfer

to the next cell sharing the contour point with the current cell. If the contour point is not a vertex, these cells share a side. In this case, the next cell is clearly determined.

However, if the contour point is a vertex, the next cell is selected by the gradient of the elevation function which is perpendicular to the contour. Namely, if the gradient is in the direction of t-axis, we select the cell adjacent to the current cell in the direction of s-axis. If the gradient is in the direction of s-axis, we select the cell adjacent to the current cell in the direction of t-axis. If the gradient is neither in the direction of s-axis nor t-axis, the next cell is selected diagonal to the current cell. Each time the next cell is determined, we conduct the coordinate transformation in Section 3.2.1.

3.3. Case study of Yakushima island

In this section, we use the JMC50 m mesh data which is a proprietary product of Japan Map Center. These data are a remake of the altitude data on 50 m mesh of Geospatial Information Authority of Japan currently out of print. In this data, the altitudes of oceanic area are expressed by −999.9 m. We change it to −1 m for the purpose of interpolation following the precedent of Goto and Shimakawa (2015).

As a case study, we apply the algorithm described above to draw the contour map of Yakushima island which is located at the south part of Kagoshima prefecture. We use an altitude data on 498 × 598 grid. The grid interval is 50 m × 50 m. The altitude of each grid point is measured by meters. The highest point of Yakushima is at an altitude of 1,935 m.

We set $m = 498$, $n = 598$, and $\delta = 50$ in expression (3) and obtain the spline function which is shown in Figure 32. As a result of the entire algorithm, the contour map is completed as shown in Figure 33 where the contours are drawn every 100 m from 0 to 1,900 m. To see the polygonal structure of the

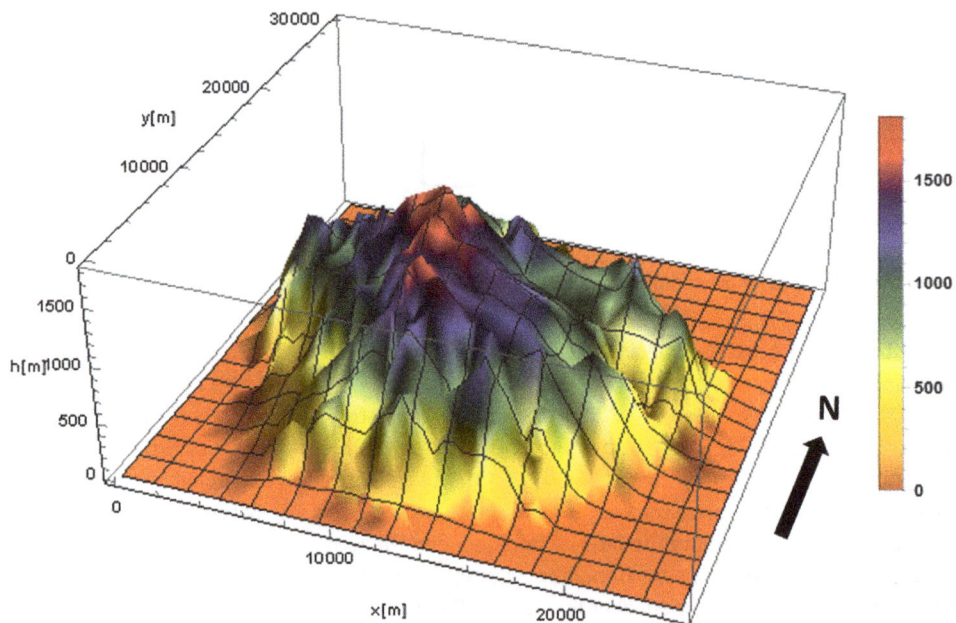

Figure 32. Topographical illustration of Yakushima island.

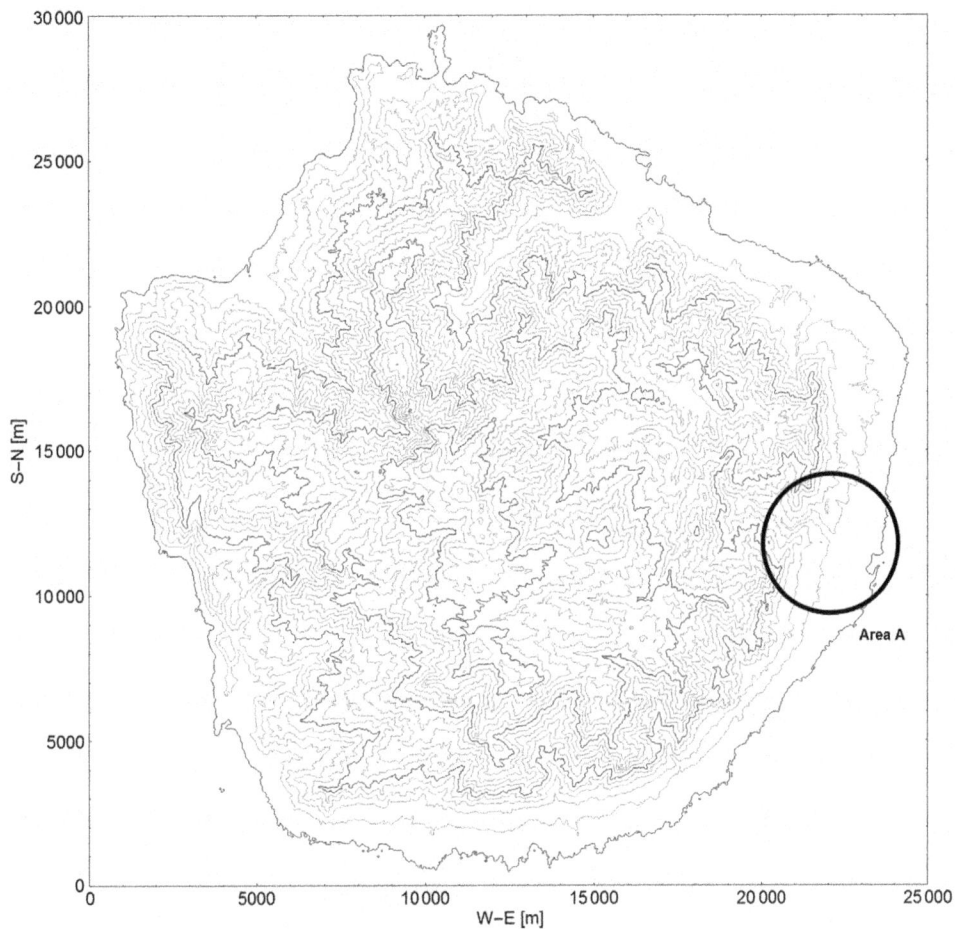

Figure 33. Contour map of Yakushima island.

contours, we focus on area A in Figure 33, and enlarge it in Figure 34, where we add the contours for every 10 m from 0 to 100 m. In this scale, the piecewise linear structure of the polygons is recognizable. Although the discussion about the superiority of bicubic splines over bilinear splines is out of scope of our research, we compare these two splines by applying them to an area in Yakushima island in Figure 35. The contour of the elevation function by bicubic spline is colored black, while the one by bilinear spline is colored gray. The central cell of Figure 35 is magnified in Figure 36. The polygons constructed by our method are delineated by black solid lines, while the one by the method of Lopes and Brodlie (1998) is drawn by gray dashed lines in Figure 37 and Figure 38. Both methods faithfully trace the contour of spline functions, respectively. As a consequence, the advantage of bicubic spline interpolation is reflected on our polygon.

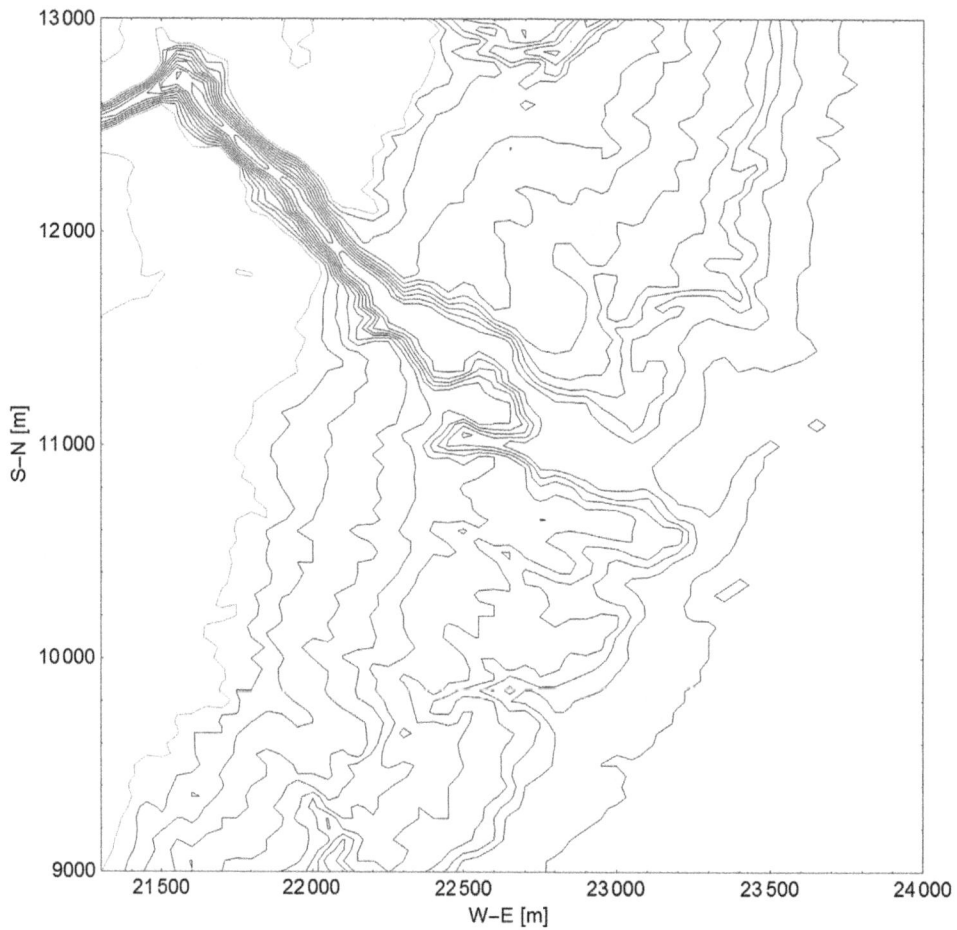

Figure 34. Contour map in area A.

We measure the time of computation for constructing each polygon as shown in Figure 39 where the level axis represents the number of points in a polygon and the vertical axis the computational time. The maximum computational time is measured 242.39 s for a polygon of 500 m consisting of 5,000 points. To compare the computational time between a cell with two contour points on its sides and a cell with more than two, let n_1 be the number of the former and n_2 be the number of the latter. The regression of computational time t on n_1 and n_2 is $t = 0.046\ n_1 + 0.108\ n_2 - 10.68$ with $R^2 = 0.993$. Although our algorithm of selecting the exit points seems complicated, the computational time is not so affected by the number of the contour points on the sides of a cell: The ratio of the regression coefficients for n_1 and n_2 above is about 1:2, while the ratios of n_1 and n_2 is around 95:5.

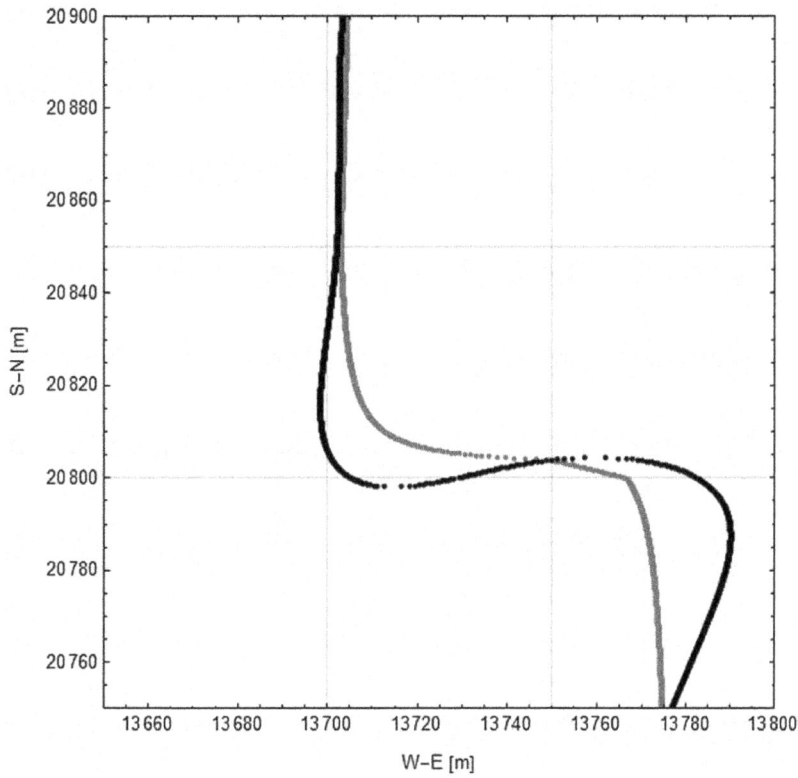

Figure 35. Contours for bilinear and bicubic splines.

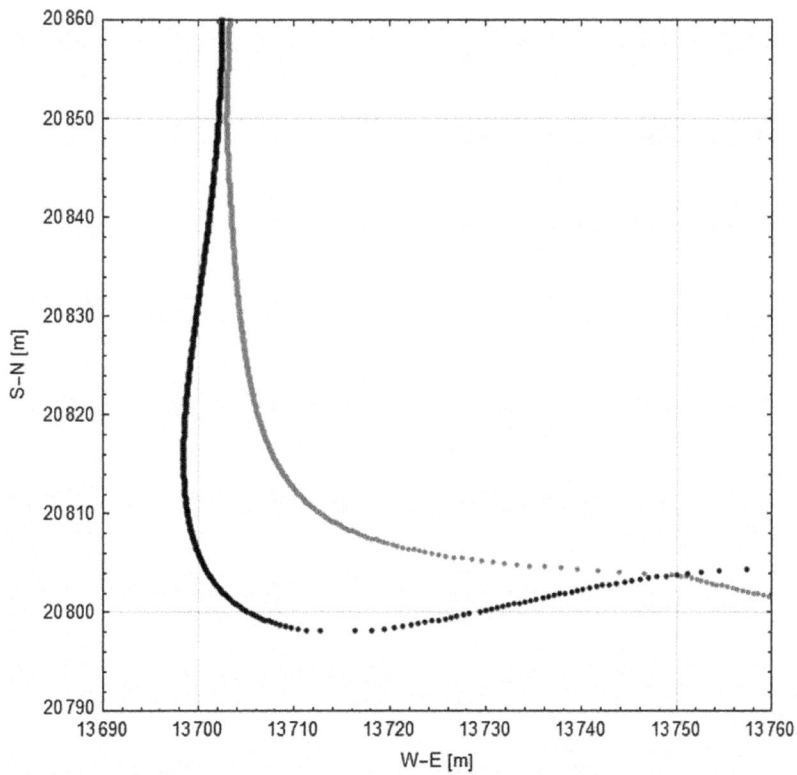

Figure 36. Contours for bilinear and bicubic splines magnified.

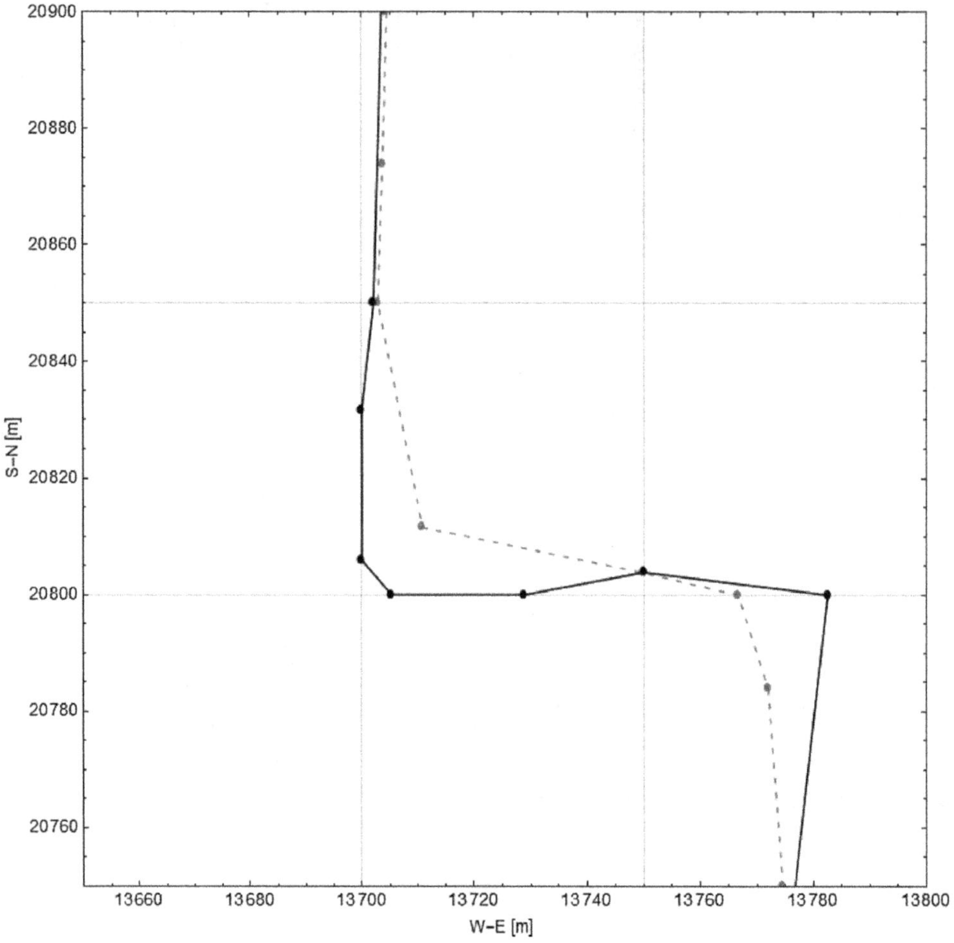

Figure 37. Polygons by our method and Lopes–Brodlie's.

The experiment above is performed in the following environment:

Machine: Aspire M3970,

CPU: Inter Core i7–3770 3.40 GHz,

OS: Windows 7 Enterprise,

Memory: 8.00 GB,

Program: Wolfram Mathematica 10.3.

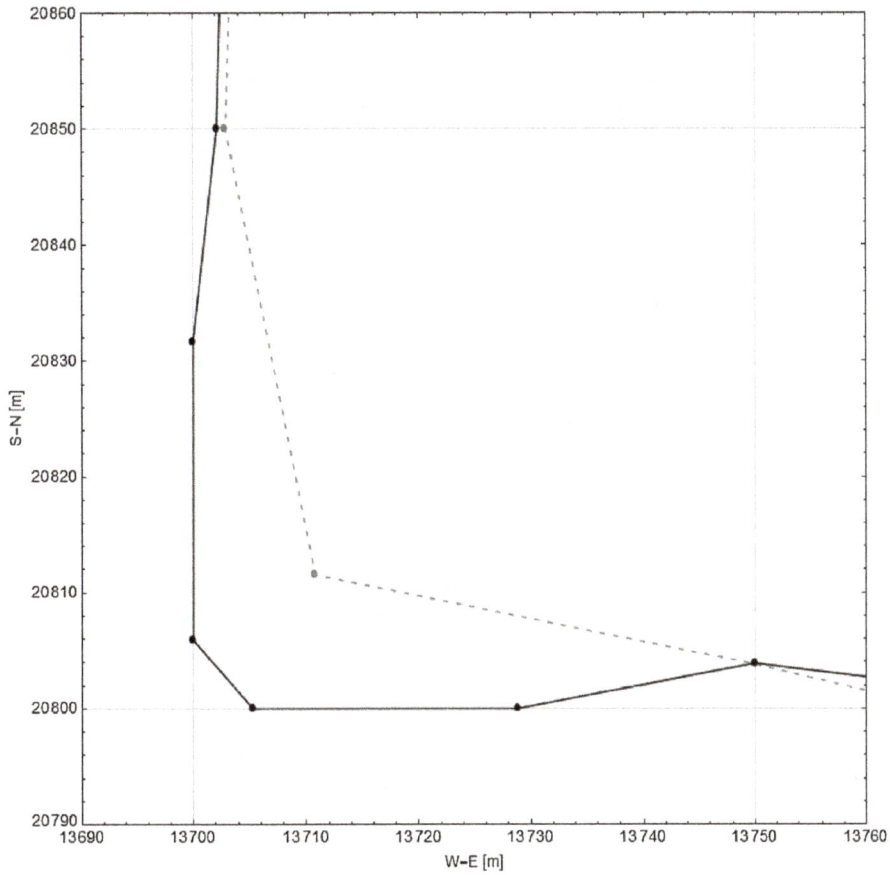

Figure 38. Polygons by our method and Lopes–Brodlie's magnified.

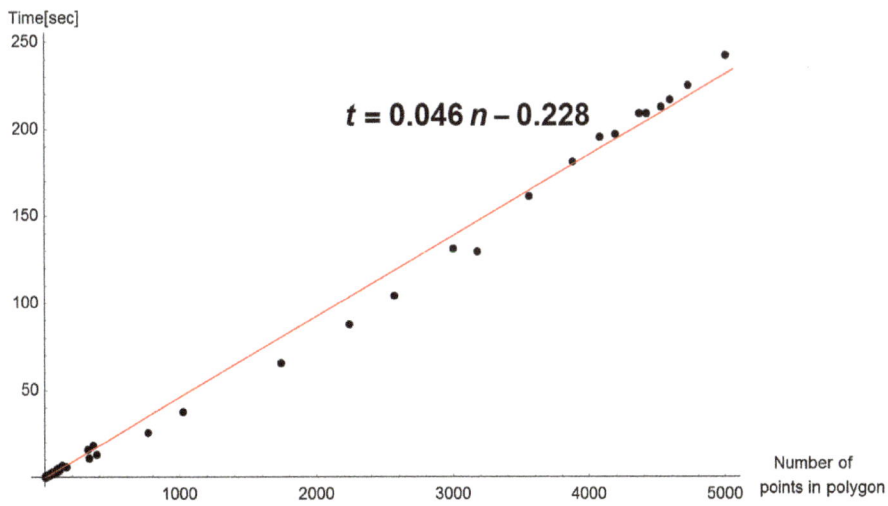

$$t = 0.046\,n - 0.228$$

Figure 39. Computational time and number of points in a polygon.

4. Conclusion

The method of contour tracing developed in this research makes full use of the fact that a bicubic function is regarded as a univariate cubic function with a parameter. The behavior of the real roots within a rectangular cell is examined to link an entry point with the corresponding exit point. This algorithm detects and traces all the connected components of a target contour which intersect the grid and gives the same polygon independent of the choice of a starting point on a connected component. In the case of complex terrain, this method shows clear benefits over the alternative methods using data on a rectangular grid in providing smooth and fitting contour representation with moderate load of implementation and computation. Tracing a closed contour contained in the interior of a cell remains as a future work.

Acknowledgments

The author is grateful to Prof Hiroyuki Goto in the Department of Industrial and System Engineering, Hosei University, Japan for introducing the author to the literature of preceding works about contour tracing including his own.

Funding

The author received no direct funding for this research.

Author details

Tatsuya Ishige[1]
E-mail: tatsuya.ishige0809@gmail.com
[1] Department of Industrial and System Engineering, Hosei University, Tokyo, Japan.

References

Bourdon, M. (1853). *Elements of algebra* (pp. 354–366). New York, NY: A. S. Barnes & .

Cottafava, G., & Moli, G. L. (1969). Automatic contour map. *Communications of the ACM, 12*, 386–391. doi:10.1145/363156.363172

Dobkin, D. P., Wilks, A. R., Levy, S. V. F., & Thurston, W. P. (1990). Contour tracing by piecewise linear approximations. *ACM Transactions on Graphics, 9*, 389–423. doi:10.1145/88560.88575

Gold, C. M., Charters, T. D., & Ramsden, J. (1977). Automated contour mapping using triangular element data structure and an interpolant over each irregular triangular domain.

In *ACM SIGGRAPH Computer Graphics, San Jose* (pp. 170–175). doi:10.1145/563858.563887

Goto, H., & Shimakawa, Y. (2015). Storage-efficient method for generating contours focusing on roundness. *International Journal of Geographical Information Science, 30*, 200–220. doi:10.1080/13658816.2015.1080830

Lopes, A., & Brodlie, K. (1998). Accuracy in contour drawing. In *Eurographics UK Conference Proceedings* (pp. 301–311), Leeds.

Maple, C. (2003). Geometric design and space planning using the marching squares and marching cube algorithms. In *2003 International conference on geometric modelling and graphics* (pp. 90–95). IEEE Computer Society. https://doi.org/10.1109/GMAG.2003.1219671

McCullagh, M. J. (1981). Creation of smooth contours over irregularly distributed data using local surface patches. *Geographical Analysis, 13*, 51–63. https://doi.org/10.1111/j.1538-4632.1981.tb00714.x

McLain, D. H. (1971). Drawing contours from arbitrary data points. *The Computer Journal, 17*, 318–324. doi:10.1093/comjnl/17.4.318

Piegl, L., & Tiller, W. (2012). *The NURBS book* (pp. 47–138). Berlin: Springer

Saux, E., Thibaud, R., Li, K.-J., & Kim, M.-H. (2004). A new approach for topographic feature-based characterization of digital elevation data. In *Workshop on Advances in Geographic Information System* (pp. 73–81). ACM Press. doi:10.1145/1032222.1032235

5

Creation of subtropical greenhouse plan for the Flora Exhibition Grounds using GIS

Zdena Dobesova[1]*

*Corresponding author: Zdena Dobesova, Faculty of Science, Department of Geoinformatics, Palacký University, 17 listopadu 50, 771 46 Olomouc, Czech Republic
E-mail: zdena.dobesova@upol.cz
Reviewing editor: Louis-Noel Moresi, University of Melbourne, Australia

Abstract: This paper describes a cartographical and Geographic Information System (GIS) work on a poster for the subtropical greenhouse. The subtropical greenhouse is part of a collection of greenhouses in Olomouc that are located in the centre of the city (Czech Republic) near Smetana Park. An overall plan exists for the collection of greenhouses and botanical gardens in the area. We have created a poster regarding the subtropical greenhouse plan. Partial plan for the subtropical greenhouse shows the detailed positions of approximately 120 plants. This plan is central information on the big presented poster (format A0). Plans arose from the cooperation of cartographers and botanists using GIS. All digital maps and plans were created using ArcGIS software after punctual measure in the field. Beside text information on the poster, there are pictures of selected plants. They are accompanied with maps of their native range. The maps of the native range are original cartographical part of the same authors. The poster also contains an old proposal for the subtropical greenhouse, which was created by a leading garden architect, I. Otruba, in 1991 before construction of the greenhouse. The book regarding species in subtropical greenhouses was issued in 2013. It contains descriptions of 33 select species. The exposition represents mainly Mediterranean flora. Each species is described with text and includes illustrations of fruits, leafs, flowers, and habitus.

ABOUT THE AUTHOR

She holds a PhD degree from Technical University of Ostrava, Faculty of Mining and Geology since 2007. Research interests are GIS, digital cartography, the visual programming language in GIS, scripting in Python for ArcGIS, spatial databases. She has several lectures for study branch Geoinformatics and Geography at Palacky University in Olomouc, Czech Republic. She is an author of 8 books and more than 70 articles from journals and conferences.

PUBLIC INTEREST STATEMENT

The design of greenhouse plan is a demonstration that the geoscience field has a wider application that only geography. The basic vector graphic was digitized using the geographical information systems. There are two most important parts of the area of greenhouses: the shapes of buildings and the positions and shapes of plants. Data could be used repetitively for different tasks. The article presents one printed output—big poster with the plan of the subtropical greenhouse. Other utilizations of the same data are also mentioned. One of them is the interactive web map of greenhouses with the detailed position of plants. Next utilization of the same data is on the plans in three botanical books about species in palm, tropical and subtropical greenhouses. The biggest advantage is using one source of digital vector data for several purposes. The results could be accomplished only by intensive cooperation specialists as geoinformatics and botanists.

Maps of the native ranges of each species are an original important part of the book. Greenhouses are open to the public, and students from Palacky University participate in botany and environmental lectures there. The visitors are tended to be intrigued by the astonishing variety of subtropical plants.

Subjects: Botany; Earth Sciences; Geography; Cartography

Keywords: GIS; biogeography; thematic map; native range; plan

1. Introduction

The collection greenhouses on the Exhibition Grounds are situated in Smetana Park, Olomouc. They have been declared to be cultural monuments. The greenhouses, together with the Palacký University botanical garden, are a part of the Union of Botanical Gardens of the Czech Republic (Figure 1).

The collection greenhouses consist of four separate greenhouses. The largest and oldest is the palm conservatory (Figure 2). The range and richness of the collections are among the largest and most interesting in the Czech Republic. The greenhouses occupy an area of approximately 4,100 m^2, of which 3,040 m^2 is open to the public. Every season, dozens of the most exotic species bud, bloom and spawn there. The greenhouses are open to visitors nearly year round.

The botanical garden belongs to Palacký University and neighbours the greenhouses. The garden area is approximately 5,600 m^2. The greenhouses and botanical garden create a unique area for spending time or learning, in addition to their classical function of conserving endangered species. Special botanical expositions are designed there as a part of the annual Flora Fairground.

The maintenance and education programs at the greenhouses and garden require the creation of detailed area plans and plant registries. Staff and students from the Department of Geoinformatics and the Department of Botany at Palacky University have created the BotanGIS information portal. Moreover, printed plans, posters and books have been produced. This article presents a poster that encompasses all of the interesting map documents regarding the subtropical greenhouse.

Figure 1. Location of Olomouc in the Czech Republic.

Figure 2. The entrance to collection greenhouses by architect J. Pelikan.

2. History and exposition of greenhouses

Before the current greenhouses, there was an orangery with beautiful wooden carvings. Its history dates back to 1886 when the wooden orangery structures were relocated from the castle park in Velká Bystřice to Smetana Park in Olomouc. It was built as the first palm greenhouse, with more than 90 types of tropical plants imported from Moravian castle greenhouses (Dančák, Šupová, Škardová, Dobesova, & Vávra, 2013b).

The conservatory expositions are divided into individual greenhouses according to the ecological requirements of particular plants, such as palm trees, cacti, tropical species and subtropical plants. Some species are represented by very old specimens, which have been grown over many years. The cacti and succulents collection is the most important and undoubtedly stands out nationwide. The subtropical crops, bromeliads and orchids collection is also of great importance. Tourists and professionals alike are attracted by the valuable cycad trees and Ceratozamia specimens, as well as many other dominant palm trees. Permanent specimens are tagged with a valid Latin name, family, distribution range and identification number (ID). This identifier can be used find information in the greenhouse plans and the BotanGIS database.

3. History and exposition of the subtropical greenhouse

The subtropical greenhouse is a typical 1980s construction. The original subtropical greenhouse project was created in 1991 by a leading garden architect, Prof. Ing. Ivar Otruba, CSc., who designed the exposition as an Israeli garden. The original plan is part of the presented poster.

The first plants (*Ceratonia siliqua*) were ceremonially planted in 1994 by the former Israeli ambassador in the Czech Republic. The current exhibition is the result of a systematic plan, which is conducted by gardeners who work for the Flora Exhibition Grounds Olomouc, JSC and cooperate with Palacký University, Olomouc. In 2012, the greenhouse received roof renovations (Dančák, Šupová, Škardová, Dobesova, & Vávra, 2013a).

The major plants in the composition come from subtropical zone across the world (e.g. the Mediterranean, Asia Minor and Australia), many of which are useful plants (Figure 3). Citrus fruits, which comprise a large portion of the exhibited plants, are spontaneous cultural hybrids of native species. In addition to commonly known species, such as mandarin, lemon and orange trees, less known citrus trees also exist, including *Citrus medica* whose fruit can grow to two kilograms and are used by the perfumery and confectionery industries.

The entrance is lined with Chinese plant *Poncirus trifoliate*, whose thorny shoots attract attention. This species is used as a rootstock for citruses. *Actinidia chinensis* is an interesting plant, with large,

Figure 3. Exposition in the subtropical greenhouse.

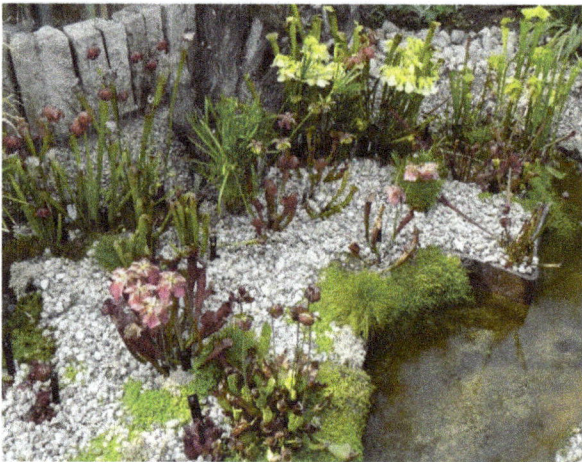

Figure 4. Carnivorous plants exposition.

felt-like leaves. Representatives of Mediterranean flora include fig trees (*Ficus carica*), olive trees *Olea europaea*) and aromatic rosemary (*Rosmarinus officinalis*). Twigs from myrtle (*Myrtus communis*), a Mediterranean shrub, have white flowers and dark fruits, making them popular choices for wedding decorations. The pomegranate tree (*Punica granatum*) is a scarlet flowering species. In the back of the greenhouse, near the lake, a very popular seasonal exhibit, carnivorous plants, is managed by the Science Faculty and the Department of Botany (Figure 4).

Tomato-like fruits, known as kaki, grow on the Japanese persimmon tree (*Diospyros kaki*). The evergreen Australian tea trees (*Melaleuca*) represent another remarkable species, from which essential oils with unique bactericidal effects are extracted. The remainder of the greenhouse is lined with strawberry trees (*Psidium cattleianum*), which produce red-purple fruits with a strawberry flavour.

4. Methods for plan creation

Some examples of digital plans for parks, and arboretums using Geographic Information System (GIS) exist. GIS technology is helping community garden managers inventory, maintain, and manage their plant collections. Shields (2010) describes the Davis Arboretum of University of California, where GIS software catalogue and map more than 30,000 plants. Arnold Arboretum of Harvard University (2011) has web map application with point positions of plants above aerial photo. Plants are labelled by ID number or by scientific or common name. There is a possibility to switch between point symbols of plants according to family, country or plant size. The crown expression by polygons is not used. Alliance for Public Gardens GIS and firm Esri published Alliance for Public Gardens GIS (2011) as a tool for creating a public garden GIS.

Some map solutions are used only a simple flash-based tree map like Kew Royal Botanical Gardens: Kew Gardens Map (2017). There is only reduced selection of visitor interesting trees. Very seldom is used the GIS for greenhouses or small detail botanical area.

Unfortunately, detailed documentation of the greenhouse collections, including the plant origins and ages, does not exist. It was destroyed by a fire in the Archives of the Exhibition Grounds, which occurred in the early 1990s. The remaining documentation was completely destroyed during the 1997 floods.

The only remaining subtropical greenhouse plan that still exists is the plan proposed by I. Otruba. This plan is presented in the poster. It contains suggestions for pavements, bridges, lakes and major plants. The contours are also mapped in the plan. The elevation difference between the entrance and rear of the greenhouse is more than 1 m. This elevated arrangement creates a more attractive exhibit. The basic arrangement of the greenhouse has remained unchanged, including elevated soils, bridges, and the lakes.

The discussion of geoinformatics-cartographers and botanists started before the creation of plans. Some problems also arose during digital plan creation. These questions were solved. What data to measure and digitize firstly, buildings and plants? What level of detail? How express the plants, by point or polygons? What is a correct shape of the polygon for plants to select, schematic or punctual? How assure the visibility of all plants, especially small plants under big plants in the plan? How many data-sets prepare in various scales for printed plan, poster, and interactive map? Next text describes the final solutions and procedures.

A systematic gathering of detailed greenhouse plans began in 2011. First, the greenhouse and botanical garden buildings were measured and digitized using ArcGIS. This resulted in an orientation plan for the entire area, depicting all greenhouses and botanical gardens without position of plants (Figure 5). Some technical building parts as heating, electricity, etc. were omitted. The main purpose

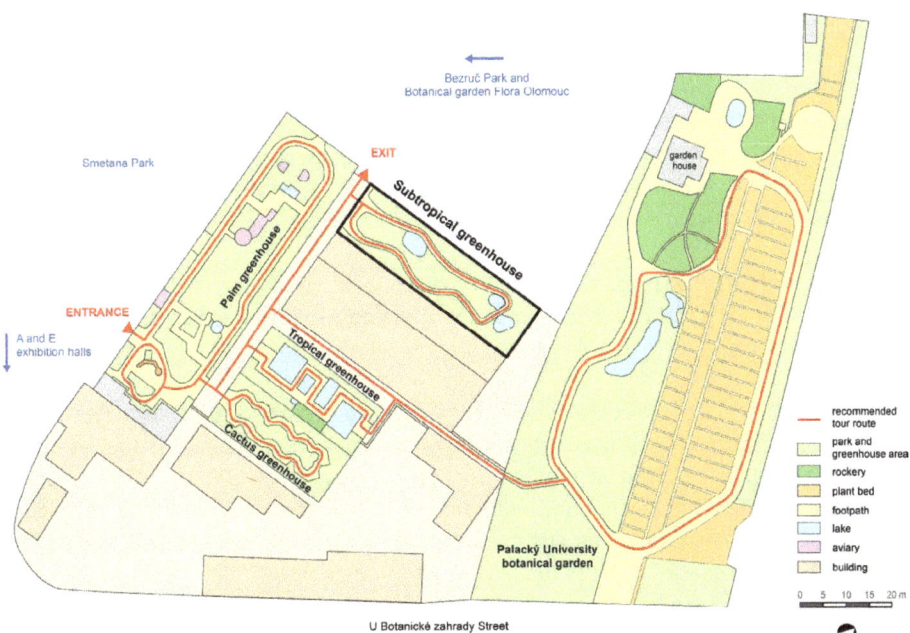

Figure 5. Collection greenhouses and Palacký University botanical garden overview plan.

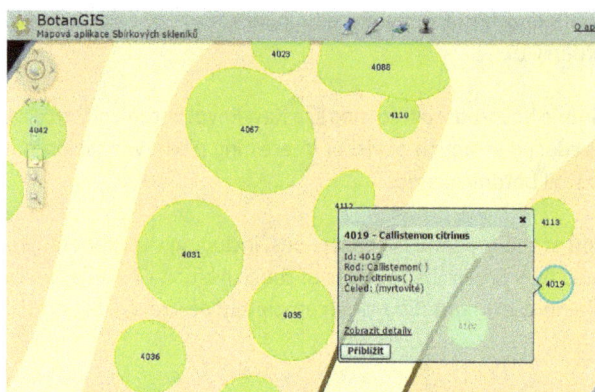

Figure 6. Interactive subtropical greenhouse plan via the portal BotanGIS with a plant callout (www.botangis.upol.cz).

of the mapping is to better imagine expositions and produce a recommended tour route for visitors. The red line denotes the recommended route, which is coloured to attract attention within the plan.

The greenhouse and botanical garden plans have similar map keys. The colour and symbol choices mainly fulfil the cartographic associativity rule. Cartographers recommend green colours for parks and greenhouse expositions. Dark green is used for six rockeries in the botanical garden. Yellow was used for footpaths in both areas. Plant beds were rendered in brown (Dobesova, Vavra, & Netek, 2013).

The format of the plan is A4. This plan is used as overview map in the presented poster. The print version is available for visitors at the entrance to the greenhouses (both in Czech language both English language).

The second part of the project collected data regarding plant position and crown size. The greenhouse plants are drawn as circles or ovals, expressing the actual plant cover of each specimen. The plant list was stored in the database. Each plant received a unique ID number that identifies the plant in the database and plan. The plan and database are both parts of the BotanGIS botanical information system (www.botangis.upol.cz). The plant database is linked to the interactive plans (Figure 6). It is possible to search for plants based on ID, genus, family and other descriptors (Nétek, Dobesova, & Vavra, 2014). The same digital data-sets were used for printed plan, poster (Appendix A) and interactive map. The differences in scales are not so big to generalize data and to create separate data sets for each scale and type of plan output. The reusability of vector data is a big advantage of creation and storing data in GIS.

Three books were issued in 2003 regarding palm, tropical and subtropical greenhouses (Dančák et al., 2013a, 2013b, 2013c). Select species are described in these books. Each book contains a detailed plan of each greenhouse.

The subtropical greenhouse book contains both the original plan from 1991 and the contemporary plan, with a complete list of plants (Dančák et al., 2013a). The book describes only 33 species. These interesting plants are denoted in the plan using a dark green colour, while others are light green. In addition, some species have more than one individual plant in the greenhouse. In the book, each species is described by three paragraphs: distribution and ecology, morphology description and interesting facts about the species. Species are illustrated using photos of fruits, leafs, flowers and habitus (Figure 7).

Moreover, each plant is depicted by a native species range map. The definition of the *native range of the species* term is based on each plant species evolving at a specific place on Earth, where it

4019

CALLISTEMON CITRINUS Skeels

Štětkovec citronový

Syn.: *Callistemon lanceolatus* Sweet
Čeled: *Myrtaceae* Juss. – myrtovité

PŮVOD, EKOLOGIE

Druh pochází z Austrálie. Roste na jejím východním pobřeží ve státech Queensland, New South Wales a Victoria.
Tento druh roste podél skalních potůčků a v příbřežních bažinách, vždy na slunném stanovišti.

POPIS

Stálezelený keř dorůstá do výšky 1–3 m. Má tence oválné až čárkovité tuhé listy, které jsou 3–7 cm dlouhé a 5–8 mm široké. Rostlina má typická hustá válcovitá květenství asi 10 cm dlouhá a 4–7 cm široká, která vynikají početnými, velmi dlouhými, červeně zbarvenými tyčinkami. Samotné květní obaly jsou nevýrazné. Plody jsou kulovitého tvaru, dřevnaté, přisedlé k větvi. Po rozemnutí voní po mentolu.

ZAJÍMAVOSTI

Callistemon citrinus je v subtropických oblastech pěstován jako oblíbený okrasný keř, vyšlechtěno bylo dokonce několik kultivarů. Rostlina obsahuje látku leptospermon, která má biocidní účinky, a na jejím základě byla syntetizována látka mezotrion, která se využívá jako součást některých herbicidů.

Figure 7. An example list of the Australian *Callistemon citrinus* species (Dančák et al., 2013a).

typically occurs at present. However, a large number of species have expanded into neighbouring and even distant areas. If they did so spontaneously, via natural processes and without human contributions, these areas are called *native ranges* (Pyšek et al., 2004; Smith, 1986; Webb, 1985).

All native range maps were created by cartographers at the Department of Geoinformatics, with help from botanists. A collection of 115 native range maps were created for the three above-mentioned greenhouse books (Dančák et al., 2013a, 2013b, 2013c). Some of them are original works. Native range maps are examples of chorochromatic maps. Chorochromatic maps only illustrate nominal data for specific areas using various colours (Kraak & Ormeling, 2003; Voženílek & Kaňok, 2011).

Experimental testing of visualization methods verified good understanding of chorochromatic maps (Pődör & Kiszely, 2014). Two types of regional delimitations were used for the native range maps. If areas were well known and detailed in the botanical literature, we used a red polygon with a detailed outline (e.g. *Actinidia chinensis* and *Arundo donax* on the poster). At times, the coastline region became discontinuous and comprised islands (Figure 8). Accurately drawing the regions proved difficult, with botanists verifying the final accuracy.

If a native range was approximated, only a red oval outline was used (e.g. *Punica granatum*, *Citrus madurensis*, *Citrus sinesis*, *Citrus limon* and *Lycianthes ranntonei* on the poster—Appendix A). Botanists were uncertain of all of the native ranges, especially for early domesticated crops. In these cases, they assumed or inferred the native ranges. Various visualization methods exist for approximating data (Brus, Vozenilek, & Popelka, 2013). Delimitation using smooth ovals is simple and comprehensive. Therefore, it is an effective generalization method.

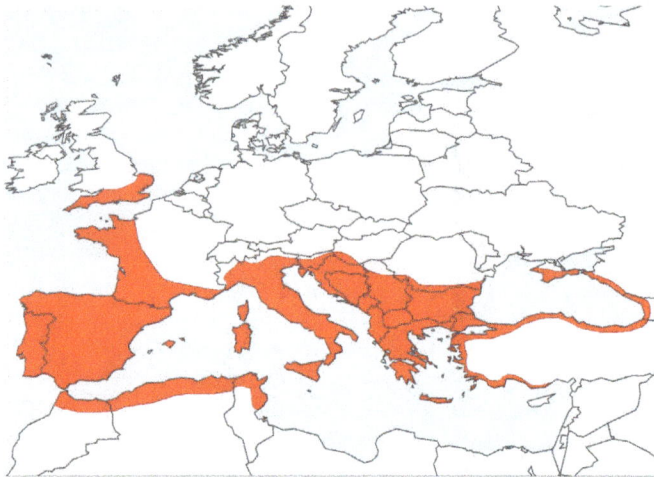

Figure 8. Europe native ranges for *Ruscus aculeatus* (Dančák et al., 2013a).

5. The contribution of cooperation

The presented utilization of GIS is a cross-disciplinary example of cooperation between botanists and geoinformatics. Chain of discussions and partial solutions of problems resulted in the series of useful map results. One of the solved problematic tasks were the recommendations and the decision how to correctly expressed suitable shape of crowns of plants from the botanical point of view. Mainly circles and ellipses were chosen. The jagged snake shapes were used only sporadically in case of three bushes. The smoothed shapes were preferred. Exact detail shapes are not necessary. The shapes could be various due to the pruning of bushes and trees in seasonal maintenance. The form of the crow expression also respects the botanic tradition, that uses only simple circles for crowns. From the cartographical point, the unsmoothed shapes also disturbed reading of plan and have non-aesthetic impression.

We also solved the expressing of cover by plant crowns to be all visible in plans. ArcGIS served very helpfully by the operation *Sort* for polygon shapes. Descending sort according to polygon area moved smaller shapes higher to be visible in the plan under bigger crowns. Without botanists were not possible to decide the correct set of important plants to be inserted to the plans. Some plants are seasonal, and they were not included in plans.

The outcome of GIS research is that data collecting could be used for mapping in very detailed scale, as the position of plants in greenhouses. The distances between plants were sometimes in centimeters in reality. The next outcome is the experience with the multiple utilizations of the same data. One source of digital vector data (shapefiles) was used for several outputs: interactive map on the web, orientation maps in greenhouses as big posters, and maps in printed books about planted species. Both administrators of greenhouses and visitors use all maps in everyday work. The map outputs are helpful also for botanical education that takes place in the area of greenhouses. All these useful experiences increase the skills and knowledge in using GIS in the fields of mapping botanical objects.

5.1. Software

The ArcGIS for Desktop 10.2 software, which is produced by Esri, was used to create the following plans: the subtropical greenhouse plan, the collection greenhouse area orientation plan and the native range maps. The AutoCAD Map 3D Raster Design 2015 software, which is produced by Autodesk, was used to clean and de-skew the scanned, original plan from 1991. This old plan was necessary to prepare before presentation on the poster. Some speckles and pencil notes were there. All these unwanted scraps were cleaned. The poster was designed and printed using AutoCAD Map 2015. This software was powerful to collect and maintained all partial plans, maps, photos and texts to one big informational poster.

5.2. Data

The source data were taken from the "Data and Maps for ArcGIS" data-set, which is available in ArcGIS (Esri, 2015). Topographic data for the continents and political boundaries were used as base layers for native range maps. The data about position and crowns of plants were measured manually in the field.

6. Conclusions

The creation of the BotanGIS information portal provided a digital data-set for the collection greenhouses and botanical garden in Olomouc. The set was used to create a series of maps with several purposes and scales. The overview map contains all of the greenhouses, buildings and botanical gardens, as well as the main tour route. The detailed plans for each greenhouse depict plant locations and crown shapes. The use of ArcGIS to store and illustrate the data allows for the publication of plans in different forms. The first form is the interactive BotanGIS portal. The second form is the plans printed in the three books. Finally, plans were used as important poster features (format A0).

The content of poster for subtropical greenhouse is depicted in this article and presented as Appendix A. Posters were placed near the greenhouse entrances. ArcGIS was also used to produce native range maps.

Funding

This work was supported by the European Social Fund (project CZ.1.07/2.3.00/20.0170) of the Ministry of Education, Youth, and Sports of the Czech Republic.

Author details

Zdena Dobesova[1]

E-mail: zdena.dobesova@upol.cz

ORCID ID: http://orcid.org/0000-0002-3989-5951

[1] Faculty of Science, Department of Geoinformatics, Palacký University, 17 listopadu 50, 771 46 Olomouc, Czech Republic.

References

Alliance for Public Gardens GIS. (2011). The ArcGIS public garden data model. Retrieved from http://publicgardensgis.ucdavis.edu/downloads/data-model/

Arnold Arboretum of Harvard University. (2011). Retrieved from http://arboretum.harvard.edu/explorer/

Brus, J., Vozenilek, V., & Popelka, S. (2013). An assessment of quantitative uncertainty visualization methods for interpolated meteorological data. In *Computational science and its applications – ICCSA, Lecture Notes in Computer Science* (vol. 7974, pp. 166–178). Ho Chi Minh City: Springer Berlin Heidelberg.

Dančák, M., Šupová, H., Škardová, P., Dobesova, Z., & Vávra, A. (2013a). *Zajímavé rostliny subtropického skleníku Výstaviště Flora Olomouc* [Interesting species in the subtropical greenhouse of flora Olomouc exhibition grounds] (48 p.). Olomouc: Palacký University. ISBN 978-80-244-3548-0.

Dančák, M., Šupová, H., Škardová, P., Dobesova, Z., & Vávra, A. (2013b). *Zajímavé rostliny palmového skleníku Výstaviště Flora Olomouc* [Interesting species in the palm greenhouse of flora Olomouc exhibition grounds] (65 p.). Olomouc: Palacký University. ISBN 978-80-244-3672-2.

Dančák, M., Šupová, H., Škardová, P., Dobesova, Z., & Vávra, A. (2013c). Zajímavé rostliny tropického skleníku Výstaviště Flora Olomouc [Interesting species in the tropical greenhouse of flora Olomouc exhibition grounds] (52 p.). Olomouc: Palacký University. ISBN 978-80-244-3885-6.

Dobesova, Z., Vavra, A., & Netek, R. (2013). Cartographic aspects of creation of plans for botanical garden and conservatories. In *SGEM 2013 13th International Multidisciplinary Scientific Geo Conference, Proceedings* (vol. I, pp.653–660). Sofia: STEF92 Technology Ltd. ISBN 978-954-91818-9-0. doi:10.5593/SGEM2013/BB2.V1/S11.006

Esri. (2015, June 6). Data and maps for ArcGIS. Retrieved from http://www.esri.com/data/data-maps/data-and-maps-dvd

Kew Royal Botanical Gardens: Kew Gardens Map. (2017). Retrieved from http://www.kew.org/visit-kew-gardens/plan-your-visit/map

Kraak, M. J., & Ormeling, F. (2003). *Cartography: Visualization of geospatial data* (2nd ed., 129 p.). Harlow: Prentice Hall.

Nétek, R., Dobesova, Z., & Vavra, A. (2014). Innovation of botany education by cloud-based geoinformatics system. In Q. Wang (Ed.), *Innovative use of online platforms for learning support and management, International Journal of Information Technology and Management* (Vol. 13, No. 1, pp. 15–31). Geneva: Inderscience Enterprises Ltd.. ISSN online: 1741-5179, ISSN print: 1461-4111. doi:10.1504/IJITM.2014.059149

Põdör, A., & Kiszely, M. (2014). Experimental investigation of visualization methods of earthquake catalogue maps. *Geodesy and Cartography, 40*, 156–162. doi:10.3846/2029 6991.2014.987451

Pyšek, P., Richardson, D. M., Rejmánek, M., Webster, G. L., Williamson, M., & Kirschner, J. (2004). Alien plants in checklists and floras: Towards better communication between taxonomists and ecologists. *Taxon, 53*, 131–143.

Shields, B. (2010). *Public gardens grow research capability with GIS, 2010*. Retrieved from http://www.esri.com/news/arcwatch/0810/uc-davis.html

Smith, P. M. (1986). Native or introduced? Problems in the taxonomy and plant geography of some widely introduced brome-grasses. *Proceedings of the Royal Society of Edinburgh. Section B. Biological Sciences, 89B*, 273–281.

Voženílek, V., & Kaňok, J. (2011). *Metody tematické kartografie: vizualizace prostorových jevů* [Methods of thematic Cartography: Visualisation of Spatial Phenomena] (216 p.). Olomouc: Palacky University. ISBN 978-80-244-2790-4.

Webb, D. A. (1985). What are the criteria for presuming native status? *Watsonia, 15*, 231–231.

6

The use of terrestrial laser scanning in monitoring and analyses of erosion phenomena in natural and anthropogenically transformed areas

Paweł B. Dąbek[1]*, Ciechosław Patrzałek[2], Bartłomiej Ćmielewski[3] and Romuald Żmuda[1]

*Corresponding author: Paweł B. Dąbek, Institute of Environmental Development and Protection, Wrocław University of Environmental and Life Sciences, pl. Grunwaldzki 24a, Wrocław 50-363, Poland

E-mail: pawel.dabek@upwr.edu.pl

Reviewing editor: Louis-Noel Moresi, University of Melbourne, Australia

Abstract: Implementation of terrestrial laser scanning creates new opportunities in the analysis of processes occurring in the environment. Terrestrial laser scanning (TLS) enables fast, precise mapping of land relief and dimensions of technical structures. TLS allows moving from traditional methods, such as point measurement, to differential analyses of a relief model. TLS allows accurate analyses of land relief changes and erosion phenomena, both in terms of the processes intensity and range, and the initial and intensifying events. The paper presents the authors' experiment with the use of the TLS technology in erosion processes monitoring and analyses. The authors present the methodology of using TLS which contributed to the acquisition of reliable results by providing optimal conditions for the research. The paper shows the effects of laser scanning in order to analyse the intensity of soil erosion in mountainous forest area with a large number of measurement stations, which resulted in a dense points cloud. Differential model method allowed to evaluate erosion phenomena in the form of rills resulting from surface flushing, which is impossible to assess by traditional methods of erosion measurements. In the paper, we showed the necessity to choose the proper coordinate system (object or scanner) for the analysis of the landslide, depending on structure and direction of the soil masses movement. The application of the object coordinate system showed greater erosion phenomena range (ca 12%) on buttress stabilizing landslide and ca 44% on natural landslide, in which the application of the object coordinate system increases the number of observations by 40%.

ABOUT THE AUTHOR

Paweł Dąbek has been awarded with a PhD degree in specialty of Water Management, Water Erosion of Soil and Forest Engineering for thesis "The water erosion processes of Soils in forest areas in the mountain catchment" in 2017 from Wrocław University of Environmental and Life Sciences, Poland, where he is currently working as a research and teaching assistant. His research interests are GIS, forest engineering, small scale water retention and soil erosion.

PUBLIC INTEREST STATEMENT

We would like to share our experience with using terrestrial laser scanning (TLS) in monitoring the land relief changes and qualitative and spatial assessment of the water erosion processes. The paper presents the conditions of using the TLS technology in erosion phenomena research. We were studied the erosion processes both occurring in natural areas locally on forest skid trails and in the large scale as a landslide, and also on technical structures. Research indicates the importance of choosing the right equipment, coordinate system and spatial resolution, which affects on the quality of measurements, time and costs, and research results.

Subjects: Environment & Agriculture; Earth Sciences; Geomorphology; Environmental Studies & Management; Engineering & Technology; Geomechanics; Remote Sensing

Keywords: terrestrial laser scanning (TLS); object coordinate system; soil erosion; skid trail; landslide

1. Introduction

The term erosion means the processes of destruction of the top layer of the ground by forces of nature, such as wind, water and gravity or by anthropogenic factors. The processes of the degrading effects include mainly the transformation of the soil and terrain and a change of water relations. Erosion can be observed and evaluated by the movements of particles and masses of soil, caused by water and wind, as well as by mass movement understood as movement of large volumes of soil materials (Prochal, 1987). All types and forms of erosion processes lead to levelling of the ground surface.

Analysis of water erosion processes requires knowledge of the intensity of the process under evaluation, the volume of transported material, and the location of erosion. In the case of spatial phenomena, such as surface erosion and mass movements such as soil creep, it is justified to determine where the erosion processes have occurred.

Traditional methods of surveying used in the assessment of erosion phenomena (Prochal, 1987; Prochal, Maślanka, & Koreleski, 2005) are based on levelling, typically in order to analyse changes in the thickness of topsoil or to determine the volume of sediment in streams and reservoirs. The use of conventional surveying tools, such as the leveller, allows only to execute point measurements. In the case of spatial phenomena, it is associated with the interpolation between the measured positions results. That leads to averaging the intensity of the phenomena. Using traditional methods for large areas is possible, but also involves a significant investment of time, work, and increases the risk of measurement errors.

Another spatial distribution analysis method is assessment by determining the amount of sediment deposition in the runoff receivers and transported erosion material (Jała & Cieślakiewicz, 2004; Pierzgalski, Janek, Kucharska, Tyszka, & Wróbel, 2007). Similar studies were conducted in the context of sediment accumulation in reservoirs (Czamara, 1992). However, one of the most common methods for measuring surface erosion (sheet erosion) and mass movements (soil creep) is to measure the volume of eroded material retained in the grippers (catchers)—these are temporary structures, usually made of impermeable synthetic material mounted on a rack, or in septic tanks, whose task is to capture the erosion material from the run-off. This methodology allows to perform an analysis of eroded material in terms of its weight and volume.

Field studies and analyses of erosion processes resulting from natural precipitation are carried out in many countries. These types of researches are similar not only in terms of the causative factor being natural, but also in the methodology of measurement and design of equipment for measuring the intensity of run-off and erosion. However, there are different reasons for undertaking the researches. Benavides-Solorio and MacDonald (2005) and Larsen, MacDonald, Brown, Rough, and Welsh (2009) conducted studies on the Colorado Front Rage (USA), where catchers were used to analyse the intensity of surface erosion in the area of a fire and the factors that cause erosion in these specific circumstances. Black and Luce (2013) undertook the analysis of erosion on forest roads, by draining run-off from the road surface with transported debris to septic tanks. Furthermore, the correlation of intensity of precipitation, runoff intensity, and volume of sediment were analysed in relation to the road area and its structure. Similar studies have also been conducted in Poland. One of them was realised by Smolska (2002) in the Suwałki Lake District. Evaluation of spatial erosion processes may also be carried out using representative experimental plots. Studies on those dedicated test stands—plots - may be conducted in the laboratory as a result of simulated precipitation, as well as in the field under natural conditions (Rejman & Usowicz, 2002; Ruiz-Sinoga,

Romero-Diaz, Ferre-Bueno, & Martínez-Murillo, 2010). Martínez-Zavala, Jordán López, and Bellinfante (2008) developed the impact of topography and vegetation on soil erosion intensity using simulated rainfall on several plots in forest areas by measuring the amount of run-off and eroded material. Similar studies were also carried out in Spain by Jordán and Martínez-Zavala (2008).

TLS technology has changed the way of analysing the natural phenomena like erosion processes. Geodesy surveying has evolved into 3D models analyses (Barbarella & Fiani, 2013; Barbarella, Fiani, & Lugli, 2015; Coppa, Guarnieri, Pirotti, Tarolli, & Vettore, 2013; Denora, Romano, & Cecaro, 2013; Guzzetti et al., 2012; Vosselman & Maas, 2010). The use of laser technology in the research of erosion phenomena is becoming more and more common in the scientific community. Such research was conducted by Dąbek in terms of surface flushing on forest roads in the Sudetes, Poland (Dąbek, Żmuda, Ćmielewski, & Szczepański, 2014), Ballesteros-Cánovas in terms of influence exposed roots on sheet erosion in experimental approach in Barranca de los Pinos, Spain (Ballesteros-Cánovas, Corona, Stoffel, Lucia-Vela, & Bodoque, 2015), Stenberg in terms of erosion of peatland forest ditches, Finland (Stenberg et al., 2016), Eltner in terms of several years soil erosion research on small, unpaved fields in Spain and Germany simultaneously showing DTM comparison obtained from TLS and photogrammetry from UAV (Eltner, Mulsow, & Maas, 2013) and many others. No homogeneous or unambiguous procedures and techniques have been developed for the investigation and monitoring of erosion processes on technical structures or in the natural environment. Such guidelines could be necessary in terms of determining the optimal density of the point cloud, the number of scanner stations relative to the research area, application the coordinate system (object or scanner), methods for interpolating the point cloud to DTM and others. The analysis of the topic and the research we conducted were the reason for preparing such a study based on our experience in the use of TLS in the research of various forms of erosion phenomena.

The authors' aim was to demonstrate the usability of modern and rapidly evolving technology—terrestrial laser scanning (TLS)—for monitoring eroded areas. The main aim of the paper was to determine the conditions for the using TLS in research of the qualitative and quantitative assessment of erosion processes, measurement intensity and range of erosion and monitoring of these phenomena. We analyse the conditions of using TLS in the terms of the adopted measurement accuracy, density of point cloud, application the coordinate system, interpolation to DTM and adopting the differential model method for further analysis. We conducted analyses in terms of erosion processes in the form of landslides on a technical structure and in the natural environment, as well as in the form of sheet and rill erosion on the example of forest areas of operation trials.

2. Methods and materials

Laser scanners allow to take inventory of an area with a high resolution. The results are presented as a point cloud (Afana, Solé-Benet, Pérez, Gilkes, & Prakongkep, 2010). This way, the collected data allow to perform quantitative and qualitative analysis of the occurring phenomenon. The economic criteria have been also discussed, such as the time and ease of measurement (a single measurement session does not exceed a few hours of field work), as well as the possibility of stepping into previously inaccessible areas of analyses.

2.1. Types of laser scanners

Two technologically different scanners are used to obtain data (Vosselman & Maas, 2010). A pulse scanner (like Leica ScanStation2 that was used in our researches) and a phase scanner (like Faro Focus3D).

The phase laser scanner measurements are based on comparing the phase shift of the emitted and returned laser light. This type of scanner usually uses a signal with a sinusoidal modulation. This system is at risk of losing one full cycle. In order to avoid this error, a plurality of frequencies is used. The low frequency of the long wavelength determines the scope, while the shortness of the wavelength determines the accuracy that can be obtained.

Measurement using a phase scanner is easy because the scanner size allows to transport the device. Due to measurement noise, the data are automatically filtered during the measurement. Consequently, many points of the object cloud are removed by filtration algorithms, making this type of scanner less useful in the measurement. Because of that there are places for which the data are inaccurate or missing. However, in other areas, it is possible to determine the areas of erosion processes.

Distance measurement method used in pulsed scanners is based on recording the time of emitting and returning reflected laser light wave from the object. Knowing this time, as well as the speed of light and atmospheric factors, it is possible to count twice the distance from the object to the scanner. Therefore, there is a need to use high-precision timers.

The way the two instruments measure the distance entails different work parameters of the devices. A phase scanner achieves a much higher measurement speed (more than one million points per second), but its range is usually twice shorter, while a pulse scanner is characterised by a much lower measurement noise (Vosselman & Maas, 2010).

2.2. Methodology of analysis
The data acquired from a terrestrial laser scanner must be filtered, especially in natural areas where vegetation can be very dense. To clean point clouds form noise, it is possible to use several algorithms. Points representing the terrain surface are located the lowest in the cloud. In this work, a programme designed to filter data from a laser scanner was used. TerraScan used the active algorithm TIN (Brodowska, 2012).

After reducing the noise, a set of points representing the surface of the terrain was analysed using geostatistical tools—Surfer v10. Algorithms implemented in the software create digital terrain models (DTM) using different interpolation methods such as kriging, triangulation, nearest neighbour method and others, allowing a thorough analysis of the data-set. The study is based on kriging because it reflects the terrain best (Chaplot et al., 2006; Czamara, 1992; Oliver & Webster, 1990). Using the primary measurement filtered data and the control one, a DTM for the study objects was developed. The primary DTM was subtracted from the control DTM, resulting in a differential model of the terrain. The obtained model allows to see the changes that have been taking place around the facility, as well as their extent. It should be reminded that the laser scanning method is one of quantitative methods and as such it determines the location of the effects of the phenomena on the entire object area. An in-depth analysis of the area models can provide data concerning phenomena intensity.

2.3. Study areas

2.3.1. Forest skid trail in a mountain area
Human influence on the natural balance of forests, including establishing networks of forest roads and trails or logging, and the effects of these treatments, negatively affects the forest soil protection function (Chang, 2003; Kusiak & Jaszczak, 2009). Changing the land use, the landform (Burley, Evans, & Youngquist, 2004; Prochal et al., 2005), and the soil's shear strength and the ability of soil to resist force working parallel to the soil surface which is the result of using the land for transport, leads to intensification of soil erosion processes. Erosion phenomena are a result of concentration run-off on forest roads and trails. Chang (2003) shows that the roads and trails network is essential for the management of forests, but as much as 90% of the erosion material deposited in streams is derived from road surface. In areas with heavy relief, erosion occurring on road scarps located on slopes could generate 70–90% soil loss from the entire network in the area. Also Croke, Hairsine, and Fogarty (2001) showed that the erosion on skid trails (trails used to transport timber from a harvest area to a storage place) is approximately 30-fold greater than erosion from the surface of harvesting. The most differences are observed 1–2 years after the logging.

Figure 1. Terrestrial laser scanning on land relief analyses. The Szklarska Poręba Forestry District.

Photo: Dąbek (2011).

Therefore, identifying the erosion and hydrologic processes and the intensity of these phenomena in the context of skid trails, as well as the effects of anthropogenic activities, is a legitimate subject. To perform such field measurements TLS technology was used (Figure 1). TLS allows the analysis of changes in local land relief in relation to the entire experimental plot, taking into account changes in the structure of the terrain in the vertical axis (depth of rills—the effect of flushing the surface and rills erosion) and in the horizontal axis (width and length of rills), as well as the spatial arrangement. This allows to analyse all the formed rills network and changes in the thickness of the topsoil. TLS also allows to analyse the line of the runoff concentration and to perform quantitative evaluation of water erosion.

The experimental area was located in the western part of the Sudetes, in the Izera Mts., Poland, on the slope of Ciemniak (GPS N: 50° 51′ 33″; E: 15° 34′ 45″). The area under study is a forest area managed by the Szklarska Poręba Forest District (The State Forests National Forest Holding). In the analysed area, timber logging was conducted until the end of 2010 according to the planned activities. For proper and efficient conduct of logging, skid trails have been designated and located in accordance with the State Forests regulations. Setting and using the trails causes topsoil degradation, changes in local land relief, and thus contributes to the intensification of soil erosion processes on the forest slopes. The increase in erosion phenomena on the trails changed the stability of the

Figure 2. Land relief changes as a result of skidding and erosion phenomena of the trail.

Photo: Dąbek (2010–2012).

topsoil (Chang, 2003). With the end of the logging, a selected section of the trail was chosen and an experimental area was created. The change of local land relief of the trail (Figure 2) was documented from the skidding time until the end of the research, focusing on the erosion effects emerging after the works.

2.3.2. Anthropogenically transformed area

The next example are erosion processes occurring on land transformed by human activity—buttress stabilising landslide (Figure 3) on the rail trail in Mazurowice, Poland (GPS N 51° 12′ 53,94″, E 16° 27′ 43,44″). Deep cuttings and high embankments favour the occurrence of mass movements, especially if they are located on cohesive soils (Figure 4).

The railway track in this area is situated on ground built with cohesive soils, including mostly Tertiary clays. The terrain forced the location of the E-30 railway line on high embankments, h > 6.0 m, and in a deep excavation, which creates a greater risk of landslides. The research area is located on the edge of the plateau of Malczyce. In the upper part of the slope, drilled geotechnical bores (3.0–7.0 m depth) revealed the Quaternary formations of Middle Polish Glaciation, in the deeper parts - sand over Tertiary formations. Cohesive soils occurring in the area, with numerous interbeds of silt, were classified as silty clays and loess-like sand over Tertiary formations. The soils in the subsurface layers (0.6–3.8 m depth) occur in a firm state (IL = 0.3), and deeper in a stiff state

Figure 3. Buttress stabilizing landslide within the railroad E-50.

Photo: Ćmielewski (2011).

Figure 4. Landslide before slope stabilisation.

Photo: Ćmielewski (2011).

(IL = 0.2). Deposits (to 2.3 m depth) were classified as embankment material due to a variety of drilled material and numerous interbeds. In cohesive soils (to 4.3 m depth), numerous interbeddings in the form of hydrated medium sands were observed. The ceiling of Tertiary clays was observed at 4.2–4.3 m depth on ~121.00 m a.s.l. These deposits are in a stiff (IL = 0.2), semi-hard (IL = 0.0) and locally in a firm state (IL = 0.3). Tertiary formations included also soils in the form of medium sand, sandy gravel, sandy clay and silt which formed local interbeddings. The landslides colluvium and the direction of rainwater runoff were determined.

2.3.3. Landslide area

In 1997, after the July flood in the Sudetes region, a landslide occurred in the area of Janowiec village, Poland, on the slope of Kurzyniec Mt. (GPS N 50° 30′ 18,53″, E 16° 44′ 57,50″). The landslide area is located on Quaternary formations, in particular on deluvial clay with a content of rubble. The oldest rocks in the area are revealed in the erosion indentations as mudstones, clays, greywacke sandstones assigned to Lower Carboniferous formation from Opolnica (Cwojdziński & Pacuła, 2009; Oberc, 1987). The development and geological position of Lower Carboniferous deposits in the Bardzkie Mts. is shown in numerous studies (Chorowska & Radlicz, 1994; Chorowska & Wajsprych, 1995; Finckh, 1932; Oberc, 1987; Wajsprych, 1995; Wajsprych, 1978, 1986).

The landslides of Quaternary formations situated directly on the Carboniferous formation created a phenomenon. The main slope is clearly defined in the morphology, numerous secondary slopes and crevices are located in the middle part of the landslide, and the tongue of the landslide is poorly formatted (Figure 5). The study area is bounded by deep erosion valleys oriented transversely to the slope (up to 8 m depth). At the bottom, strongly fractured Carboniferous shales are exposed. In the eastern canyon, due to backward erosion, a sub-notch is developing in the SW-NE direction. This direction is analogous to the slopes. It may reflect the neotectonic movements related with a system of the Sudetes Marginal Fault that occur along the sub-block faults (Badura et al., 2003; Krzyszkowski, Migoń, & Sroka, 1995).

The landslide moves directly to the bottom of the Nysa Kłodzka valley (Figure 6). The area of the landslide is about 1.1 ha (130 m length, 120 m width), and it reaches about 20–25 m above the level of the river. However, this did not result in flow limitation in river Nysa Kłodzka, due to the small volume of moving mass. The landslide is located from 295 m a.s.l. to the bottom of the Nysa Kłodzka valley (225 m a.s.l.). The landslide was founded on the northern slope of the Kurzyniec Mt. In the upper part of the investigated area the slope does not exceed 5°. On the edge of the Holocene valley of the Nysa Kłodzka, following the Sudetes fault, the slope attains even 30°. The area of the landslide forehead is covered mostly with deciduous trees, the upper part of the landslide is agriculturally used (Figure 7).

One of important elements in research on landslides is to determine the area and micromorphology of the phenomenon. Several methods can be used for surface measurements, like tachymetry, GNSS, hand-held GPS loggers. It is also important that the data are intended to be used at a later stage of deformation research. One of the methods—modern and relatively fast allowing for the accurate reproduction of micromorphology as well as allowing for deformation analyses is Airborne laser scanning (ALS) (Pawłuszek & Borkowski, 2016) or close range photogrammetry from unmanned aerial vehicle (UAV). Landslides located in the mountainous region are situated in areas usually covered dense vegetation where those methods do not allow for capture occurring changes of terrain morphology because of occurrence a large covering of high vegetation. With TLS method we can work below the vegetation, directly from the ground, we also obtain lots of additional information while scanning, like vegetation. Using filtration data can be classified into classes of vegetation (low, medium and high) and the ground class.

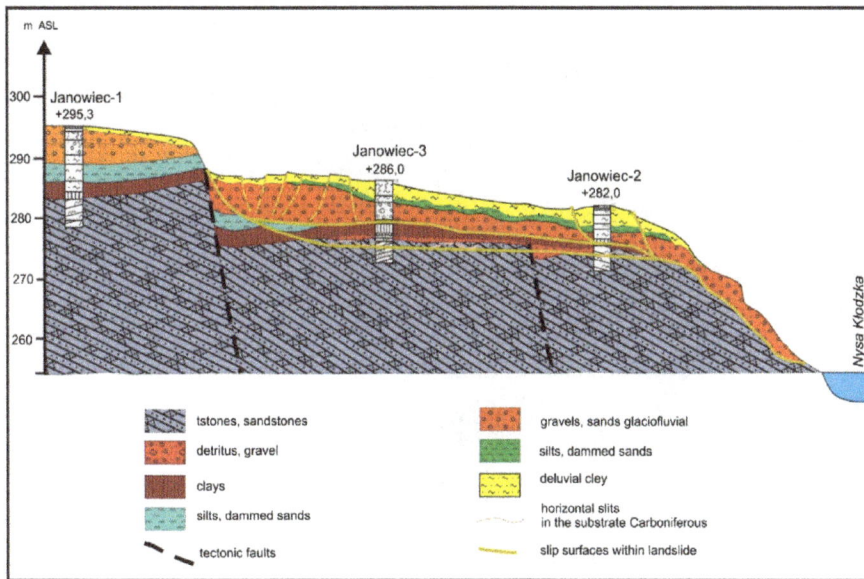

Figure 5. Geological cross-section of the Janowiec landslide.

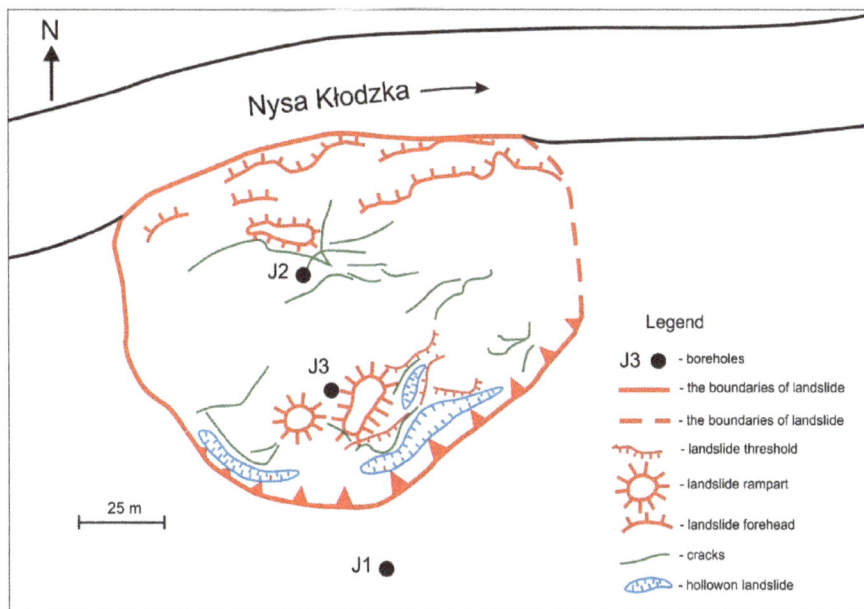

Figure 6. Mapping of Janowiec landslide by SOPO guidelines (Różański and Urbański).

3. Results

3.1. Forest skid trail in the mountain area

Applying TLS made it possible to automatically measure and to produce measurable and fully objective models of the terrain in a short time. Scanning resolution was scheduled and set as a grid of 3 × 4 mm at a distance of 10 m. The accuracy of scan registration and adjustment for a single measurement campaign did not exceed ±3 mm. The average density of the point cloud after filtration in different sectors of skid trail on 1 m² is shown in Figure 8. Data prepared in such a way were imported into Surfer10, where the interpolations of the point cloud and the basic analyses were done. GRID

Figure 7. The ortophotomap of the Janowiec landslide mad from a UAV.

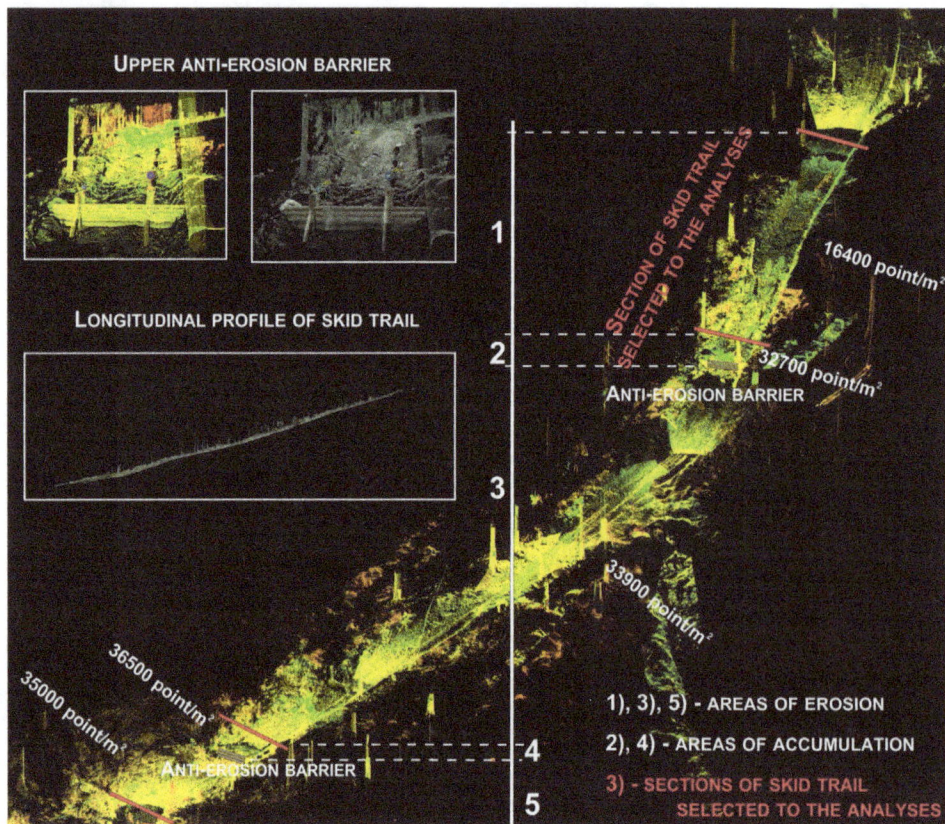

Figure 8. Profile of skid trail on research area, density of the point cloud from TLS.

with a density 1 × 1 cm was created. As a result of the measurements and conducted analyses, the areas of soil erosion occurrence and the areas of eroded material deposition were determined. Analysis of DTM created from laser scanning also allowed to calculate precisely the intensity of the erosion processes. Interpolation of point cloud (obtained from the scanning) to resolution of 1 × 1 cm allowed accurate assessment of erosion rills, even for their minimal depth. In comparison, data interpolation with lower resolution or using traditional methods such as levelling would not show such small changes in local land relief. As a result of inaccurate interpolation or measurement errors, the magnitude of erosion processes may be underestimated or excessively averaged.

To present the research results, the lower section of the analysed trail, which ends with an anti-erosion barrier, was chosen. The area of the selected section is 131 m² in total, with an average width of the trail at about 2 m. The section is characterised by an average decrease of about 20%.

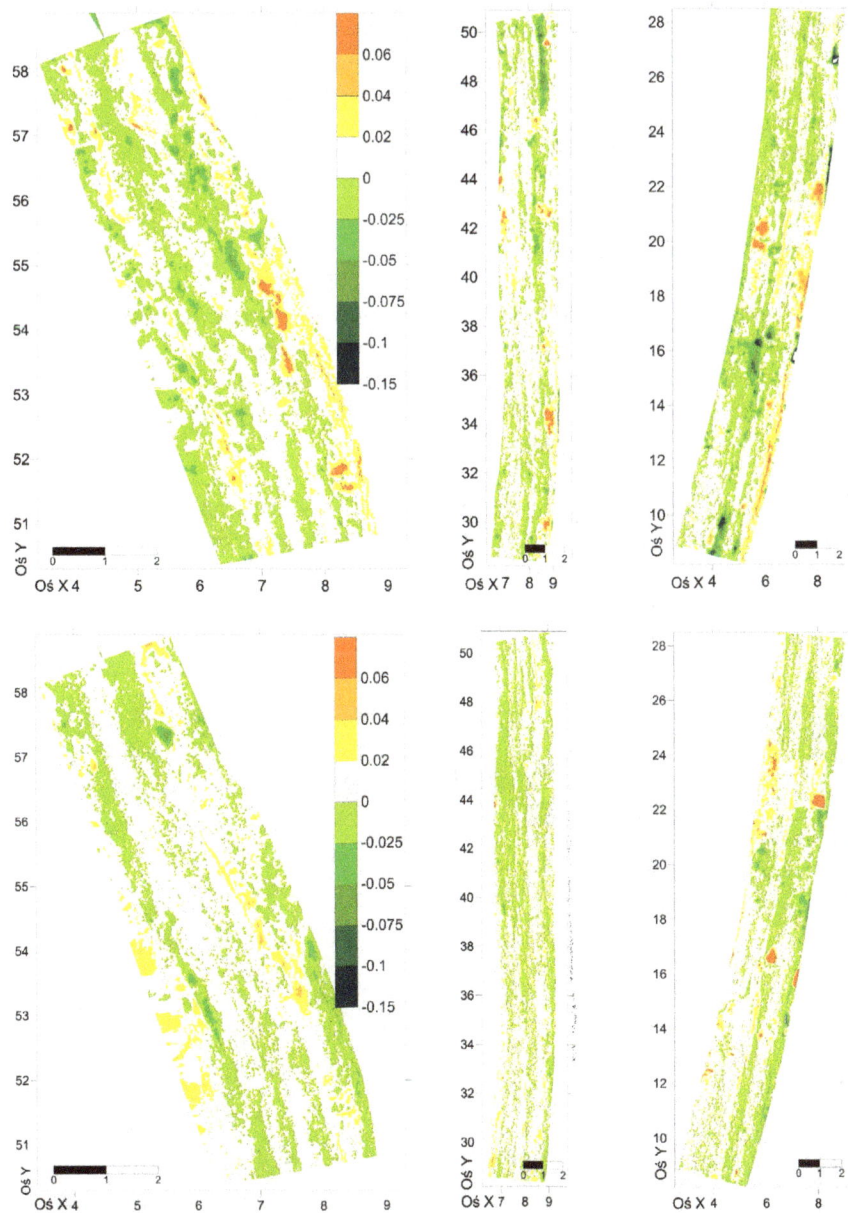

Figure 9. Contour map of the differential model of the skid trail section (the scanner coordinate system), model of June–August.

The density of the point cloud from TLS for the selected trail section is above 33,000 points per m^2 (Figure 8). This accuracy of data allowed to create DTMs of this area, to carry out the analyses and to interpret the results.

The analysis of changes in local land relief was obtained by comparing the subsequent models. The data from scanning were acquired in June, August and November, 2011. The effects of water erosion during that period are shown in Figure 9.

Local land relief changes, resulting from erosion processes, occurred in more than 38% of the analysed section between June and August, and in about 33% between August and November, 2011. Run-off concentration lines were formed in the ruts, caused by the movement of heavy forestry tractors and timber hauling. Erosion rills, with an average depth of 1–10 cm, were formed in those lines. At the same time, terrain elevation outside the ruts was reduced by approximately 15 cm. Analysis of the DTMs also shows intensive erosion occurring on trail scarps. The dimensions of the formed rills and their location on the trail area indicate that the intensity of erosion depends on local land relief of the trail, local slope, and varying topsoil stability. During the research period, the estimated volume of eroded material in the analysed area was approximately 0.9 m^3. A half of this volume was eroded in the first part of the study. This value was calculated as the volume of all the rills formed on the analysed trail section.

3.2. Anthropogenically transformed area

Two measurement campaigns were performed, with an interval of about one year. Both measurements referred to the local survey matrix, located outside of the object impact. The result of the inferred overlapping point clouds is shown in Figure 10. Registration of the objects in a common spatial arrangement and graphical presentation shows the scale of the phenomenon. The sectors of evenly coloured penetrations (point cloud) suggest a low intensity of the phenomenon. Intensification of one colour indicates greater intensity of erosion.

The analysis of the deformation processes was done in two ways: in the scanner coordinate system (Z-axis directed along the vector of gravity but in the opposite direction) and in the object coordinate system (Z-axis directed perpendicularly to the plane fitted to the research object). The aim of this was to improve the accuracy of the analysis by better interpolation in connection with the extension of the object area (Figure 11).

Figure 10. Point cloud of output (green colour) and control measurement (red colour) of the Mazurowice landslide.

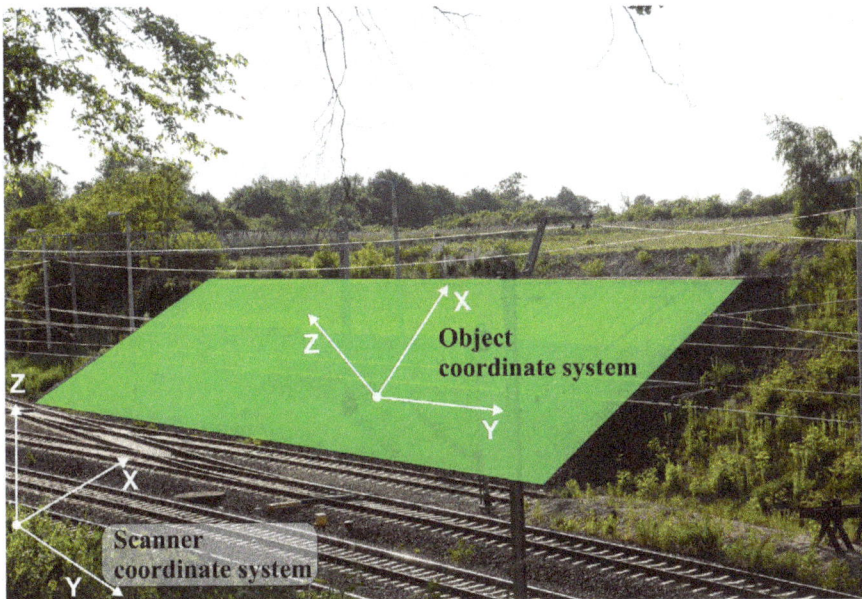

Figure 11. Methods of coordinate system arrangement.

As a result of the calculations the DTMs were developed. Next, the differential models were created (control DTM was subtracted from the output DTM). This procedure was performed for the two cases—the scanner and the object coordinate system (Figure 12). The results are presented in Figure 13.

Both the differential models show object deformation, however, it can be assumed that displacements are not associated with the object movements but with erosion processes—ablation. The differential model coordinate systems show (Figure 13) that the largest changes of land relief occur in the upper part of the object, which is due to the design of the buttresses—in the lower parts it is

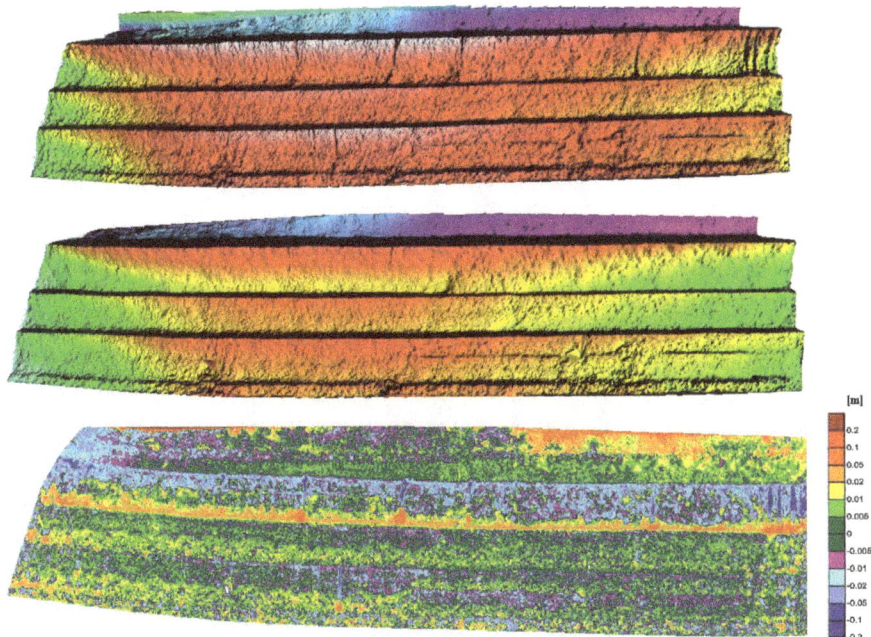

Figure 12. Escarpment of excavation—the object coordinate system: DTM from the control measurement (top, January 2012), DTM from the output measurement (middle).

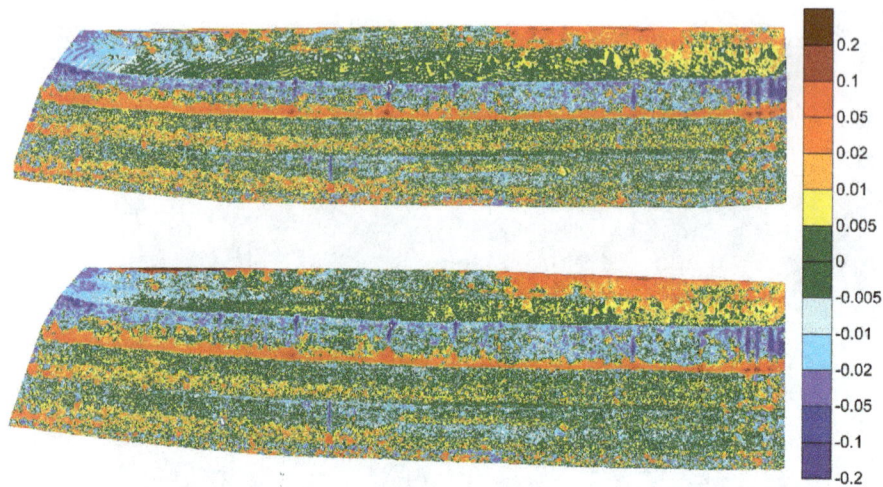

Figure 13. Comparison of differential models from the scanner coordinate system (top) and the object coordinate system (bottom).

made of crushed stones and the upper part is made of sand. The sand was eroded and deposited on the shelf formed by the first layer of the crushed stone.

To examine the distribution of displacements occurring on the object (Figure 14), two histograms were created. While comparing the two histograms, the greater frequency of erosion phenomena (about 12%) can be observed in the object coordinate system. Such a manner of presentation of the changes taking place on the area allows a more detailed analysis of local land relief, especially on very steep slopes.

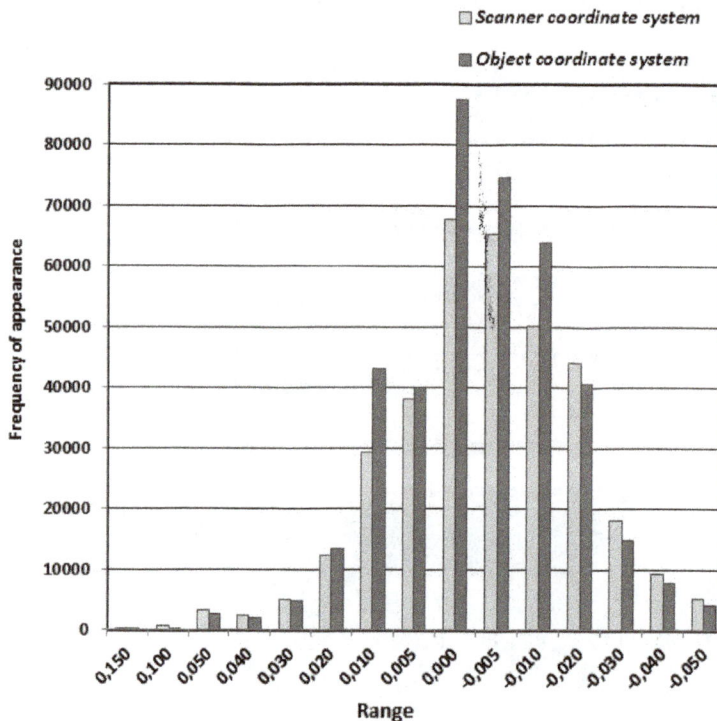

Figure 14. Histogram of differential models for the scanner and object coordinate system.

3.3. Landslide

In the opinion of the authors, the best way to inventory such an area is to perform a large number of scans at various positions, with a short measurement time. That reduces the amount of shade, does not take a lot of time, and allows to represent the terrain micromorphology accurately. After filtration of the data obtained from TLS, the points of the vegetation class are ca 75% and those of the ground class are ca 25% of the total acquired information. Using TLS allows to develop a digital terrain model representing both the area of the landslide and its micromorphology (Figure 15). Analysis of the deformation of the landslide obtained from TLS is possible after control measurement. The results of the measurement campaigns should be compared with each other by performing the differential model. In our presentation of the method, we focus primarily on the head of the landslide as representing the best the occurring transformation of the landslide.

As a result of subsequent measurements which were filtered and oriented to a common reference system, we obtained DTMs of the scanner and the object coordinate system (Figures 16 and 17). As a result it was possible to determine the areas where the mass movement phenomenon is intensified, and to compute the volume of the eroding material.

Comparing these two representations of occurring phenomenon it can be seen that using the object coordinate system the area of deformation increases by ca 44%. Analysing the histograms (Figure 18) of the differential models, the number of observations also increases (ca 40%). In both approaches the net volume of eroding material is positive, which suggests bloating of the landslide. In the authors' opinion, more reliable results are obtained using the object coordinate system,

Figure 15. DTM of the Janowiec landslide forehead.

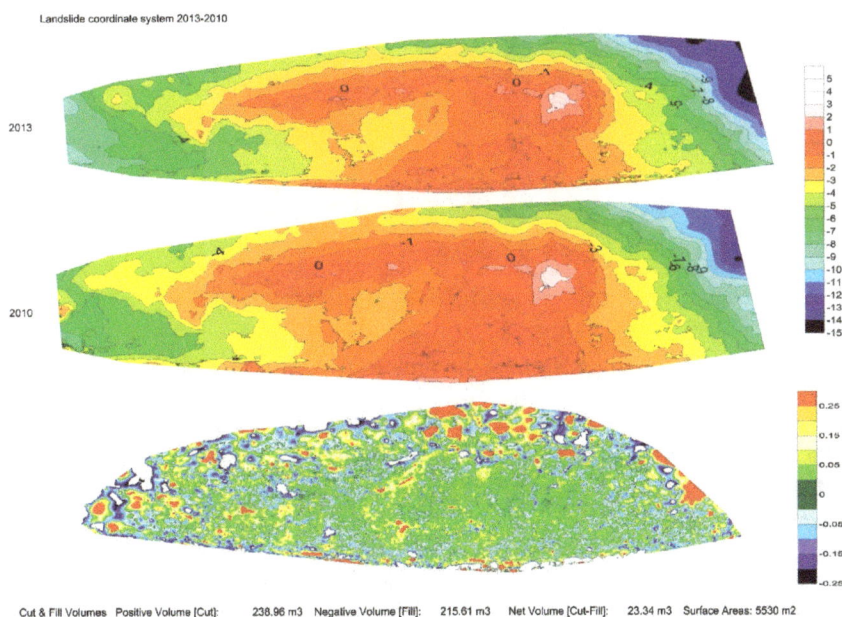

Landslide coordinate system 2013-2010

Cut & Fill Volumes   Positive Volume [Cut]:   238.96 m3   Negative Volume [Fill]:   215.61 m3   Net Volume [Cut-Fill]:   23.34 m3   Surface Areas: 5530 m2

Figure 16. Forehead of the landslide—the object coordinate system: DTM from the control measurement (top, 2013), DTM from the output measurement (middle, 2010).

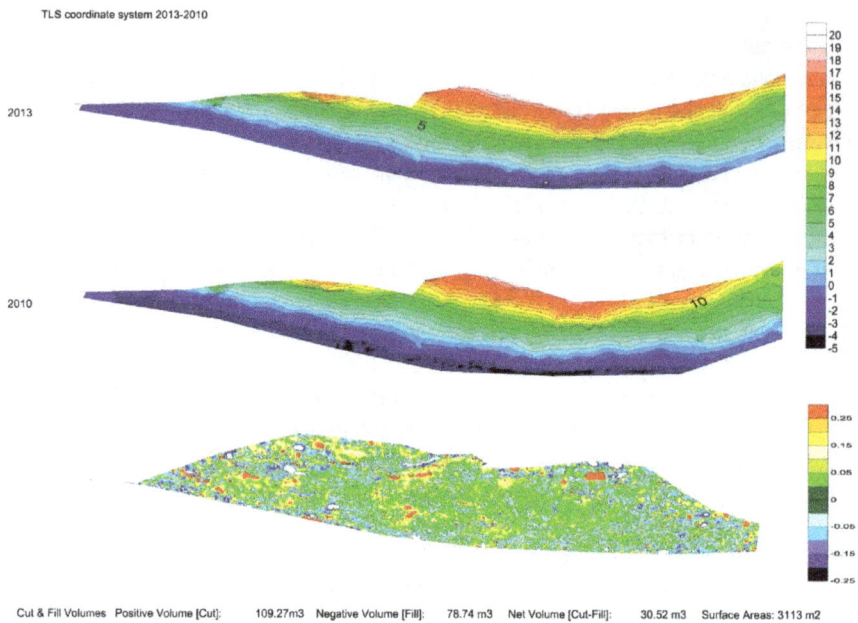

Figure 17. Forehead of the landslide—the scanner coordinate system: DTM from the control measurement (top, 2013), DTM from the output measurement (middle, 2010).

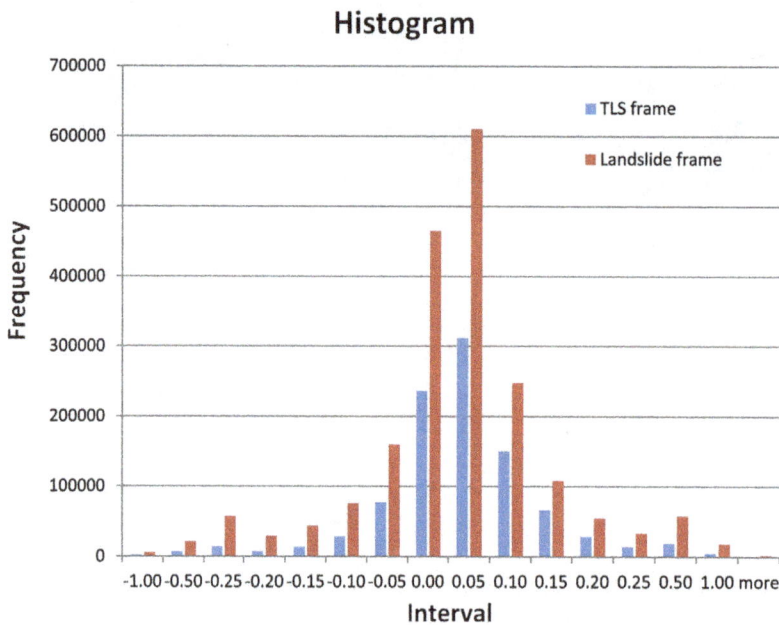

Figure 18. Histogram of the differential models from the scanner and object coordinate system.

because the observed landslide area is more perpendicular to the Z axis. The use of the object coordinate system improves the quality of interpretation of the occurring phenomenon, which significantly affects the reliability of the method.

4. Conclusions

The use of modern techniques and technologies for measuring and processing data in research on erosion phenomena now seems to be natural, easy, faster and giving better and more accurate results than traditional methods. The use of measurement techniques such as TLS, ALS or photogrammetry in combination with geostatistical programmes and GIS gives almost unlimited possibilities, as well as better recognition of the characteristic of erosion processes, their assessment and monitoring. However, the use of modern technologies, measurement techniques and data processing methods requires the adoption of certain assumptions and conditions that need to be defined.

The application of terrestrial laser scanning allows to perform accurate evaluation of the terrain condition, in natural areas (landslide) as well as in anthropogenically transformed ones, such as forest trails and roads or railway embankments and cuttings. The area of erosion processes may be accurately located by means of control measurements. The adopted method allowed rapid assessment of the intensity of these processes and the calculation of the erosion material volume. In comparison with the traditional methods, field measurements are relatively quick and easy to conduct. The methodology has another important advantage—it is possible to identify precisely the areas where intensive erosion processes occur. Changing the scanner coordinate system to the object one increases the accuracy of determining the area of the phenomenon and its intensity on steep surfaces.

The applied automatic laser measurement technology allows to perform modelling of the erosion phenomena that take place on skid trails in mountain forest areas. The traditional measuring method based on catchers allows to evaluate the volume of erosion material. However, this method limits the trails usability. The TLS method does not limit the functionality of the terrain, and the applied resolution gives very accurate results of the erosion processes in terms of volume, variability in time and spatial range of the phenomenon. The study results concerning water erosion in forests are crucial for managing such areas, including damage in forests and small-scale retention, as well as for the cognitive and scientific aspects of erosion.

Erosion processes of landslide (on anthropogenically transformed and natural objects) in the top layer of buttresses built of loose soil are well represented. On the rest of the analysed object, possible processes of water erosion are within the range of scanner measurement error. By transformation of the scanner coordinate system to the object coordinate system, the number of observations increased, making it possible to accurately identify the areas of occurrence of intensive erosion processes.

Funding
The authors received no direct funding for this research.

Author details
Paweł B. Dąbek[1]
E-mail: pawel.dabek@upwr.edu.pl
ORCID ID: http://orcid.org/0000-0003-0203-3116
Ciechosław Patrzałek[2]
E-mail: ciechoslaw.patrzalek@upwr.edu.pl
Bartłomiej Ćmielewski[3]
E-mail: bartlomiej.cmielewski@pwr.wroc.pl
Romuald Żmuda[1]
E-mail: romuald.zmuda@upwr.edu.pl

[1] Institute of Environmental Development and Protection, Wrocław University of Environmental and Life Sciences, pl. Grunwaldzki 24a, Wrocław 50-363, Poland.
[2] Department of Landscape Management, Wrocław University of Environmental and Life Sciences, ul. Grunwaldzka 53, Wrocław 50-363, Poland.
[3] Institute of History of Architecture, Wrocław University of Technology, LabScan3D, ul. Bolesława Prusa 53/55, Wrocław 50-317, Poland.

References
Afana, A., Solé-Benet, A., Pérez, J. C., Gilkes, R. J., & Prakongkep, N. (2010, August 1–6). Determination of soil erosion using laser scanners. In *Proceedings of the 19th World Congress of Soil Science: Soil solutions for a changing world* (pp. 39–42). Brisbane. Published on DVD.

Badura, J., Zuchiewicz, W., Górecki, A., Sroka, W., Przybylski, B., & Żyszkowska, M. (2003). Morphotectonic properties of the Sudetic Marginal Fault, SW Poland. *Acta Montana A, 24*(131), 21–49.

Ballesteros-Cánovas, J. A., Corona, C., Stoffel, M., Lucia-Vela, A., & Bodoque, J. M. (2015). Combining terrestrial laser scanning and root exposure to estimate erosion rates. *Plant and Soil, 394*(1–2), 127–137. doi:10.1007/s11104-015-2516-3

Barbarella, M., & Fiani, M. (2013). Monitoring of large landslides by Terrestrial Laser Scanning techniques: Field data collection and processing. *European Journal of Remote Sensing, 45*, 126–151. doi:10.5721/EuJRS20134608

Barbarella, M., Fiani, M., & Lugli, A. (2015). Landslide monitoring using multitemporal terrestrial laser scanning for ground displacement analysis. *Geomatics, Natural Hazards and Risk, 6*(5–7), 398–418. doi:10.1080/19475705.2013.863808

Benavides-Solorio, J. D. D., & MacDonald, L. H. (2005). Measurement and prediction of post-fire erosion at the hillslope scale, Colorado Front Range. *International Journal of Wildland Fire, 14*(4), 457–474. doi:10.1071/WF05042

Black, T. A., & Luce, C. H. (2013). *Measuring water and sediment discharge from a road plot with a settling basin and tipping bucket* (Gen. Tech. Rep. RMRS-GTR-287, 38 pp.). Fort

Collins, CO: U.S. Department of Agriculture, Forest Service, Rocky Mountain Research Station. https://doi.org/10.2737/RMRS-GTR-287

Brodowska, P. (2012). Comparison of TIN active model and prediction of linear algorithms for terrain points segmentation. *Archiwum Fotogrametrii, Kartografii i Teledetekcji, 24*, 63–71 (in Polish).

Burley, J., Evans, J., & Youngquist, J. A. (Eds.). (2004). *Encyclopedia of forest sciences* (2061 pp). Cambridge, MA: Academic Press.

Chang, M. (2003). *Forest hydrology: An introduction to water and forests* (p. 373). Boca Raton, FL: CRC Press LLC.

Chaplot, V., Darboux, F., Bourennane, H., Leguédois, S., Silvera, N., & Phachomphon, K. (2006). Accuracy of interpolation techniques for the derivation of digital elevation models in relation to landform types and data density. *Geomorphology, 77*(1–2), 126–141. doi:10.1016/j.geomorph.2005.12.010

Chorowska, M., & Radlicz, K. (1994). Revision of the age of the Lower Carboniferous deposits of the northern part of the Góry Bardzkie (Sudetes). *Geological Quarterly, 38*(2), 249–288.

Chorowska, M., & Wajsprych, B. (1995). Góry Bardzkie Mts. Unit. In: The Carboniferous System in Poland. *Prace Pafstwowego Instytutu Geologicznego, 148*, 137–139.

Coppa, U., Guarnieri, A., Pirotti, F., Tarolli, P., & Vettore, A. (2013). Comparing data acquisition methodologies for DTM production. *International Archives of the Photogrammetry, Remote Sensing and Spatial Information Sciences - ISPRS Archives, 40*(5W3), 59–62. doi:10.5194/isprsarchives-XL-5-W3-59-2013

Croke, J., Hairsine, P., & Fogarty, P. (2001). Soil recovery from track construction and harvesting changes in surface infiltration, erosion and delivery rates with time. *Forest Ecology and Management, 143*, 3–12. doi:10.1016/S0378-1127(00)00500-4

Cwojdziński, S., & Pacuła, J. (2009). *Objaśnienia do arkusza Ząbkowice Śląskie (869) Szczegółowej mapy geologicznej Polski*. Warszawa: Narodowe Archiwum Geologiczne (in Polish).

Czamara, W. (1992). Prognoza erozji powierzchniowej w zlewniach zbiorników wodnych zlokalizowanych na przedgórzu sudeckim. *Zeszyty Naukowe Akademii Rolniczej w Krakowie, 271*, 79–89 (in Polish).

Dąbek, P., Żmuda, R., Ćmielewski, B., & Szczepański, J. (2014). Analysis of water erosion processes using terrestrial laser scanning. *Acta Geodynamica et Geomaterialia, 11*(1, 173), 45–52. doi:10.13168/AGG.2013.0054

Denora, D., Romano, L., & Cecaro, G. (2013). Terrestrial laser scanning for the Montaguto landslide (Southern Italy). In C. Margottini, P. Canuti, & K. Sassa (Eds.), *Landslide science and practice* (Vol. 2, pp. 33–38). doi:10.1007/978-3-642-31445-2_4

Eltner, A., Mulsow, C., & Maas, H. G. (2013). Quantitative measurement of soil erosion from tls and uav data. *ISPRS - International Archives of the Photogrammetry, Remote Sensing and Spatial Information Sciences, XL-1/W2*(2), 119–124. https://doi.org/10.5194/isprsarchives-XL-1-W2-119-2013

Finckh, L. (1932). Erlaeuterungen zur geologischen Karte von Preussen und benachbarten deutschen Laendern 1:25 000. Blatt Frankenstein. *Preussische Geologische Landesanstalt*, 343.

Guzzetti, F., Mondini, A. C., Cardinali, M., Fiorucci, F., Santangelo, M., & Chang, K.-T. (2012). Landslide inventory maps: New tools for an old problem. *Earth-Science Reviews, 112*(1–2), 42–66. doi:10.1016/j.earscirev.2012.02.001

Jała, Z., & Cieślakiewicz, D. (2004). Potential erosion of soils in the Karkonosze National Park. In Geoekologicke problemy Krkonoš. Mez. Věd. Konf. November 2003, Szklarska Poręba. *Opera Corcontica, 41*, 66–73 (in Polish).

Jordán, A., & Martínez-Zavala, L. (2008). Soil loss and runoff rates on unpaved forest roads in southern Spain after simulated rainfall. *Forest Ecology and Management, 255*(3–4), 913–919. doi:10.1016/j.foreco.2007.10.002

Krzyszkowski, D., Migoń, P., & Sroka, W. (1995). Neotectonic quaternary history of the Sudetic marginal fault, SW Poland. *Folia Quartern, 66*, 73–98.

Kusiak, W., & Jaszczak, R. (2009). *Propedeutyka leśnictwa* (143 pp.). Wyd. Uniwersytetu Przyrodniczego w Poznaniu (in Polish).

Larsen, I. J., MacDonald, L. H., Brown, E., Rough, D., & Welsh, M. J. (2009). Causes of post-fire runoff and erosion: Water repellency, cover, or soil sealing? *Soil Science Society of America Journal, 73*(4), 1393–1407. doi:10.2136/sssaj2007.0432

Martínez-Zavala, L., Jordán López, A., & Bellinfante, N. (2008). Seasonal variability of runoff and soil loss on forest road backslopes under simulated rainfall. *Catena, 74*(1), 73–79. doi:10.1016/j.catena.2008.03.006

Oberc, J. (1987). Struktura bardzka jako reper rozwoju waryscydów wschodniej części Sudetów Zachodnich i ich przedpola. *Przew, 58*, 165–180 (in Polish).

Oliver, M. A., & Webster, R. (1990). Kriging: A method of interpolation for geographical information systems. *International journal of geographical information systems, 4*(3), 313–332. doi:10.1080/02693799008941549

Pawłuszek, K., & Borkowski, A. (2016, July 12–19). Landslides identification using airborne laser scanning data derived topographic terrain attributes and support vector machine classification. In *The International Archives of the Photogrammetry, Remote Sensing and Spatial Information Sciences, XXIII ISPRS Congress* (Volume XLI-B8). Prague. doi:10.5194/isprsarchives-XLI-B8-145-2016

Pierzgalski, E., Janek, M., Kucharska, K., Tyszka, J., & Wróbel, M. (2007). *Badania hydrologiczne w leśnych zlewniach sudeckich* (84 pp.). Sękocin: IBL (in Polish).

Prochal, P. (Eds.). (1987). Podstawy melioracji rolnych. *PWRiL, Warszawa, 2*, 419 (in Polish).

Prochal, P., Maślanka, K., & Koreleski, K. (2005). *Ochrona środowiska przed erozją wodną* (126 pp). Kraków (in Polish).

Rejman, J., & Usowicz, B. (2002, October). Ocena erozji wodnej gleb lessowych w oparciu o pomiary poletkowe. In *Proccedings of conference: "Erozja gleb i transport rumowiska rzecznego"*. Zakopane (in Polish).

Ruiz-Sinoga, J. D., Romero-Diaz, A., Ferre-Bueno, E., & Martínez-Murillo, J. F. (2010). The role of soil surface conditions in regulating runoff and erosion processes on a metamorphic hillslope (Southern Spain). *Catena, 80*(2), 131–139. doi:10.1016/j.catena.2009.09.007

Smolska, E. (2002, October). Erozja powierzchniowa gleb na Pojezierzu Suwalskim i niektóre jej uwarunkowania klimatyczno-topograficzne. In *Proccedings of conference "Erozja gleb i transport rumowiska rzecznego"*. Zakopane (in Polish).

Stenberg, L., Tuukkanen, T., Finér, L., Marttila, H., Piirainen, S., Kløve, B., & Koivusalo, H. (2016). Evaluation of erosion and surface roughness in peatland forest ditches using pin meter measurements and terrestrial laser scanning. *Earth Surface Processes and Landforms, 41*, 1299–1311. doi:10.1002/esp.3897

Vosselman, G., & Maas, H. G. (2010). *Airborne and terrestrial laser scanning* (p. 336). Dunbeath, Scotland: Whittles Publishing.

Wajsprych, B. (1978). Allochthonous Paleozoic rocks in the Visean of the Bardzkie Mts. (Sudetes). *Rocz Pol Tow Geol, 38*(1), 99–128 (in Polish).

Wajsprych, B. (1986). Sedimentary record of tectonic activity on a Devonian—Carboniferous continental margin (Sudetes). In *IAS 7th European Meeting. Kraków - Poland*. (pp. 141–164). Wrocław: Excursion Guide Book.

Wajsprych, B. (1995). The Bardo Mts. rock complex: The Famennian – Lower Carboniferous preflysch (platform)-to-flysch (foreland) basin succession, the Sudetes. In O. Kumpera, M. Paszkowski, B. Wajsprych, J. Foldyna, J. Malec, & H. Zakowa (Eds.), *Transition of the early carboniferous pelagic sedimentation into synorogenic flysch. An example from the Silesian-Moravian zone and adjacent areas. XIII International Geological Congress on Carboniferous-Permian* (Vol. B2, pp. 23–42). Kraków: Guide to Excursion.

7

Review of preprocessing techniques used in soil property prediction from hyperspectral data

S. Minu[1*], Amba Shetty[1] and Binny Gopal[2]

*Corresponding author: S. Minu, Department of Applied Mechanics and Hydraulics, National Institute of Technology Karnataka, Surathkal, Mangalore 575025, Karnataka, India

E-mail: minu.s88@gmail.com

Reviewing editor: Lachezar Hristov Filchev, Space Research and Technology Institute - Bulgarian Academy of Sciences (SRTI-BAS), Bulgaria

Abstract: Soil properties are neither static nor homogenous with space and time. Capturing the spatial variation of soil properties through conventional methods is a difficult task. Hyperspectral remote sensing data provide rich source of information produced in the form of spectrum at each pixel which can be used to identify surface materials. Airborne and spaceborne narrowband hyperspectral sensors have come to the fore which provides spectral information across large area. Thus, it is a promising tool for studying soil properties and can be used as an alternative to conventional method. But atmospheric attenuation and low signal to noise ratio are major problems with this type of data. Preprocessing of hyperspectral airborne/spaceborne data is required to extract soil properties. This paper reviews previous studies on prediction of soil properties from hyperspectral airborne and satellite data during the past years and the preprocessing techniques used in these predictions.

Subjects: Earth Sciences; Engineering & Technology; Environment Agriculture

Keywords: airborne; hyperspectral; prediction model; preprocessing techniques; soil properties; spaceborne

1. Introduction

Remotely sensed hyperspectral satellite data have great potential for quantitative assessment of soil and vegetation parameter at spatial scale. The development of methods to map soil properties using optical remote sensing data in combination with field measurements has been the objective of several studies during the last decade (Ben-Dor et al., 2009). Also it has been a challenge to find the most appropriate technique for studying soil properties from optical data and thus reducing the time and effort involved in field sampling and laboratory analysis.

Soil reflectance in the visible near-infrared and mid-infrared regions has been widely used in many studies. Some of the soil properties predicted from reflectance data were organic matter (OM), soil

ABOUT THE AUTHORS

Our group works on application of Hyperspectral data for soil and vegetation discrimination applications. In the process of applying this data to any application, major issue to be addressed is to account for the effects of atmosphere on the hyperspectral data and to account for it appropriately. Though there are several algorithms are available to address this, there is no guideline on the application of them. One of the issues that we wish to address is to how best to account for the effect of atmosphere so that proper signal of the targets is extracted for further analysis.

PUBLIC INTEREST STATEMENT

The present review paper would be very useful in the process of digital soil mapping mission from satellite data. As soil is a precious non-renewable resource, it has to be examined periodically. Prediction from satellite data provides a continuous method of monitoring soil quality. The accuracy of prediction depends on the quality of satellite data. The methods to improve quality of data are reviewed in this paper.

organic carbon (SOC), total nitrogen (TN), pH, moisture content (MC), electrical conductivity (EC), phosphorous (P), potassium (K), calcium (Ca), magnesium (Mg), sodium (Na), manganese (Mn), zinc (Zn), and iron (Fe) with various levels of prediction accuracy. Various prediction models such as multiple linear regression (MLR), principal components regression (PCR), stepwise multiple linear regression (SMLR), partial least squares regression (PLSR), artificial neural networks (ANN), etc. were used. These models work well with signals obtained under laboratory conditions, with minimal source of noise. Thus, performance of these models on remotely sensed airborne or spaceborne data is influenced by atmospheric interference and the occurrence of spectral noises. At this juncture, the role of preprocessing techniques on the prediction accuracy of soil properties from remotely sensed data needs to be studied.

Preprocessing techniques consist of atmospheric correction algorithms as well as spectral pretreatment and smoothening methods. Over the years, atmospheric correction algorithms have evolved from applied math approach to ways supported on rigorous radiative transfer (RT) modeling (Minu & Shetty, 2015). Noise and unwanted spectral signals are removed by spectral pretreatment and smoothening methods. Only good-quality data with better signal-to-noise ratios can be conveniently used for the purpose.

Minu and Shetty (2015) review different hyperspectral atmospheric correction algorithms developed during the past years. Internal average reflectance approach (Kruse, Raines, & Watson, 1985), flat field approach (Roberts, Yamaguchi, & Lyon, 1986), empirical line (EL) method (Roberts, Yamaguchi, & Lyon, 1985), QUick atmospheric correction (Bernstein et al., 2005) etc. are empirical or semi-empirical atmospheric correction methods. RT codes try to simulate the transfer process of an electromagnetic wave in the atmosphere. The normally used RT codes are LOWTRAN (Kneizys et al., 1988), MODTRAN (Berk, Bernstein, & Robertson, 1989), 5S (Tanré, Deroo, Duhaut, Herman, & Morcrette, 1990), and 6S (Vermote et al., 1997). There are a range of software programs available to model the atmosphere including ATmospheric REMoval algorithm (ATREM) (Gao, Heidebrecht, & Goetz, 1993), ATmospheric CORrection (ATCOR) (Richter, 1996), Fast Line-of-sight Atmospheric Analysis of Spectral Hypercubes (FLAASH) (Adler-Golden et al., 1998), Imaging Spectrometer Data Analysis System (ISDAS) (Staenz, Szeredi, & Schwarz, 1998), High-accuracy ATmosphere Correction for Hyperspectral data (HATCH) (Qu, Goetz, & Heidbrecht, 2001), Atmospheric CORrection Now (ACORN) (ACORN 4.0, 2002) etc. Hybrid methods include combinations of empirical approaches and radiative modeling for the derivation of surface reflectance from hyperspectral imaging data. Each preprocessing technique is made of its own assumptions. So there is a need to analyze limitations of different preprocessing techniques and to come up with a universal method.

2. Prediction of soil properties from airborne/spaceborne hyperspectral data

Hyperspectral sensors operate with more than hundreds of bands with good spatial and spectral resolution producing continuous spectra. With the progress and maturity of technology, hyperspectral remote sensing has found a wide range of applications in mapping soil types and quantifying soil constituents. Review papers by Ben-Dor et al. (2009); Ge, Thomasson, and Sui (2011); Mulder, de Bruin, Schaepman, and Mayr (2011), etc. point toward it. Airborne sensors provide high spatial resolution (2–20 m), high spectral resolution (10–20 nm), and high SNR (>500:1) data. Even though satellite hyperspectral imageries have become available since 2000, only few attempts have been made to use them for mapping soil properties. This may be due to their low signal to noise ratio. Tables 1 and 2 summarize previous studies carried out using airborne and satellite hyperspectral imageries to predict soil properties. The preprocessing techniques used are also mentioned in the table.

It is seen that RT models are mainly used in preprocessing of airborne imagery. It may be due to the fact that more information on atmospheric conditions are available in the case of airborne sensors, so that modeling of atmosphere can be done precisely and it can be removed to obtain pure signal. Whereas semi-empirical models like FLAASH are mainly used in hyperspectral imageries. Comparison of different models are still lacking in this field. Also EL method which also requires ground information gives good results. But it is limited only to the areas where ground information

Table 1. Summary of soil properties prediction, using airborne hyperspectral imagery							
Soil property	Platform/ spectral range/spatial resolution	Field nature	Country	Preprocessing method	Prediction tech.	R^2 value	Author
Fe	AVIRIS (400–2,500 nm) (20 m)	Pasture and seasonal crops	Brazil	MODTRAN-based (Green, Conel, & Roberts, 1993)	Regression equations	0.83	Galvão, Pizarro, and Epiphanio (2001)
TiO$_2$						0.74	
Al$_2$O$_3$						0.68	
OM	DAIS-7915 (400–2,500 nm) (5 m)	Agriculture fields	Israel	Minimum noise fraction (MNF) (Green, Berman, Switzer, & Craig 1988) for noise reduction; EL technique	Visible and NIR analysis	0.827	Ben-Dor, Patkin, Banin, and Karnieli (2002)
MC						0.647	
EC						0.665	
EC	RDACS/H3 (471–828 nm) (1 m)	Bare soil of corn–soybean rotation field	Missouri	Calibrated with chemically treated reference traps with known reflectance	SMLR	0.66	Hong et al. (2002)
pH						0.68	
Mg						0.67	
K						0.59	
OM						0.55	
OM	CASI (408.73–947.07 nm) (2 m)	Corn field with clay-loam soil	Canada	CAM5S model O'Neill et al. (1997)	SMLR	0.49	Uno et al. (2005)
					ANN	0.592	
Iron oxide	CASI-A (400–1,000 nm) (3 m)	Sand dunes	Israel	EL technique	Spectral indices based model	0.59	Ben-Dor et al. (2006)
Gravel coverage %	DAIS-7915 (400–2,500 nm) (5 m)	Alluvial fan	Negev desert, Israel	MNF technique for noise reduction and EL technique	Ferric absorption feature depth(AFD) model	0.83	Crouvi, Ben-Dor, Beyth, Avigad, and Amit (2006)
					Al–OH AFD model	0.67	
					Carbonate AFD model	0.57	
SOC	HyMap (450–2,500 nm) (3.5 m)	Agriculture fields	Germany	ATCOR Richter & Schläpfer, (2002; Schläpfer & Richter, 2002)	MLR	0.9	Selige, Böhner, and Schmidhalter (2006)
TN						0.92	
Sand						0.95	
Clay						0.71	
SOC					PLSR	0.86	
TN						0.87	
Sand						0.87	
Clay						0.65	
EC	HyMap (420–2,480 nm) (6 m)	Wetland	Western Australia	Corrected for atmospheric effects and multiplicative signal correction (MSC) techniques	PLSR	0.86	Farifteh, Van der Meer, Atzberger, and Carranza (2007)
					ANN	0.86	
EC	AVNIR (429–1,010 nm) (1.2 m)	Cotton field	California	Atmospheric calibrated with black and gray reference panels	MLR	0.6696	De Tar, Chesson, Penner, and Ojala (2008)
Ca						0.6188	
Mg						0.582	
Na						0.6224	
Cl						0.7376	
Clay	HYMAP (400–2,500 nm) (5 m)	Area is mainly devoted to vineyards	France	ATCOR4, Savitzky–Golay filter	PLSR	0.64	Gomez, Lagacherie, and Coulouma (2008)
CaCO$_3$						0.77	
Clay					Continuum removal	0.58	
CaCO$_3$						0.47	

(Continued)

Table 1. (*Continued*)							
Soil property	Platform/ spectral range/spatial resolution	Field nature	Country	Preprocessing method	Prediction tech.	R^2 value	Author
MC	HyMap (440–2,470 nm) (4 m)	Sandy substrates and low vegetation cover area	Germany	MODTRAN4 based ACUM algorithm	Normalized soil moisture index (NSMI) model	0.819	Haubrock, Cha-brillat, Kuhnert, Hostert, and Kaufmann (2008)
Clay	HYMAP (400–2,500 nm) (5 m)	Area is devoted to vineyards	France	ATCOR4 code for airborne sensors	Continuum re-moval analysis	0.58	Lagacherie, Baret, Feret, Ma-deira Netto, and Robbez-Masson (2008)
CaCO$_3$						0.47	
SOC	AHS-160 sensor (430 nm–2540 nm) (2.6 m)	Agriculture fields	Belgium	MODTRAN-4 embedded with ATCOR-4 (Richter, Schläpfer, & Müller, 2006)	PLSR	RPD = 1.47	Stevens et al. (2008
Clay	HYMAP (400–2,500 nm) (5 m)	Area is devoted to vineyards	France	ATCOR4 code for airborne sensors	PLSR	0.64	Lagacherie, Go-mez, Bailly, Baret, and Coulouma (2010)
CaCO$_3$						0.77	
SOC	AHS-160 sensor (430 nm–2,540 nm) (2.6 m)	Cropland	Luxem-bourg	MODTRAN4-based algorithm; (Richter, 2005; Rodger & Lynch, 2001)	PLSR	0.71	Stevens et al. (2010)
					PSR	0.75	
					SVMR	0.69	
C	HyperSpecTIR (400–2,450 nm) (2.5 m)	Tilled agricul-tural fields	MD, USA	Imagery processing by ENVI 4.7; & different signal smoothening methods	PLSR	0.65	Hively et al. (2011)
Al						0.76	
Fe						0.75	
Silt						0.79	
Clay	MIVIS (430–1,270 nm) (4.8 m)	Maize field, but the crop had not emerged	Central Italy	MODTRAN4-based model (Vermote, Tanre, Deuze, Herman, & Morcette, 1997)	PLSR	0.78	Casa, Castaldi, Pascucci, Palom-bo, and Pignatti (2013)
Silt						0.56	
Sand						0.81	
SOC	CASI 1500 (380–1,050 nm) (0.2 m)	Compost added soil	Italy	EL calibration with asphalt spectral signatures	Correlation between the second deriva-tive value and SOC	0.85	Matarrese et al. (2014)

is available. Also it is seen that prediction of SOC gives good results compared to other properties. This may be because the soil reflectance curve is affected more by presence of OM.

3. Inference

Several surface soil properties were modeled from remotely sensed hyperspectral imagery. Since soil is a more heterogeneous material, more careful spectral manipulations need to be done in as-sessing its properties from spectral data. For the best performance of any prediction system, the key influencing factors are to be identified and optimized. Although there are many soil properties pre-diction models, the prediction accuracy is found to be still very low.

The noises should be removed from the hyperspectral imagery in order to utilize it to the best. The signal to noise ratio should be maximum. Several spectral pre-processing methods are employed in various studies to improve the performance and robustness of the prediction models. Even though the preproc-essing techniques affect the prediction model considerably, it was not given that much importance. So to develop a good model there is a need to perform a better preprocessing. In this percept, different preproc-essing techniques used in various studies are listed in this review paper. Hybrid methods which combine physical model and image statistics need to be promoted. There is a need to give guidelines on selection of suitable preprocessing technique for the prediction of soil chemical properties.

Table 2. Summary of soil properties prediction using satellite remote sensing techniques

Soil Prop	Platform	Field characteristics	Country	Preprocessing method	Prediction tech.	R^2 values	Author
SOC	EO1 Hyperion (400–2,500 nm) (30 m)	Cotton crops and pasture. Field size = 100 × 500 m²	Australia	Algorithm based on ATREM and 5S code.	PLSR	0.5	Gomez, Viscarra Rossel, and Mc-Bratney (2008))
OM	EO1 Hyperion (400–2,500 nm) (30 m)	Row-crop agriculture field	Central Indiana, USA	ENVI FLAASH module	PLSR	0.74	Zheng (2008)
TN						0.72	
TP						0.67	
TN	EO1 Hyperion (400–2,500 nm) (30 m)	Arid regions; 4,332 km².	Shanxi, China	EL atmospheric correction	Linear regression model	0.84	Wu, Liu, Chen, Wang, and Chai (2009)
MC	EO1 Hyperion (400–2,500 nm) (30 m)	Bare field	Central Indiana, USA	ACORN	PLSR	0.79	Zhang, Li, and Zheng (2009)
OM						0.89	
TN						0.7	
TP						0.69	
TC						0.86	
Clay						0.49	
OM	EO1 Hyperion (400–2,500 nm) (30 m)	Agriculture-pasture mixed area.	Hengshan County, China	Internal average relative reflectance	Land degradation spectral response units (DSRU) model	0.722	Wang, He, Lv, Chen, and Jian (2010)
Clay	CHRIS-PROBA (415–1,050 nm) (17 m)	Maize field, but the crop had not emerged, 12 and 17 ha plots	Central Italy	FLAASH	PLSR	0.6	Casa et al. (2013)
Silt						0.3	
Sand						0.62	
OM	EO1 Hyperion (400–2,500 nm) (30 m)	Wheat and potato fields. Field size = 90 × 90 m².	China	FLAASH	PLSR	0.63	Lu, Wang, Niu, Li, and Zhang (2013)
pH						0.68	
P						0.62	
N	EO1 Hyperion (400–2,500 nm) (30 m)	Scattered paddy fields, 47 km²	Karnataka India	FLAASH,	PLSR	0.63	Gopal, Shetty, and Ramya (2014)
				Moving average			
				Savitzky–Golay		0.63	
POM	EO1 Hyperion (400–2,500 nm) (30 m)	Coastal soils densely covered with vegetation	Florida, USA	FLAASH, MNF filter	PLSR	0.67	Anne, Abd-Elrah-man, Lewis, and Hewitt (2014)
MAOM						0.74	
labile C						0.93	
labile N						0.96	

Funding
The authors received no direct funding for this research.

Author details
S. Minu[1]
E-mail: minu.s88@gmail.com
Amba Shetty[1]
E-mail: amba_shetty@yahoo.co.in
Binny Gopal[2]
E-mail: binnycoorg07@gmail.com

[1] Department of Applied Mechanics and Hydraulics, National Institute of Technology Karnataka, Surathkal, Mangalore 575025, Karnataka, India.

[2] Department of Agronomy, University of Agricultural and Horticultural Sciences, Navile, Shimoga, Karnataka 577225, India.

References
ACORN 4.0. (2002). *"User's Guide", Analytical Imaging and Geophysics.* Boulder, CO: LLC.
Adler-Golden, S. M., Berk, A., Bernstein, L. S., Richtsmeier, S., Acharya, P. K., Matthew, M. W., ... Chetwynd, J. (1998). FLAASH, A MODTRAN4 atmospheric correction package for hyperspectral data retrievals and simulations. *Proceedings of the 7th Annual JPL Airborne Earth Science Workshop* (Vols. 97–21, pp. 9–14). CA: JPL Publication Pasadena.
Anne, N. J. P., Abd-Elrahman, A. H., Lewis, D. B., & Hewitt, N. A. (2014). Modeling soil parameters using hyperspectral

image reflectance in subtropical coastal wetlands. *International Journal of Applied Earth Observation and Geoinformation, 33*, 47–56. http://dx.doi.org/10.1016/j.jag.2014.04.007

Ben-Dor, E., Patkin, K., Banin, A., & Karnieli, A. (2002). Mapping of several soil properties using DAIS-7915 hyperspectral scanner data—A case study over clayey soils in Israel. *International Journal of Remote Sensing, 23*, 1043–1062. http://dx.doi.org/10.1080/01431160010006962

Ben-Dor, E., Levin, N., Singer, A., Karnieli, A., Braun, O., & Kidron, G. J. (2006). Quantitative mapping of the soil rubification process on sand dunes using an airborne hyperspectral sensor. *Geoderma, 131*(1–2), 1–21. http://dx.doi.org/10.1016/j.geoderma.2005.02.011

Ben-Dor, E., Chabrillat, S., Demattê, J. A. M., Taylor, G. R., Hill, J., Whiting, M. L., & Sommer, S. (2009). Using imaging spectroscopy to study soil properties. *Remote Sensing of Environment, 113*, S38–S55. http://dx.doi.org/10.1016/j.rse.2008.09.019

Berk, A., Bernstein, L. S., & Robertson, D. C. (1989). MODTRAN: A moderate resolution model for LOWTRAN7. Final report, GL-TR-89-0122, AFGL, Hanscom AFB, MA 42 pp.

Bernstein, L. S., Adler-Golden, S. M., Sundberg, R. L., Levine, R. Y., Perkins, T. C., Berk, A., ... Hoke, M. L. (2005). A new method for atmospheric correction and aerosol optical property retrieval for VIS-SWIR multi- and hyperspectral imaging sensors: QUAC (QUick Atmospheric Correction). *Geoscience and Remote Sensing Symposium, IEEE International, 5*, 3552.

Casa, R., Castaldi, F., Pascucci, S., Palombo, A., & Pignatti, S. (2013). A comparison of sensor resolution and calibration strategies for soil texture estimation from hyperspectral remote sensing. *Geoderma, 197–198*, 17–26. http://dx.doi.org/10.1016/j.geoderma.2012.12.016

Crouvi, O., Ben-Dor, E., Beyth, M., Avigad, D., & Amit, R. (2006). Quantitative mapping of arid alluvial fan surfaces using field spectrometer and hyperspectral remote sensing. *Remote Sensing of Environment, 104*, 103–117. http://dx.doi.org/10.1016/j.rse.2006.05.004

De Tar, W. R., Chesson, J. H., Penner, J. V., & Ojala, J. C. (2008). Detection of soil properties with airborne hyperspectral measurements of bare fields. *American Society of Agricultural and Biological Engineers, 51*, 463–470.

Farifteh, J., Van der Meer, F. D., Atzberger, C. G., & Carranza, E. J. M. (2007). Quantitative analysis of salt-affected soil reflectance spectra: A comparison of two adaptive methods (PLSR and ANN). *Remote Sensing of Environment, 110*, 59–78. http://dx.doi.org/10.1016/j.rse.2007.02.005

Galvão, L. S., Pizarro, M. A., & Epiphanio, J. C. N. (2001). Variations in reflectance of tropical soils. *Remote Sensing of Environment, 75*, 245–255. http://dx.doi.org/10.1016/S0034-4257(00)00170-X

Gao, B. C., Heidebrecht, K. B., & Goetz, A. F. H. (1993). Derivation of scaled surface reflectances from AVIRIS data. *Remote Sensing of Environment, 44*, 165–178. http://dx.doi.org/10.1016/0034-4257(93)90014-O

Ge, Y., Thomasson, A., & Sui, R. (2011). Remote sensing of soil properties in precision agriculture: A review. *Frontiers of Earth Science, 5*, 229–238.

Gomez, C., Lagacherie, P., & Coulouma, G. (2008). Continuum removal versus PLSR method for clay and calcium carbonate content estimation from laboratory and airborne hyperspectral measurements. *Geoderma, 148*, 141–148. http://dx.doi.org/10.1016/j.geoderma.2008.09.016

Gomez, C., Viscarra Rossel, R. A., & McBratney, A. B. (2008). Soil organic carbon prediction by hyperspectral remote sensing and field vis-NIR spectroscopy: An Australian case study. *Geoderma, 146*, 403–411. http://dx.doi.org/10.1016/j.geoderma.2008.06.011

Gopal, B., Shetty, A., & Ramya, B. J. (2014). Prediction of topsoil nitrogen from spaceborne hyperspectral data. *Geocato International, 30*, 82–92. doi:10.1080/1010604 9.2014.894585

Green, A. A., Berman, M., Switzer, P., & Craig, M. D. (1988). A transformation for ordering multispectral data in terms of image quality with implications for noise removal. *IEEE Transactions on Geoscience and Remote Sensing, 26*, 65–74. http://dx.doi.org/10.1109/36.3001

Green, R. O., Conel, J. E., & Roberts, D. A. (1993). Estimation of aerosol optical depth and additional atmospheric parameters for the calculation of the reflectance from radiance measured by the Airborne Visible/Infrared Imaging Spectrometer. *In Summaries of the Forth Annual JPL Airborne Geoscience Workshop, JPL Publication, 93-26*, 73–76.

Haubrock, S.-N., Chabrillat, S., Kuhnert, M., Hostert, P., & Kaufmann, H. (2008). Surface soil moisture quantification and validation based on hyperspectral data and field measurements. *Journal of Applied Remote Sensing., 2*(023552), 1–26.

Hively, W. D., McCarty, G. W., Reeves, J. B., Lang, M. W., Oesterling, R. A., & Delwiche, S. R. (2011). Use of airborne hyperspectral imagery to MAP soil properties in tilled agricultural fields. *Applied and Environmental Soil Science*, Article ID 358193, 13.

Hong, S. Y., Sudduth, K. A., Kitchen, N. R, Drummond, S. T., Palm, H. L., & Wiebold, W. J. (2002). Estimating within-field variations in soil properties from airborne hyperspectral images. In *Pecora 15/Land Satellite Information IV/ISPRS Commission I/FIEOS 2002 Conference Proceedings.* Denver, CO.

Kneizys, F. X., Shettle,E .P., Abreau, L. W., Chetwynd, J. H., Anderson, G. P., Gallery, W. O, ... Clough, S. A. (1988). Users guide to LOWTRAN-7. In *AFGL-TR-8-0177 Air Force Geophysics Laboratories*, Bedford, MA.

Kruse, F. A., Raines, G .I, & Watson, K. (1985). Analytical techniques for extracting geologic information from multichannel airborne spectroradiometer and airborne imaging spectrometer data. In *Proceedings of the 4th thematic conference on remote sensing for exploration geology.* Ann Arbor, MI.

Lagacherie, P., Baret, F., Feret, J.-B., Madeira Netto, J. M., & Robbez-Masson, J. M. (2008). Estimation of soil clay and calcium carbonate using laboratory, field and airborne hyperspectral measurements. *Remote Sensing of Environment, 112*, 825–835. http://dx.doi.org/10.1016/j.rse.2007.06.014

Lagacherie, P., Gomez, C., Bailly, J. S., Baret, F., & Coulouma, G. (2010). The use of hyperspectral imagery for digital soil mapping in Mediterranean areas. *Digital Soil Mapping, Progress in Soil Science, 2*, 93–102. http://dx.doi.org/10.1007/978-90-481-8863-5

Lu, P., Wang, L., Niu, Z., Li, L., & Zhang, W. (2013). Prediction of soil properties using laboratory VIS–NIR spectroscopy and Hyperion imagery. *Journal of Geochemical Exploration, 132*, 26–33. http://dx.doi.org/10.1016/j.gexplo.2013.04.003

Matarrese, R., Ancona, V., Salvatori, R., Muolo, M. R., Uricchio, V. F., & Vurro, M. (2014). Detecting soil organic carbon by CASI hyperspectral images. In *Geoscience and Remote Sensing Symposium* (INSPEC Accession Number: 14716443, pp. 3284–3287). IEEE Conference Publications. doi:10.1109/IGARSS.2014.694718

Minu, S., & Shetty, A. (2015). Atmospheric correction algorithms for hyperspectral imageries: A review. *International Research Journal of Earth Sciences, 3*, 14–18.

Mulder, V. L., de Bruin, S., Schaepman, M. E., & Mayr, T. R. (2011). The use of remote sensing in soil and terrain mapping—A review. *Geoderma, 162*, 1–19. http://dx.doi.org/10.1016/j.geoderma.2010.12.018

O'Neill, N. T., Zagolski, F., Bergeron, M., Royer, A., Miller, J. R., & Freemantle, J. (1997). Atmospheric correction validation of casi images acquired over the Boreas Southern study area. Canadian Journal of Remote Sensing, 23, 143–162. http://dx.doi.org/10.1080/07038992.1997.10855196

Qu, Z., Goetz, A. F. H., & Heidbrecht, K. B. (2001). High accuracy atmosphere correction for hyperspectral data (HATCH). Proceedings of the Ninth JPL Airborne Earth Science Workshop JPL-Pub, 00–18, 373–381.

Richter, R. (1996). A spatially adaptive fast atmosphere correction algorithm. International Journal of Remote Sensing, 11, 159–166.

Richter, R. (2005). Atmospheric/topographic correction for airborne imagery (DLR report, DLR-IB 562-02/05, p. 107). Wesseling.

Richter, R., & Schläpfer, D. (2002). Geo-atmospheric processing of airborne imaging spectrometry data. Part 2: Atmospheric/topographic correction. International Journal of Remote Sensing, 23, 2631–2649. http://dx.doi.org/10.1080/01431160110115834

Richter, R., Schläpfer, D., & Müller, A. (2006). An automatic atmospheric correction algorithm for visible/NIR imagery. International Journal of Remote Sensing, 27, 2077–2085. http://dx.doi.org/10.1080/01431160500486690

Roberts, D. A., Yamaguchi, Y., & Lyon, R. J .P. (1985). Calibration of airborne imaging spectrometer data to percent reflectance using air borne imaging spectrometer data to percent reflectance using field spectral measurements. In Proceedings of the Nineteenth International Symposium on Remote Sensing of the Environment (pp. 21–25). Michigan, 21–25 October 1985.

Roberts, D. A., Yamaguchi, Y., & Lyon, R. (1986). Comparison of various techniques for calibration of AIS data. In Proceedings of the 2nd Airborne Imaging Spectrometer Data Analysis Workshop (Vols. 86–35, pp. 21–30). Pasadena, CA: JPL Publication Laboratory.

Rodger, A., & Lynch, M. J. (2001). Determining atmospheric column water vapour in the 0.4–2.5 μm spectral region. In Proceedings of the AVIRIS Workshop 2001. Pasadena, CA.

Schläpfer, D., & Richter, R. (2002). Geo-atmospheric processing of airborne imaging spectrometry data: Part 1. Parametric orthorectification. International Journal of Remote Sensing, 23, 2609–2630. http://dx.doi.org/10.1080/01431160110115825

Selige, T., Böhner, J., & Schmidhalter, U. (2006). High resolution topsoil mapping using hyperspectral image and field data in multivariate regression modeling procedures. Geoderma, 136, 235–244. http://dx.doi.org/10.1016/j.geoderma.2006.03.050

Staenz, K., Szeredi, T., & Schwarz, J. (1998). ISDAS—A system for processing/analyzing hyperspectral data. Canadian Journal of Remote Sensing, 24, 99–113. http://dx.doi.org/10.1080/07038992.1998.10855230

Stevens, A., van Wesemael, B., Bartholomeus, H., Rosillon, D., Tychon, B., & Ben-Dor, E. (2008). Laboratory, field and airborne spectroscopy for monitoring organic carbon content in agricultural soils. Geoderma, 144, 395–404. http://dx.doi.org/10.1016/j.geoderma.2007.12.009

Stevens, A., Udelhoven, T., Denis, A., Tychon, B., Lioy, R., Hoffmann, L., & van Wesemael, Bv (2010). Measuring soil organic carbon in croplands at regional scale using airborne imaging spectroscopy. Geoderma, 158, 32–45. http://dx.doi.org/10.1016/j.geoderma.2009.11.032

Tanré, D., Deroo, C., Duhaut, P., Herman, M., & Morcrette, J. J. (1990). Technical note description of a computer code to simulate the satellite signal in the solar spectrum: The 5S code. International Journal of Remote Sensing, 11, 659–668. http://dx.doi.org/10.1080/01431169008955048

Uno, Y., Prasher, S. O., Patel, R. M., Strachan, I. B, Pattey, E., & Karimi, Y. (2005). Development of field-scale soil organic matter content estimation models in Eastern Canada using airborne hyperspectral imagery. Canadian Biosystems Engineering, 47, 1.9–1.14.

Vermote, E. F., Tanre, D., Deuze, J. L., Herman, M., & Morcette, J. J. (1997). Second simulation of the satellite signal in the solar spectrum, 6S: An overview. IEEE Transactions on Geoscience and Remote Sensing, 35, 675–686. http://dx.doi.org/10.1109/36.581987

Vermote, E. F., El Saleous, N., Justice, C. O., Kaufman, Y. J., Privette, J. L., Remer, L., ... Tanré, D. (1997). Atmospheric correction of visible to middle-infrared EOS-MODIS data over land surfaces: Background, operational algorithm and validation". Journal of Geophysical Research, 102, 131–141.

Wang, J., He, T., Lv, C., Chen, Y., & Jian, W. (2010). Mapping soil organic matter based on land degradation spectral response units using Hyperion images. International Journal of Applied Earth Observation and Geoinformation, 12, S171–S180. http://dx.doi.org/10.1016/j.jag.2010.01.002

Wu, J., Liu, Y., Chen, D., Wang, J., & Chai, X. (2009). Quantitative mapping of soil nitrogen content using field spectrometer and hyperspectral remote sensing. IEEE International Conference on Environmental Science and Information Application Technology, 2, 379–382. doi:10.1109/ESIAT.2009.296

Zhang, T., Li, L., Zheng, B. (2009). Partial least squares modeling of Hyperion image spectra for mapping agricultural soil properties. Proceedings of SPIE—The International Society for Optical Engineering, 7454, 74540P-1–74540P-12.

Zheng, B. (2008). Using satellite hyperspectral imagery to map soil organic matter, total nitrogen and total phosphorus (MSc thesis, pp. 1–81). Department of Earth Science, Indiana University.

8

Assessing the potential of remote sensing to discriminate invasive *Asparagus laricinus* from adjacent land cover types

Bambo Dubula[1]*, Solomon Gebremariam Tesfamichael[1] and Isaac Tebogo Rampedi[1]

*Corresponding author: Bambo Dubula, Department of Geography, Environmental Management and Energy Studies, University of Johannesburg, Johannesburg, South Africa
E-mail: bambodubula@gmail.com

Reviewing editor: Louis-Noel Moresi, University of Melbourne, Australia

Abstract: The utility of remote sensing technique to discriminate *Asparagus laricinus* from adjacent land cover types using a field spectrometer data was explored in this study. Analysis made use of original spectra and spectra simulated based on Landsat and SPOT 5 bands. Comparisons were made at individual and plot levels using original spectra, and individual and group level using simulated spectra. The near-infrared region showed consistent significant differences between *A. laricinus* and adjacent land cover types at the individual level analysis. In particular, Landsat- and SPOT 5-simulated spectra showed significant differences in only the NIR band. The findings suggest the potential of upscaling field-based data into airborne or spaceborne remote sensing techniques with more emphasis on the NIR band. However, more studies need to be undertaken that will make up for the shortcomings encountered in this study. In this regard, improvements can be made using large number of samples, stratifying target plants according to phenologies, and taking spectral measurements at ideal times as much as possible. Furthermore, laboratory measurements would help in drawing up conclusive statements on the discriminability of the species.

Subjects: Biodiversity; Botany; Earth Sciences

Keywords: *Asparagus laricinus*; remote sensing; spectral reflectance; spectral bands; field spectrometer

1. Introduction

Invasive alien plants are a growing global concern (Richardson & Van Wilgen, 2004; Rouget, Hui, Renteria, Richardson, & Wilson, 2015; Schor, Farwig, & Berens, 2015; Vicente et al., 2013). These

ABOUT THE AUTHORS

Bambo Dubula is a postgraduate student in environmental management, Solomon Gebremariam Tesfamichael is a senior lecturer in GIS and remote sensing, and Isaac Tebogo Rampedi is a senior lecturer in environmental management in the Department of Geography, Environmental Management and Energy Studies, University of Johannesburg. We are conducting research work on plant invasions in the Kliprivierberg Nature Reserve and we aim toward developing distribution maps of invasive plant infestation that will be helpful in the design of better management strategies by the reserve managers.

PUBLIC INTEREST STATEMENT

This study gives insights into remote sensing capabilities in discriminating *A. laricinus* from adjacent land cover types. The information will be helpful in developing spatial distribution maps of the species. Such maps are valuable to land managers for the timely information they provide on the distribution of the species, thus aid in development of better management strategies of the species.

plants hold special characters that make them outcompete and replace indigenous vegetation, and have the potential of spreading to other areas (Bradley & Marvin, 2011; Mgidi et al., 2007; Van Wilgen, 2006). As a result, they compromise ecosystem stability, delivery of ecosystem goods and services, and threaten biodiversity and economic productivity (van Wilgen, Reyers, Le Maitre, Richardson, & Schonegevel, 2008; van Wilgen et al., 2012; Van Wilgen, 2006). Mitigating these effects is costly; South Africa, for example, spends considerable amounts of money in programs such as the Working for Water (WfW) which is mandated to control invasive alien plants.

Most invasive plant control measures focus primarily on established invasions and less attention is given to new infestations (Mgidi et al., 2007). The success of this practice is unsatisfactory, since an effective management of invasive alien plants should depend on early detection and eradication (Mgidi et al., 2007). One method of achieving early detection of plant invasions is through the use of spatial and temporal maps that show the distribution of invasive plants (Dorigo, Lucieer, Podobnikar, & Čarni, 2012). Traditional methods can be used to provide spatial and temporal distribution of invasive plants, but the methods often rely on field inventories that are limited in spatial coverage, time-consuming, and relatively expensive (Dewey, Price, & Ramsey, 1991; Dorigo et al., 2012; Rodgers, Pernas, & Hill, 2014).

Remote sensing methods make up for most inefficiencies of the traditional mapping methods and are used to characterize the spatial and temporal distribution of plants (Alparone, Aiazzi, Baronti, & Andrea, 2015; Campbell & Wynne, 2011; Galvão, Epiphanio, Breunig, & Formaggio, 2011; Jensen, 2014; Lillesand, Kiefer, & Chipman, 2015). Remote sensing is the science of deriving information from electromagnetic energy reflected from objects on the ground (Alparone et al., 2015; Campbell & Wynne, 2011; Jensen, 2014). The method differentiates earth features using varying sensitivity of ground objects to electromagnetic radiation often acquired within the visible, infrared, and microwave regions of the spectrum (Campbell & Wynne, 2011; Lillesand et al., 2015). Several studies have used a variety of remote sensing techniques to study invasive alien plants (e.g. Abdel-Rahman, Mutanga, Adam, & Ismail, 2014; Adam & Mutanga, 2009; Adelabu, Mutanga, Adam, & Sebego, 2014; Bentivegna et al., 2012; Berg, Kotze, & Beukes, 2013; Manevski, Manakos, Petropoulos, & Kalaitzidis, 2011; Martín, Barreto, & Fernández-Quintanilla, 2011; Mirik et al., 2014; Narumalani, Mishra, Wilson, Reece, & Kohler, 2009; Prasad & Gnanappazham, 2014).

Plants have been mapped using multispectral remote sensing techniques in a number of studies (e.g. Dronova, Gong, Wang, & Zhong, 2015; Johansen, Phinn, & Witte, 2010; Laba et al., 2008; Lemke, Hulme, Brown, & Tadesse, 2011; Vancutsem, Pekel, Evrard, Malaisse, & Defourny, 2009). This method is good particularly for large spatial area mapping purposes (Azong, Malahlela, & Ramoelo, 2015; Cuneo, Jacobson, & Leishman, 2009; Dronova et al., 2015; Vancutsem et al., 2009). In comparison, hyperspectral remote sensing offers better accuracy levels of vegetation characterization due to the high spectral resolution and continuous hyperspectral bands they possess (Alparone et al., 2015; Carroll, Glaser, Hunt, & Sappington, 2008; Gavier-pizarro, Kuemmerle, & Stewart, 2012; Huang & Asner, 2009; Jensen, 2014). For example, Bentivegna et al. (2012) detected invasive cutleaf teasel (*Dipsacus laciniatus* L.) in Missouri, USA using high spatial resolution (1 m) hyperspectral images (63 bands in visible to near-infrared spectral region). Mirik et al. (2013) explored the ability of hyperspectral imagery for mapping infestation of musk thistle (*Carduus nutans*) on a native grassland during the pre-and peak-flowering stages using support vector machine classifier in Friona, Parmer County, USA. Ouyang et al. (2013) used a field spectrometer data to find the most appropriate period for mapping invasive *Spartina alterniflora* by measuring its community and major victims at different phonological stages in Chongming Island, China. Similarly, Rudolf, Lehmann, Große-stoltenberg, Römer, and Oldeland (2015) developed a classification model to spectrally discriminate between invasive shrub *Acacia longifolia* from other non-native and native species using field-based spectra and condensed leaf tannin content in Portuguese dune ecosystems, Portugal.

However, discrimination of plant species using hyperspectral data often places emphasis on identification of the optimal specific bands for discrimination. These bands are narrow and cannot be separated from within the broader bandwidth of multispectral data. Hyperspectral remote sensing has grown significantly in the past few decades. However, its application in operational characterization is rather limited. Although there is a promise to translate research efforts of hyperspectral remote sensing into operational tools, current advances in data availability show that multispectral remote sensing remains the most important source of information in vegetation monitoring. For example, DeVries, Verbesselt, Kooistra, and Herold (2015) monitored small-scale forest disturbances in a tropical montane forest of southern Ethiopia using Landsat time series. Gu and Wylie (2015) developed a 30-m grassland productivity estimation map for central Nebraska in USA using 250 m MODIS and 30 m Landsat 8 observations, United States. Johansen, Phinn, and Taylor (2015) mapped woody vegetation clearing in Queensland, Australia from Landsat imagery using the Google Earth Engine. Kennedy et al. (2015) described factors attributing to disturbance change from Landsat time-series in support of habitat monitoring in the Puget Sound region, USA. Therefore, research efforts involving hyperspectral remote sensing analysis need to consider extending the technique into multispectral remote sensing techniques.

This study uses a continuum of hyperspectral bands to identify best wavelength regions for discriminating *Asparagus laricinus* from adjacent land cover types. As such, it focuses on spectral regions rather than identifying individual bands in an attempt to simulate multispectral remote sensing systems. Specific objectives of the study are (1) determining whether or not *A. laricinus* can be differentiated from adjacent land cover types using a field spectrometer data and (2) to investigate the performance of spectra simulated according to Landsat and SPOT 5 images in discriminating *A. laricinus* from adjacent land cover types. There have been little or no studies that focused on discriminating *A. laricinus* from other vegetation or land cover types. *A. laricinus* is a plant belonging to the Asparagaceae family and occurs in different parts of South Africa. However, the plant is not indigenous in South Africa and has a status of "list concern" in the South African National Biodiversity Institute (SANBI) national Red List of South African plants (Foden & Potter, 2005). Knowledge on the spectral and spatial characteristics of the species assists the development of better management strategies in areas where it invades. Such maps can also help traditional health practitioners and pharmaceutical industries to locate stands of the plant for medicinal purposes, as it also has medicinal uses (Fuku, Al-Azzawi, Madamombe-Manduna, & Mashele, 2013; Mashele & Kolesnikova, 2010; Ntsoelinyane & Mashele, 2014).

2. Methods

2.1. Study area

The study was conducted in the Klipriviersberg Nature Reserve, in Johannesburg, South Africa (Figure 1). It covers an area of approximately 680 hectares in extent and is managed by the City of Johannesburg. The reserve lies in the Klipriviersberg area, a transition zone between the grass land and the savannah biome in the northern edge of the Highveld (Faiola & Vermaak, 2014). Climatic conditions experienced in the reserve vary from warm to hot summer (17–26°C) and cool to cold winter (5–7°C) (Kotze, 2002). Three geology types occur in the reserve, namely basalt and andesite volcanic rocks that underlay the reserve; quartzites and conglomerates of the upper Witwatersrand system underneath the lavas in north of the reserve; and dolomites of the Transvaal system south of the reserve (Kotze, 2002). The flora of the reserve is categorized into two broad vegetation types, the Andesite Mountain Bushveld and a section of Tsakane Clay Grassland at its flatter southern end (Faiola & Vermaak, 2014). There is relatively rich biodiversity with approximately 650 indigenous plant species, 215 bird species, 16 reptile species, and 32 butterfly species. Mammals that occur in the reserve include lesser spotted genet, African civet, zebra, red hartebeest, blesbok, springbok, duiker, black wildebeest, porcupines, meerkats, and otters (Faiola & Vermaak, 2014).

2.2. Field data

Field surveys were conducted between the 2 and 14 December 2014 during summer season of the area with the aim of characterizing the vegetation under relatively high vigor condition. *A. laricinus* is found extensively in one part of the reserve, while other occurrences are scattered in small spatial

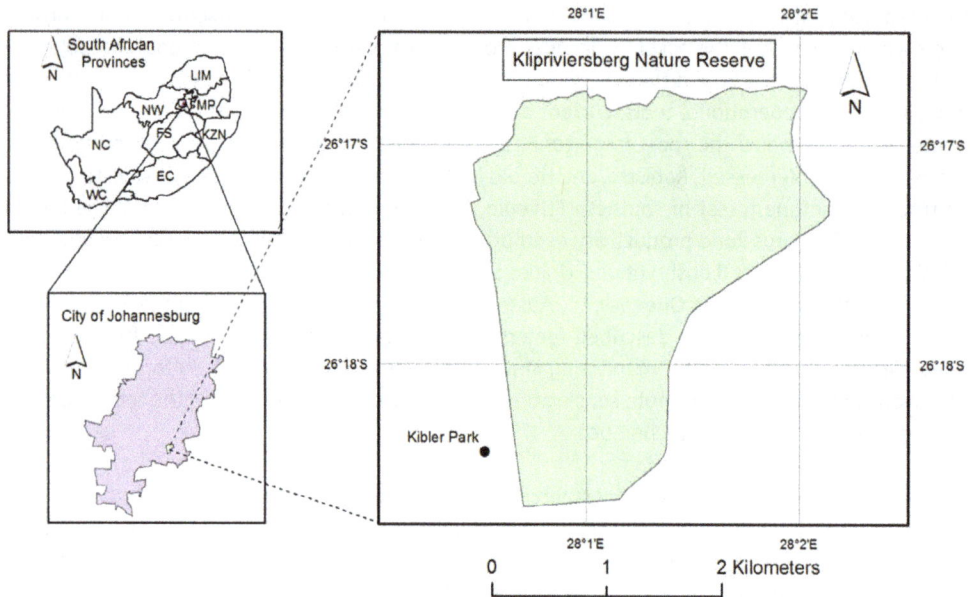

Figure 1. Map showing the Klipriviersberg Nature Reserve.

extents. Such a rather limited distribution resulted in delineation of 10 plots of 15 m radius each (Figure 2). The plot size was chosen with the anticipation of extending the investigation to space-borne remote sensing techniques. Each plot therefore accommodates at least one pixel of Landsat imagery (30 m resolution) and a number of SPOT 5 imagery pixels (1.5–10 m resolutions). The center of each plot was recorded using GPS (Garmin GPSmap® 76) within 3 m accuracy. A total of 13 sample plants were taken randomly within the 15 m radius plot area. Although random, sampling was attempted to follow a systematic design as shown in Figure 2. Therefore, samples were taken at 5-m intervals along perpendicular transects that intersect at the center of plot (Appendix A). However, this was rarely achieved as it was difficult walking through the thorny and dense stands of *A. laricinus*, prompting use of random sampling. *A. laricinus* individuals varied between six and eight plants within each plot.

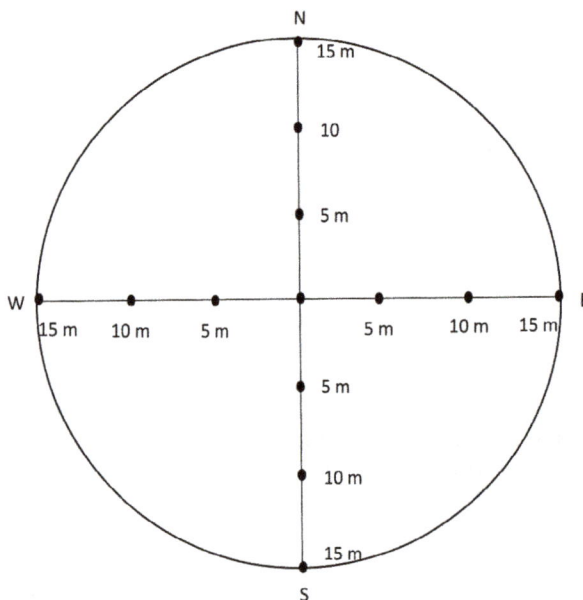

Figure 2. A layout of sampling design for spectral measurements of individual target plant.

Spectral data were collected using Spectral Evolution® SR-3500 Remote Sensing Portable Spectroradiometer (Spectral Evolution Inc., Lawrence, MA, USA). The spectrometer has a 1.6-nm spectral resolution ranging between 340 and 2,503 nm. Target radiance in energy unit was converted into percent reflectance using a white reference measurement (Prospere, McLaren, & Wilson, 2014). Three spectral measurements were taken for each *A. laricinus* plant from different leaf canopy parts of the plant with all measurements taken at 5 cm above leaf canopy to mimic a remotely sensed data (airborne and spaceborne) viewpoint. The three spectral measurements were averaged to represent the reflectance spectra of each sample plant. Spectral measurements from adjacent land cover types were taken in a similar manner. These measurements should ideally be taken when the sun is overheard to acquire electromagnetic radiation reflectance optimally (Cho, Sobhan, Skidmore, & de Leeuw, 2008; Fernandes, Aguiar, Silva, Ferreira, & Pereira, 2013; Mansour, 2013; Olsson, van Leeuwen, & Marsh, 2011; Rudolf et al., 2015). However, time constraints did not necessarily allow the application of this protocol for all measurements.

2.3. Analysis of spectral reflectance per region

Analysis was limited to the regions that showed consistent spectral differences between *A. laricinus* and adjacent land cover types. In order to identify these regions an average spectrum was computed from the three spectral measurements taken from each target (*A. laricinus* and adjacent land cover type, respectively). The resultant average values were pooled per land cover type and averaged to generate "global" spectral curves representing *A. laricinus* and each adjacent land cover type in the study area as illustrated (Figure 3). The global spectra of *A. laricinus* was compared against each adjacent land cover types, as illustrated in an example comparing *A. laricinus* and grass in Figure 4. Please note, not all global comparisons are presented in here for the sake of brevity. The global spectra of adjacent land cover types were computed to determine the potential discrimination of *A. laricinus* from them, since the species can co-exist with a mixture of land cover types in a natural environment. Comparison using global pairs is deemed a better representation of the study area than comparison of individual pairs that most likely yields results that are unable to converge to a compromise generic conclusion.

A visual assessment of the global spectra was used to determine regions that were considered unnecessary for differentiating *A. laricinus* and adjacent land cover types. Two rules were used to determine these regions. The first rule included regions that returned random reflectance properties commonly known as atmospheric noise (*A. laricinus* vs. Grass: 1,873–1,954 and 2,351–2,503 nm; *A. laricinus* vs. Acacia: 1,821–1,956 nm and 2,282–2,503 nm; *A. laricinus* vs. Herbaceous: 1,838–1,942 nm and 2,272–2,503 nm; *A. laricinus* vs. mixture of herbaceous and bare ground: 1,831–1,970 nm and 2,351–2,503 nm). The second rule included regions that did not show spectral reflectance difference between *A. laricinus* and adjacent land cover types (*A. laricinus* vs. Grass: 340–343, 684–750 nm and 1,350–1,824 nm; *A. laricinus* vs. Acacia: 650–749 and1,331–1,448 nm: *A. laricinus* vs. Herbaceous: 340–387, 641–748 nm and 1,316–1,448 nm; *A. laricinus* vs. Mixture of herbaceous and bare ground: 340–467, 685–745 nm and 1357–1,455 nm). These exclusions resulted in four discontinuous regions (Table 1, Figure 5) based on which spectra of individual targets (individuals of *A. laricinus* and adjacent land cover types) were used in further analyses.

Analysis involved comparison of reflectance between *A. laricinus* and adjacent land cover types at two levels, namely individual and plot levels. Individual level comparison was made between *A. laricinus* and adjacent land cover type at a sampling point within each plot. On the other hand, plot level comparison was made between plot level mean reflectance of *A. laricinus* against plot level mean reflectance of dominant adjacent land cover type. Differences at both levels were assessed graphically and using statistical tests such as the analysis of variance (ANOVA) and *t*-test (Weiss, 2012). All the tests were calculated using 95% confidence level ($\alpha = 0.05$).

2.4. Simulation of Landsat and SPOT 5 imagery bands

Wavelength regions corresponding to Landsat and SPOT 5 bands were extracted from the original reflectance spectra for all *A. laricinus* and adjacent land cover types. This was an initial step to

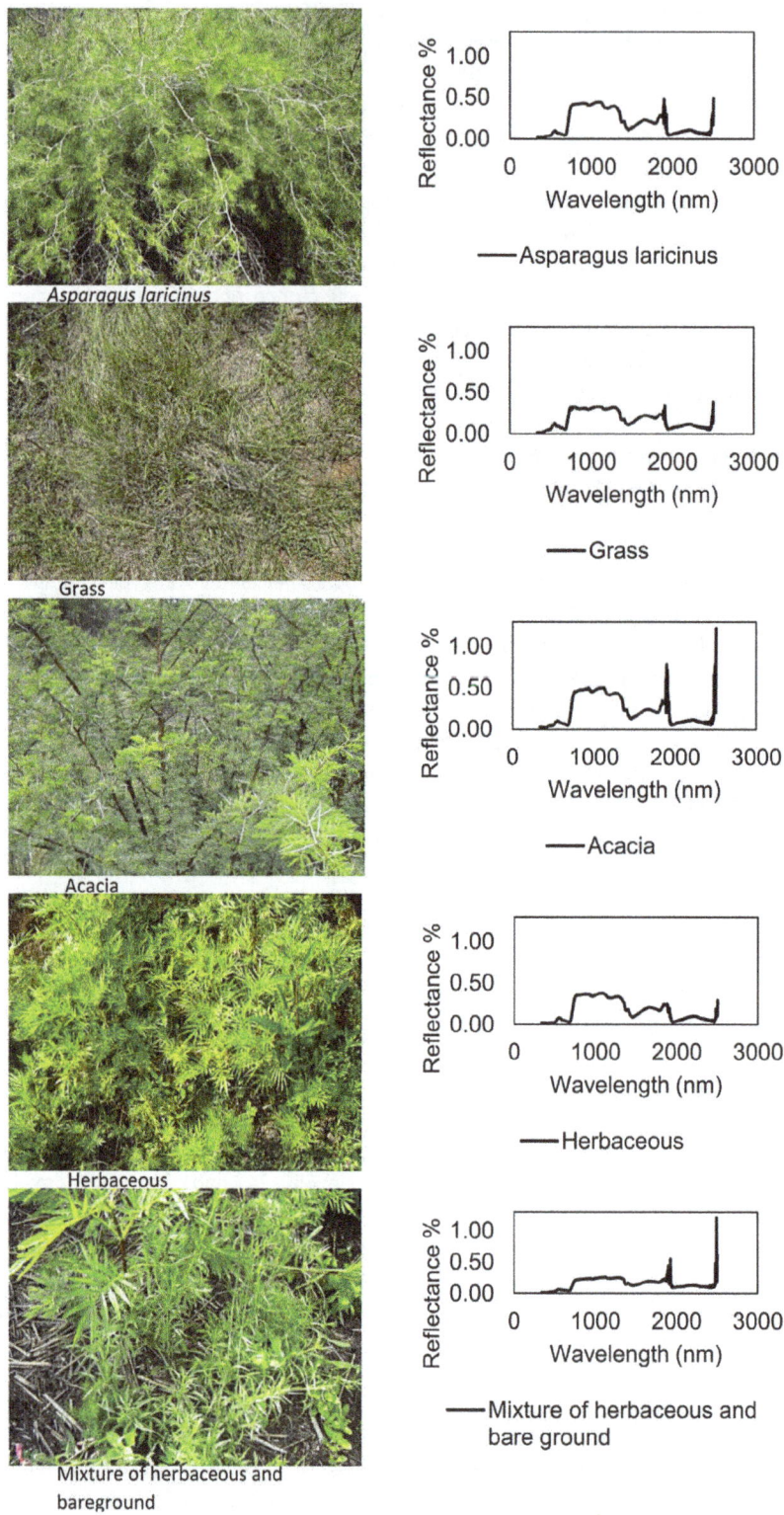

Figure 3. Global spectra of *A. laricinus* and adjacent land cover types.

testing the potential of upscaling field-based remote sensing information to airborne or satellite-based remote sensing. Only blue, green, red, and NIR bands were simulated or Landsat, while green, red, and NIR spectral bands were simulated for SPOT 5 imagery. These elected bands are widely used

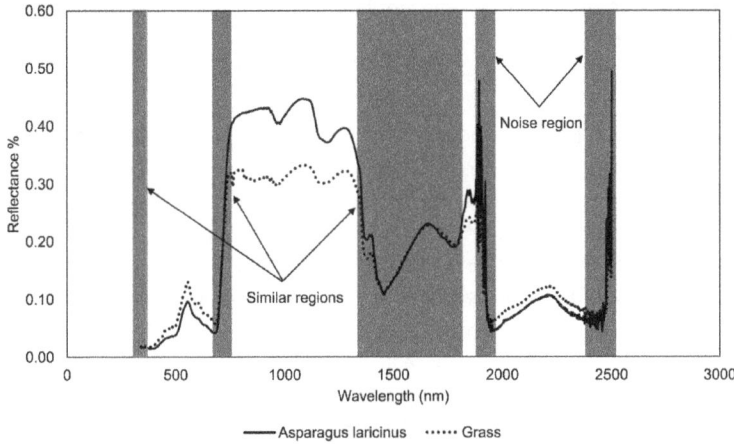

Figure 4. Global reflectance of *A. laricinus* and grass across the full spectrum.

Note: Highlighted regions show spectral parts excluded from further analysis.

Table 1. Wavelength regions used for analysis				
Comparison pairs	**Wavelength regions**			
	Region 1 (Ultraviolet & Visible), nm	**Region 2 (NIR), nm**	**Region 3 (NIR & SWIR), nm**	**Region 4 (SWIR), nm**
A. laricinus vs. Grass	345–683	752–1,346	1,828–1,872	1,956–2,349
A. laricinus vs. Acacia	340–648	750–1,327	1,452–1,817	1,959–2,279
A. laricinus vs. Herbaceous	389–640	749–1,312	1,452–1,835	1,945–2,269
A. laricinus vs. Mixture of herbaceous and bare ground	468–684	747–1,354	1,459–1,828	1,973–2,349

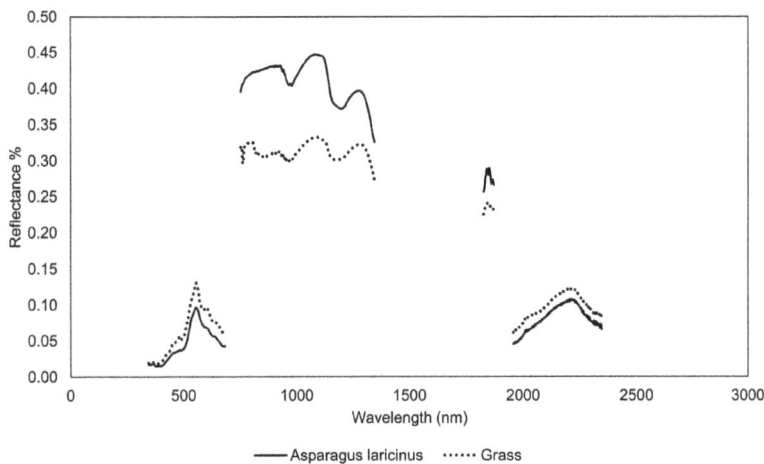

Figure 5. An example of global reflectance of *A. laricinus* and grass, representing wavelength regions used in further analysis.

in the assessment of vegetation characteristics (e.g. Manevski et al., 2011; Mirik, Ansley, et al., 2013; Mirik, Emendack, et al., 2014). Five separate pools representing *A. laricinus*, grass, acacia, herbaceous, and mixture of herbaceous and bare ground were created. Reflectance comparisons were done at individual and group level. Individual level compared the pool of *A. laricinus* against separate pools of grass, acacia, herbaceous, and mixture of herbaceous and bare ground. The group level compared *A. laricinus* pool against combined pool of adjacent land cover types. Spectral differences were assessed using ANOVA and *t*-test.

3. Results

Individual-level comparisons between *A. laricinus* and adjacent land cover types resulted in an overall significant difference in all plots for each spectral region, based on ANOVA results. However, separate reflectance comparisons of each of the individuals per plot showed inconsistent significant differences. Distinct spectral separability between *A. laricinus* and adjacent land cover types was observed mostly in the NIR region (region 2), with seven of 10 plots. In contrast, only two in the ultraviolet–visible (region 1), three in the NIR–SWIR (region 3), and five in the SWIR (region 4) regions showed clear separation. These differences are illustrated in Figure 6 which show spectral reflectance differences between *A. laricinus* and grass of one plot. Significant differences are presented using different letters, whereas same letters represent insignificant differences. Distinct separation between *A. laricinus* and adjacent land cover types in the NIR region (region 2) is shown by higher reflectance of *A. laricinus* than other land cover types (Figure 6).

Grasses represented majority of land cover types at plot level analysis (7 of 10 plots), while Herbaceous, Acacia, and Mixture of ground and herbaceous were dominant in each of the remaining plots. Comparisons at this level resulted in significant differences in all plots based on *t*-test results as illustrated in Figure 7. In most cases, *A. laricinus* had higher reflectance than adjacent land cover types in the NIR region (region 2), with 8 of 10 plots. The species had higher reflectance in five plots in the ultraviolet–visible (region 1), six plots in the NIR–SWIR (region 3) and five plots in the SWIR region (region 4).

3.1. Landsat simulation

Comparisons between *A. laricinus* and adjacent land cover types at the individual level resulted in an overall significant difference in all Landsat simulated bands (blue, green, red, and NIR), based on the ANOVA results. Individual pair comparisons using least significance difference (LSD) resulted in significant difference between *A. laricinus* and all land cover types, in most cases (Figure 8). Similarities were, however, observed between *A. laricinus* and grass in the blue and red bands, and between *A. laricinus* and herbaceous vegetation in the green band (Figure 8). *A. laricinus* had higher reflectance than other adjacent land cover types with exceptions of Acacia in the blue band, Acacia and herbaceous in the green and NIR bands, and Acacia and grass in the red band.

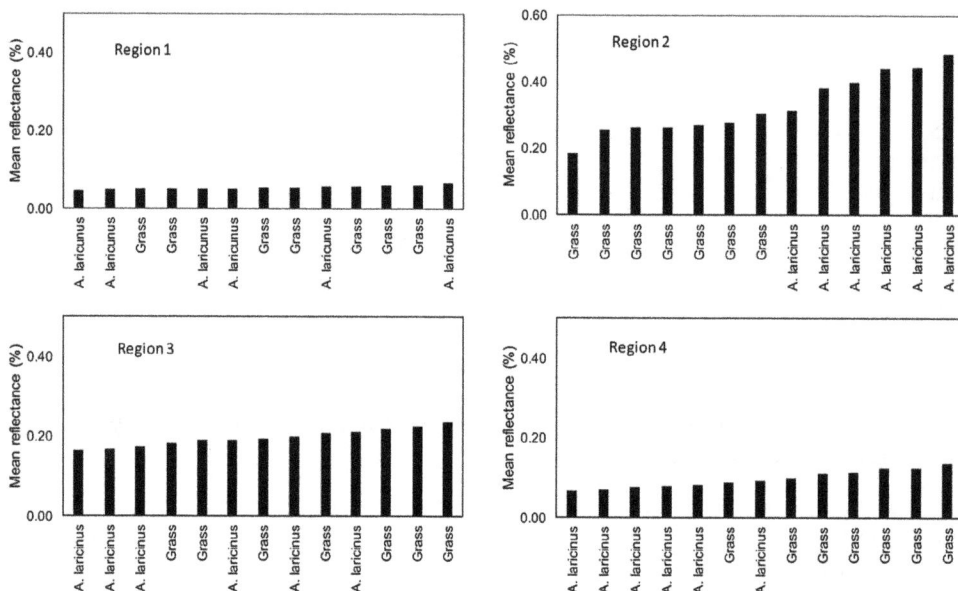

Figure 6. Reflectance of the regions used for analysis at individual plant level for a typical plot.

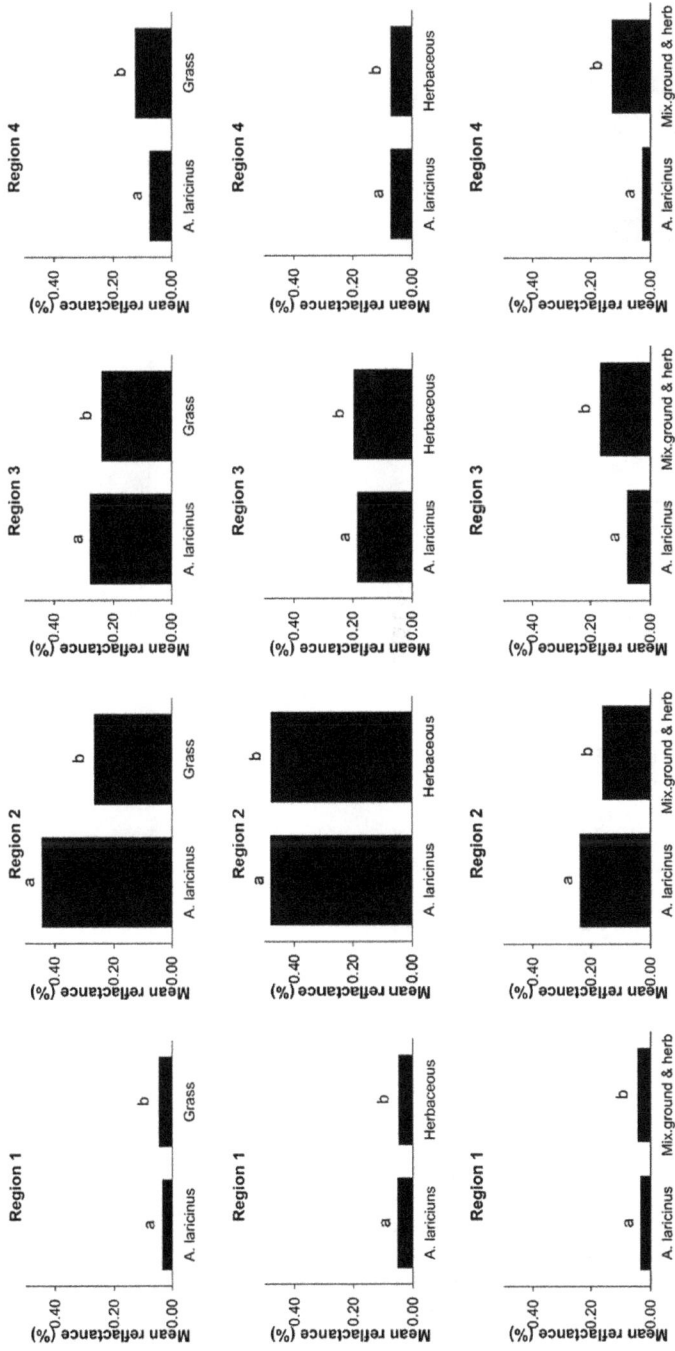

Figure 7. Plot-level mean reflectance of A. *laricinus* and adjacent land cover types.

Note: The comparisons are per region and per plot. (Mix. Ground & herb = Mixture of herbaceous and bare ground).

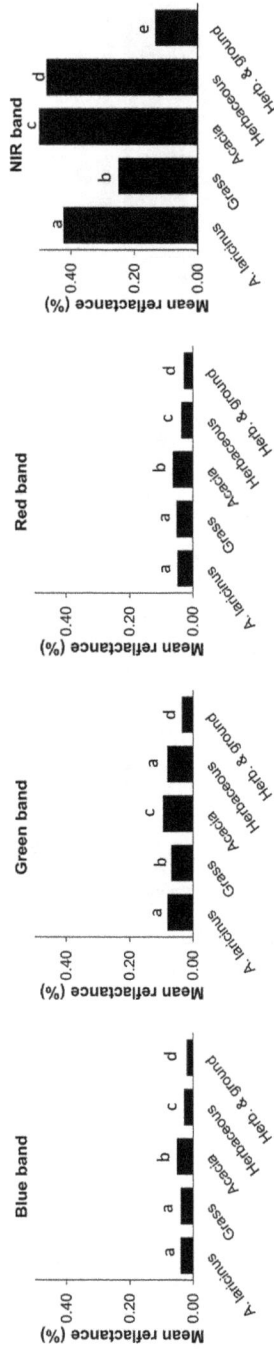

Figure 8. Mean reflectance of simulated Landsat bands per land cover type (individual level).

Note: The comparison is per spectral band. (Herb. & ground = Mixture of herbaceous and bare ground).

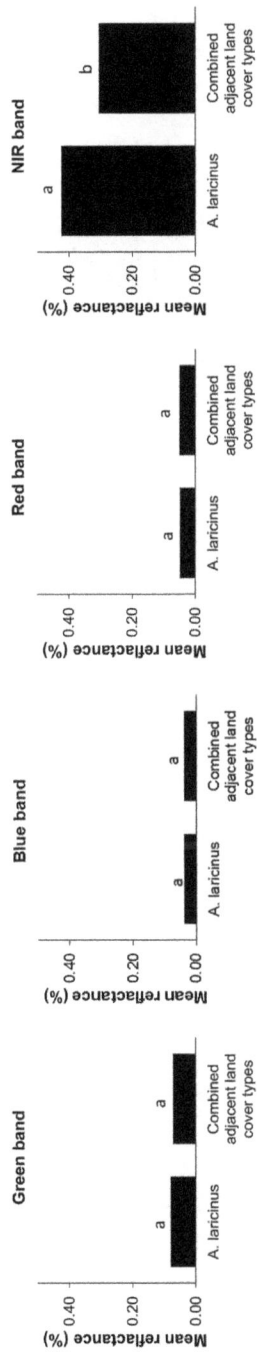

Figure 9. Mean reflectance of simulated Landsat bands per land cover type (Group level).

Note: The comparison is per spectral band.

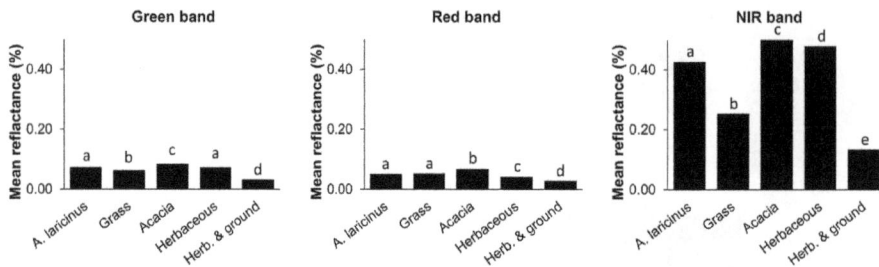

Figure 10. Mean reflectance of simulated SPOT 5 bands per land cover type (individual level).

Note: The comparison is per spectral band. (Herb. & ground = Mixture of herbaceous and bare ground).

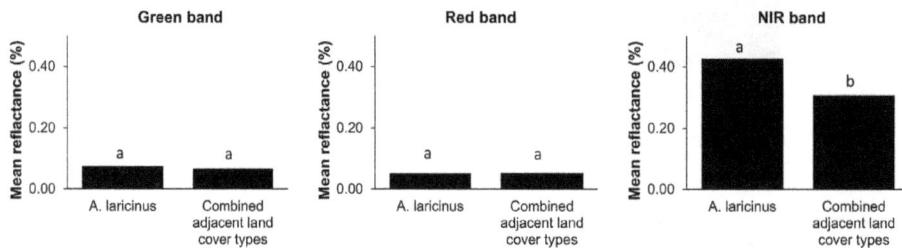

Figure 11. Mean reflectance of simulated SPOT 5 bands per land cover type (Group level).

Note: The comparison is per spectral band.

Comparison of reflectance at the group level between *A. laricinus* and combined adjacent land cover types resulted in insignificant difference in the blue, green, and red bands, while the difference was significant in the NIR (Figure 9). *A. laricuns* had higher reflectance than combined adjacent land cover types in the green and NIR band, while it had lower reflectance in the blue and red bands (Figure 9).

3.2. SPOT 5 simulation

Reflectance comparisons of SPOT 5 simulated bands resulted in overall significant differences in all bands, based on ANOVA. Individual pair comparisons using LSD showed significant differences between *A. laricinus* and adjacent land cover types in all bands, except for comparison between *A. laricinus* and herbaceous vegetation in the green band as well as between *A. laricinus* and grass in the red band (Figure 10). *A. laricinus* had a relatively high reflectance in all bands. However, it had lower reflectance than Acacia plants in all bands and herbaceous vegetation in the green and NIR bands, and grass in the red band (Figure 10).

Group-level comparisons between *A. laricinus* and combined adjacent land cover types showed significant difference in only the NIR band (Figure 11). *A. laricinus* had higher reflectance than combined adjacent land cover types in the green and NIR bands, while it had negligible reflectance in the red band (Figure 11).

4. Discussion

The utility of a field-based spectral data to discriminate *A. laricinus* from adjacent land cover types was investigated in this study. Investigations were made using original spectra and spectra simulated based on bands of Landsat and SPOT 5 images. These simulations were intended to assess the potential of upscaling the technique to spaceborne remote sensing techniques. Analyses were done at individual and plot levels using original spectra, and individual and group level for the simulated spectra. Visual comparisons using global pair reflectance of *A. laricinus* and each adjacent land cover type showed differentiation in the ultraviolet–visible (region 1), NIR (region 2), NIR–SWIR (region 3), and SWIR (region 4) spectral regions, but the difference was considerable in the NIR region (e.g. Figure 5). *A. laricinus* had high reflectance in NIR (region 2) and NIR–SWIR (region 3) and low reflectance in ultraviolet–visible region (region 1) and SWIR region (region 4) when compared with grass. *A. laricinus* reflectance was high in all regions when compared with herbaceous, while it was high in ultraviolet–visible (region 1), NIR (region 2), and NIR–SWIR (region 3) when compared with mixture of bare ground and herbaceous plants, but it was low in all regions when compared with Acacia. All these wavelength regions are considered best at characterizing vegetation types (e.g. Manevski et al., 2011; Mirik, Ansley, et al., 2013; Mirik, Emendack, et al., 2014). The far-SWIR region on the other hand is considered best at discriminating between photosynthetic, non-photosynthetic vegetation components, and ground due to spectral absorption attributable to presence of cellulose in healthy vegetation (Daughtry et al., 2006; Guerschman et al., 2009; Nagler, Daughtry, & Goward, 2000; Serbin, Daughtry, Hunt, Reeves, & Brown, 2009).

The overall significant differences observed for individual-level comparisons per plot are not attributable to reflectance difference between *A. laricinus* and adjacent land cover types. This is because significant differences were observed even within individuals of same land cover types, based on pairwise comparisons using LSD. There were further inconsistent significant differences when comparing individuals per plot separately. As such, distinct separation between *A. laricinus* and adjacent land cover types was mostly achieved in the NIR region, for 7 of 10 plots, while only a few plots showed clear separation in the ultraviolet–visible region, NIR–SWIR, and SWIR regions (Figure 6). Consistent significant difference observed in the NIR region was somewhat expected, given the distinct reflectance differences between *A. laricinus* and adjacent land cover types from the global spectra comparisons (e.g. Figure 5).

The plot-level differences between *A. laricinus* and dominant adjacent land cover types were considerable particularly between *A. laricinus* and grass as well as *A. laricinus* and mixture of herbaceous vegetation and bare ground (e.g. Figure 7). The differences were somewhat expected given different global reflectance patterns of *A. laricinus*, grass, and mixture of herbaceous vegetation and bare ground (Figure 3). In contrast, the differences between *A. laricinus* and herbaceous were lower, although they were significant in the visible, NIR, and lower end of SWIR regions. This can as well be explained by the global reflectance resemblance of *A. laricinus* and herbaceous (Figure 3). Another noteworthy observation at the plot level was the fact that the magnitude of reflectance of *A. laricinus* was greater than for herbaceous vegetation in the ultraviolet–visible (regions 1), NIR region (region 2), and SWIR (region 4), and smaller in NIR–SWIR (region 3). This is the opposite of what were observed in comparisons between *A. laricinus* and grass as well as *A. laricinus* and a mixture of herbaceous vegetation and bare ground. This dissimilarity can be attributed to the relatively heterogeneous species composition of herbaceous plants within a plot. In contrast, grass and bare ground can be comparatively considered homogenous land cover types, respectively, having marked spectral difference with *A. laricinus*.

The significant difference between *A. laricinus* and adjacent land cover types using the Landsat- and SPOT 5-simulated bands achieved at the individual-level analysis (Figures 8–11) was anticipated, given the distinct homogeneous setup of *A. laricinus* and adjacent land cover types. This setting does, however, occur rarely in an ideal natural environment where plant of different species co-exist. Unlike individual level analysis which showed significant differences in all bands (Figure 8 and 9), only the NIR band showed significant difference at group level (Figure 10 and 11). These results showed the potential of discriminating *A. laricinus* from adjacent land cover types using this band which is available in most remotely sensed data. This agrees with a study that classified *Asparagus officinalis* (a species that belongs to the same family as *A. laricinus*) successfully using Landsat imagery (Tatsumi, Yamashiki, Canales Torres, & Taipe, 2015).

The NIR band was most useful in discriminating between *A. laricinus* and adjacent land cover types. This is not surprising as the band has been widely used in discriminating between plant species in a number of studies. For example, *A. officinalis* was successfully identified using NIR reflectance spectroscopy by Perez and Sanchez (2001). This region was used in studies on plants not related to *A. laricinus*, too, such as by Xu, Yu, Fu, and Ying (2009) who successfully discriminated between two tomato varieties in China using visible–near-infrared reflectance spectroscopy. Thenkabail et al. (2013) on the other hand identified individual bands that included the visible–NIR bands as well as vegetation indices that best characterize, classify, model, and map the world's main agricultural crops. Bentivegna et al. (2012) detected cutleaf teasel (*D. laciniatus*) with hyperspectral imagery using visible–NIR spectral region along Missouri Highway, USA. Calvini, Ulrici, and Amigo (2015) tested sparse methods for classifying Arabica and Robusta coffee species using near-infrared hyperspectral images.

5. Conclusion

This study aimed at determining the potential of discriminating between *A. laricinus* and adjacent land cover types in the Klipriviersberg Nature Reserve using a field spectrometer data. Analysis of spectral reflectance was done at individual and plot levels using the original spectra. Although different spectral wavelength regions showed the ability to differentiate the species from other land cover types, the NIR region was found to be the most consistent of all. This finding is in line with other vegetation studies, although such studies on asparagus are rare.

A comparative similarity between *A. laricinus* and herbaceous plants was noteworthy. This similarity can make identification of the plant challenging in such co-existence. In contrast, the species can be discriminated from grass and mixed land cover (ground and herbaceous vegetation) at relative ease. The separability from grass is particularly important if the species favors to co-exist more with grass than with other species (7 of 10 plots were dominated by *A. laricinus* and grass in this study). The ability to discriminate these pecies from mixed land cover types that include bare ground, among others, is useful since it enables early detection in sparsely vegetated areas. Further studies are however needed to determine the relative contribution of different land cover types in the mixture to spectral reflectance.

Analysis of spectra simulated based on Landsat and SPOT 5 imagery bands showed the NIR to be consistent in discriminating *A. laricinus* from other land cover types. This finding is encouraging in that it shows the potential of upscaling the application to airborne and spaceborne remote sensing that mostly include the NIR region of electromagnetic energy. This study, however, used limited number of samples and thus should rather be considered a preliminary indicator that needs further studies. Future studies should attempt to utilize large number of samples. Such sample size can be achieved with the use of small sampling units and high spatial resolution imagery (e.g. SPOT 5, 6/7), particularly in areas where the spatial extent of invasion is small relative to imagery with lower spatial resolution (e.g. Landsat). In addition, limiting spectral measurements within ideal time frames when there is enough illumination would need to be considered. Furthermore, it is vital to profile the biochemical contents of the species so that relationships can be built between the inherent contents of the plant and their effects on spectral signatures. In connection to this, it is important to take into consideration spectral properties at different phenological stages of the species.

Acknowledgements
The authors are thankful to the Klipriviersberg Nature Reserve for permitting use of the reserve as a study site. The authors also acknowledge Bongeka Wendy Mbatha and Minenhle Gumede for helping in data collection.

Funding
This research was funded by the University of Johannesburg.

Author details
Bambo Dubula[1]
E-mail: bambodubula@gmail.com
ORCID ID: http://orcid.org/0000-0002-4718-8026
Solomon Gebremariam Tesfamichael[1]
E-mail: gtesfamichael@uj.ac.za
Isaac Tebogo Rampedi[1]

E-mail: isaacr@uj.ac.za
[1] Department of Geography, Environmental Management and Energy Studies, University of Johannesburg, Johannesburg, South Africa.

References
Abdel-Rahman, E. M., Mutanga, O., Adam, E., & Ismail, R. (2014). Detecting Sirex noctilio grey-attacked and lightning-struck pine trees using airborne hyperspectral data, random forest and support vector machines classifiers. *ISPRS Journal of Photogrammetry and Remote Sensing, 88,* 48–59. doi:10.1016/j.isprsjprs.2013.11.013

Adam, E., & Mutanga, O. (2009). Spectral discrimination of papyrus vegetation (*Cyperus papyrus* L.) in swamp wetlands using field spectrometry. *ISPRS Journal of Photogrammetry and Remote Sensing, 64,* 612–620. doi:10.1016/j.isprsjprs.2009.04.004

Adelabu, S., Mutanga, O., Adam, E., & Sebego, R. (2014). Spectral discrimination of insect defoliation levels in mopane woodland using hyperspectral data. *IEEE Journal of Selected Topics in Applied Earth Observations and Remote Sensing, 7,* 177–186. doi:10.1109/JSTARS.2013.2258329

Alparone, L., Aiazzi, B., Baronti, S., & Andrea G. (2015). *Remote sensing image fusion.* CRC Press. doi:10.1201/b18189-9

Azong, M., Malahlela, O., & Ramoelo, A. (2015). Assessing the utility WorldView-2 imagery for tree species mapping in South African subtropical humid forest and the conservation implications: Dukuduku forest patch as case study. *International Journal of Applied Earth Observations and Geoinformation, 38,* 349–357. doi:10.1016/j.jag.2015.01.015

Bentivegna, D. J., Smeda, R. J., Wang, C., Bentivegna, D. J., Smeda, R. J., & Wang, C. (2012). Detecting cutleaf teasel (*Dipsacus laciniatus*) along a Missouri highway with hyperspectral imagery. *Invasive Plant Science and Management, 5,* 155–163. doi:10.1614/IPSM-D-10-00053.1

Berg, V. Den, Kotze, I., & Beukes, H. (2013). Detection, quantification and monitoring of prosopis in the Northern Cape Province of South Africa using remote sensing and GIS. *South African Journal of Geomatics, 2,* 68–81.

Bradley, B. A., & Marvin, D. C. (2011). Using expert knowledge to satisfy data needs: Mapping invasive plant distributions in the western United States. *Western North American Naturalist, 71,* 302–315. doi:10.3398/064.071.0314

Calvini, R., Ulrici, A., & Amigo, J. (2015). Practical comparison of sparse methods for classification of Arabica and Robusta coffee species using near infrared hyperspectral imaging. *Chemometrics and Intelligent Laboratory Systems, 146,* 503–511. doi:10.1016/j.chemolab.2015.07.010

Campbell, J. B., & Wynne, R. H. (2011). *Introduction to remote sensing* (5th ed.). New York, NY: The Guilford Press.

Carroll, M. W., Glaser, J. A., Hunt, T. E., & Sappington, T. W. (2008). Use of spectral vegetation indices derived from airborne hyperspectral imagery for detection of European Corn Borer infestation in Iowa Corn plots. *Journal of Economic Entomology, 101,* 1614–1623.

Cho, M., Sobhan, I., Skidmore, A., & de Leeuw, J. (2008). Discriminating species using hyperspectral indices at leaf and canopy scales. *The International Archives of the Photogrammetry, Remote Sensing and Spatial Information Sciences, 37,* 369–376.

Cuneo, P., Jacobson, C., & Leishman, M. (2009). Landscape-scale detection and mapping of invasive African olive (*Olea europaea* L. ssp. cuspidata Wall ex G. Don Ciferri) in SW Sydney, Australia using satellite remote sensing. *Applied Vegetation Science, 12,* 145–154. http://dx.doi.org/10.1111/avsc.2009.12.issue-2

Daughtry, C. S. T., Doraiswamy, P. C., Hunt, E. R., Stern, A. J., McMurtrey, J. E., & Prueger, J. H. (2006). Remote sensing of crop residue cover and soil tillage intensity. *Soil and Tillage Research, 91,* 101–108. doi:10.1016/j.still.2005.11.013

DeVries, B., Verbesselt, J., Kooistra, L., & Herold, M. (2015). Robust monitoring of small-scale forest disturbances in a tropical montane forest using Landsat time series. *Remote Sensing of Environment, 161,* 107–121. doi:10.1016/j.rse.2015.02.012

Dewey, S. A., Price, K. P., & Ramsey, D. (1991). Satellite remote sensing to predict potential distribution of Dyers wood (*Isatis tinctoria*). *Weed Technology, 5,* 479–484.

Dorigo, W., Lucieer, A., Podobnikar, T., & Čarni, A. (2012). Mapping invasive Fallopia japonica by combined spectral, spatial, and temporal analysis of digital orthophotos. *International Journal of Applied Earth Observation and Geoinformation, 19,* 185–195. doi:10.1016/j.jag.2012.05.004

Dronova, I., Gong, P., Wang, L., & Zhong, L. (2015). Mapping dynamic cover types in a large seasonally flooded wetland using extended principal component analysis and object-based classification. *Remote Sensing of Environment, 158,* 193–206. doi:10.1016/j.rse.2014.10.027

Faiola, J., & Vermaak, V. (2014). Klipriviersberg. *Veld and Flora, 100,* 68–71.

Fernandes, M. R., Aguiar, F. C., Silva, J. M. N., Ferreira, M. T., & Pereira, J. M. C. (2013). Spectral discrimination of giant reed (*Arundo donax* L.): A seasonal study in riparian areas. *ISPRS Journal of Photogrammetry and Remote Sensing, 80,* 80–90. doi:10.1016/j.isprsjprs.2013.03.007

Foden, W., & Potter, L. (2005). Asparagus laricinus Burch. National assessment: Red list of South African plants version 2015.1. Retrieved September 18, 2015, from http://redlist.sanbi.org/species.php?species=728-59

Fuku, S., Al-Azzawi, A., Madamombe-Manduna, I., & Mashele, S. (2013). Phytochemistry and free radical scavenging activity of Asparagus laricinus. *International Journal of Pharmacology, 9,* 312–317.

Galvão, L., Epiphanio, J., Breunig, F., & Formaggio, A. (2011). Crop type discrimination using hyperspectral data. In *Hyperspectral remote sensing of vegetation* (pp. 397–422). CRC Press. doi:10.1201/b11222-25

Gavier-pizarro, G. I., Kuemmerle, T., & Stewart, S. I. (2012). Monitoring the invasion of an exotic tree (*Ligustrum lucidum*) from 1983 to 2006 with Landsat TM/ETM + satellite data and support vector machines in Córdoba, Argentina. *Remote Sensing of Environment,* 1–12.

Gu, Y., & Wylie, B. K. (2015). Developing a 30-m grassland productivity estimation map for central Nebraska using 250-m MODIS and 30-m Landsat-8 observations. *Remote Sensing of Environment, 171,* 291–298. doi:10.1016/j.rse.2015.10.018

Guerschman, J. P., Hill, M. J., Renzullo, L. J., Barrett, D. J., Marks, A. S., & Botha, E. J. (2009). Estimating fractional cover of photosynthetic vegetation, non-photosynthetic vegetation and bare soil in the Australian tropical savanna region upscaling the EO-1 Hyperion and MODIS sensors. *Remote Sensing of Environment, 113,* 928–945. doi:10.1016/j.rse.2009.01.006

Huang, C., & Asner, G. P. (2009). Applications of remote sensing to alien invasive plant studies. *Sensors, 9,* 4869–4889. doi:10.3390/s90604869

Jensen, J. R. (2014). *Remote sensing of the environment: An earth resource perspective* (2nd ed.). Harlow: Pearson.

Johansen, K., Phinn, S., & Witte, C. (2010). Mapping of riparian zone attributes using discrete return LiDAR, QuickBird and SPOT-5 imagery: Assessing accuracy and costs. *Remote Sensing of Environment, 114,* 2679–2691. doi:10.1016/j.rse.2010.06.004

Johansen, K., Phinn, S., & Taylor, M. (2015). Mapping woody vegetation clearing in Queensland, Australia from landsat imagery using the Google Earth Engine. *Remote Sensing Applications: Society and Environment, 1,* 36–49. doi:10.1016/j.rsase.2015.06.002

Kennedy, R. E., Yang, Z., Braaten, J., Copass, C., Antonova, N., Jordan, C., & Nelson, P. (2015). Attribution of disturbance change agent from landsat time-series in support of habitat monitoring in the Puget Sound region, USA. *Remote Sensing of Environment, 166,* 271–285. doi:10.1016/j.rse.2015.05.005

Kotze, P. (2002). *The ecological integrity of the Klip River and the development of a sensitivity weighted fish index of biotic integrity (SIBI). ujdigispace.* Johannesburg: University of Johannesburg.

Laba, M., Downs, R., Smith, S., Welsh, S., Neider, C., White, S., ... Baveye, P. (2008). Mapping invasive wetland plants in the Hudson River National Estuarine Research Reserve using quickbird satellite imagery. *Remote Sensing of Environment, 112,* 286–300. doi:10.1016/j.rse.2007.05.003

Lemke, D., Hulme, P. E., Brown, J. A., & Tadesse, W. (2011). Distribution modelling of Japanese honeysuckle (Lonicera japonica) invasion in the Cumberland Plateau and

Mountain Region, USA. *Forest Ecology and Management, 262*, 139–149. doi:10.1016/j.foreco.2011.03.014

Lillesand, T., Kiefer, R., & Chipman, J. (Eds.). (2015). *Remote sensing and image interpretation* (7th ed.). Hoboken, NJ: John Wiley & Sons.

Manevski, K., Manakos, I., Petropoulos, G. P., & Kalaitzidis, C. (2011). Discrimination of common Mediterranean plant species using field spectroradiometry. *International Journal of Applied Earth Observation and Geoinformation, 13*, 922–933. doi:10.1016/j.jag.2011.07.001

Mansour, K. (2013). Comparing the new generation world view-2 to hyperspectral image data for species discrimination. *International Journal of Development Research, 3*, 8–13.

Martín, M., Barreto, L., & Fernández-Quintanilla, C. (2011). Discrimination of sterile oat (*Avena sterilis*) in winter barley (*Hordeum vulgare*) using QuickBird satellite images. *Crop Protection, 30*, 1363–1369. doi:10.1016/j.cropro.2011.06.008

Mashele, S., & Kolesnikova, N. (2010). In vitro anticancer screening of Asparagus laricinus extracts. *Pharmacologyonline, 2*, 246–252.

Mgidi, T. N., Le Maitre, D. C., Schonegevel, L., Nel, J. L., Rouget, M., & Richardson, D. M. (2007). Alien plant invasions—Incorporating emerging invaders in regional prioritization: A pragmatic approach for Southern Africa. *Journal of Environmental Management, 84*, 173–187. doi:10.1016/j.jenvman.2006.05.018

Mirik, M., Ansley, R. J., Steddom, K., Jones, D. C., Rush, C. M., Michels, G. J., & Elliott, N. C. (2013). Remote distinction of a noxious weed (Musk Thistle: Carduus Nutans) using airborne hyperspectral imagery and the support vector machine classifier. *Remote Sensing, 5*, 612–630. doi:10.3390/rs5020612

Mirik, M., Emendack, Y., Attia, A., Chaudhuri, S., Roy, M., Backoulou, G. F., & Cui, S. (2014). Detecting musk thistle (*Carduus nutans*) infestation using a target recognition algorithm. *Advances in Remote Sensing, 3*, 95–105. http://dx.doi.org/10.4236/ars.2014.33008

Nagler, P. L., Daughtry, C. S. T., & Goward, S. N. (2000). Plant litter and soil reflectance. *Remote Sensing of Environment, 71*, 207–215. doi:10.1016/S0034-4257(99)00082-6

Narumalani, S., Mishra, D. R., Wilson, R., Reece, P., & Kohler, A. (2009). Detecting and mapping four invasive species along the floodplain of North Platte River, Nebraska. *Weed Technology, 23*, 99–107. doi:10.1614/WT-08-007.1

Ntsoelinyane, P. H., & Mashele, S. (2014). Phytochemical screening, antibacterial and antioxidant activities of asparagus laricinus leaf and stem extracts. *Bangladesh Journal of Pharmacology, 9*, 10–14. doi:10.3329/bjp.v9i1.16967

Olsson, A. D., van Leeuwen, W. J. D., & Marsh, S. E. (2011). Feasibility of invasive grass detection in a desertscrub community using hyperspectral field measurements and landsat TM imagery. *Remote Sensing, 3*, 2283–2304. doi:10.3390/rs3102283

Ouyang, Z.-T., Gao, Y., Xie, X., Guo, H.-Q., Zhang, T.-T., & Zhao, B. (2013). Spectral discrimination of the invasive plant spartina alterniflora at multiple phenological stages in a saltmarsh wetland. *PLoS ONE, 8*, e67315. doi:10.1371/journal.pone.0067315

Perez, D., & Sanchez, M. (2001). Authentication of green asparagus varieties by near-infrared reflectance spectroscopy. *Journal of Food Science, 66*, 323–327. doi:10.1111/j.1365-2621.2001.tb11340.x

Prasad, K., & Gnanappazham, L. (2014). Discrimination of mangrove species of Rhizophoraceae using laboratory spectral signatures. In *IEEE Geoscience and Remote Sensing Symposium*, Quebec City.

Prospere, K., McLaren, K., & Wilson, B. (2014). Plant species discrimination in a tropical wetland using in situ hyperspectral data. *Remote Sensing, 6*, 8494–8523. doi:10.3390/rs6098494

Richardson, D. M., & Van Wilgen, B. W. (2004). Invasive alien plants in South Africa: How well do we understand the ecological impacts? *South African Journal of Science, 100*, 45–52.

Rodgers, L., Pernas, T., & Hill, S. D. (2014). Mapping invasive plant distributions in the Florida everglades using the digital aerial sketch mapping technique. *Invasive Plant Science and Management, 7*, 360–374. doi:10.1614/IPSM-D-12-00092.1

Rouget, M., Hui, C., Renteria, J., Richardson, D. M., & Wilson, J. R. U. (2015). Plant invasions as a biogeographical assay: Vegetation biomes constrain the distribution of invasive alien species assemblages. *South African Journal of Botany*. doi:10.1016/j.sajb.2015.04.009

Rudolf, J., Lehmann, K., Große-stoltenberg, A., Römer, M., & Oldeland, J. (2015). Field spectroscopy in the VNIR-SWIR region to discriminate between Mediterranean native plants and exotic-invasive shrubs based on leaf tannin content. *Remote Sensing, 7*, 1225–1241. doi:10.3390/rs70201225

Schor, J., Farwig, N., & Berens, D. G. (2015). Intensive land-use and high native fruit availability reduce fruit removal of the invasive Solanum mauritianum in South Africa. *South African Journal of Botany, 96*, 6–12. doi:10.1016/j.sajb.2014.11.004

Serbin, G., Daughtry, C. S. T., Hunt, E. R., Reeves, J. B., & Brown, D. J. (2009). Effects of soil composition and mineralogy on remote sensing of crop residue cover. *Remote Sensing of Environment, 113*, 224–238. doi:10.1016/j.rse.2008.09.004

Tatsumi, K., Yamashiki, Y., Canales Torres, M. A., & Taipe, C. L. R. (2015). Crop classification of upland fields using Random forest of time-series Landsat 7 ETM+ data. *Computers and Electronics in Agriculture, 115*, 171–179. doi:10.1016/j.compag.2015.05.001

Thenkabail, P. S., Mariotto, I., Gumma, M. K., Middleton, E. M., Landis, D. R., & Huemmrich, K. F. (2013). Selection of hyperspectral narrowbands (HNBs) and composition of hyperspectral twoband vegetation indices (HVIs) for biophysical characterization and discrimination of crop types using field reflectance and hyperion/EO-1 data. *IEEE Journal of Selected Topics in Applied Earth Observations and Remote Sensing, 6*, 427–439. http://dx.doi.org/10.1109/JSTARS.2013.2252601

Van Wilgen, B. (2006). Invasive alien species—An important aspect of global change. *CSIR Science Scope, 1*, 8–11.

van Wilgen, B. W., Forsyth, G. G., Le Maitre, D. C., Wannenburgh, A., Kotzé, J. D. F., van den Berg, E., & Henderson, L. (2012). An assessment of the effectiveness of a large, national-scale invasive alien plant control strategy in South Africa. *Biological Conservation, 148*, 28–38. doi:10.1016/j.biocon.2011.12.035

van Wilgen, B., Reyers, B., Le Maitre, D., Richardson, D., & Schonegevel, L. (2008). A biome-scale assessment of the impact of invasive alien plants on ecosystem services in South Africa. *Journal of Environmental Management, 89*, 336–349. doi:10.1016/j.jenvman.2007.06.015

Vancutsem, C., Pekel, J. F., Evrard, C., Malaisse, F., & Defourny, P. (2009). Mapping and characterizing the vegetation types of the Democratic Republic of Congo using SPOT VEGETATION time series. *International Journal of Applied Earth Observation and Geoinformation, 11*, 62–76. doi:10.1016/j.jag.2008.08.001

Vicente, J. R., Fernandes, R. F., Randin, C. F., Broennimann, O., Gonçalves, J., Marcos, B., ... Honrado, J. P. (2013). Will climate change drive alien invasive plants into areas of high protection value? An improved model-based regional assessment to prioritise the management of invasions. *Journal of Environmental Management, 131*, 185–195. doi:10.1016/j.jenvman.2013.09.032

Weiss, N. A. (2012). *Introductory statistics* (9th ed.). Boston, MA: Pearson Education.

Xu, H., Yu, P., Fu, X., & Ying, Y. (2009). On-site variety discrimination of tomato plant using visible-near infrared reflectance spectroscopy. *Journal of Zhejiang University SCIENCE B, 10*, 126–132. doi:10.1631/jzus.B0820200

Appendix A

Center coordinates of sample plots used in the analysis

Plots	Latitude	Longitude
1	−26.30169	28.01205
2	−26.30117	28.01164
3	−26.30085	28.01121
4	−26.30076	28.01141
5	−26.3002	28.01127
6	−26.30018	28.01058
9	−26.30063	28.01058
8	−26.30148	28.01138
9	−26.30257	28.01096
10	−26.30291	28.01106

GEO-CEOS stage 4 validation of the Satellite Image Automatic Mapper lightweight computer program for ESA Earth observation level 2 product generation - Part 1: Theory

Andrea Baraldi[1,2,3,4]*, Michael Laurence Humber[2], Dirk Tiede[3] and Stefan Lang[3]

*Corresponding author: Andrea Baraldi, Department of Agricultural Sciences, University of Naples Federico II, Portici 80055, Italy
E-mail: andrea6311@gmail.com
Reviewing Editor: Louis-Noel Moresi, University of Melbourne, Australia

Abstract: ESA defines as Earth Observation (EO) Level 2 information product a single-date multi-spectral (MS) image corrected for atmospheric, adjacency and topographic effects, stacked with its data-derived scene classification map (SCM), whose legend includes quality layers cloud and cloud-shadow. No ESA EO Level 2 product has ever been systematically generated at the ground segment. To fill the information gap from EO big data to ESA EO Level 2 product in compliance with the GEO-CEOS stage 4 validation (*Val*) guidelines, an off-the-shelf Satellite Image Automatic Mapper (SIAM) lightweight computer program was validated by independent means on an annual 30 m resolution Web-Enabled Landsat Data (WELD) image composite time-series of the conterminous U.S. (CONUS) for the years 2006–2009. The SIAM core is a prior knowledge-based decision tree for MS

ABOUT THE AUTHOR

Andrea Baraldi (*Laurea* in Electronic Engineering, Univ. Bologna, 1989; Master in Software Engineering, Univ. Padua, 1994; PhD in Agricultural Sciences, Univ. Naples Federico II, 2017) has held research positions at the Italian Space Agency (2018–to date); Dept. Geoinformatics, Univ. Salzburg, Austria (2014–2017); Dept. Geographical Sciences, Univ. Maryland, College Park, MD (2010–2013); European Commission Joint Research Centre (2000–2002; 2005–2009); International Computer Science Institute, Berkeley, CA (1997–1999); European Space Agency Research Institute (1991–1993); Italian National Research Council (1989, 1994–1996, 2003–2004). In 2009, he founded Baraldi Consultancy in Remote Sensing. He was appointed with a Senior Scientist Fellowship at the German Aerospace Center, Oberpfaffenhofen, Germany (February 2014) and was a visiting scientist at the Ben Gurion Univ. of the Negev, Sde Boker, Israel (February 2015). His main interests center on the research and technological development of automatic near real-time spaceborne/airborne image pre-processing and understanding systems in operating mode, consistent with human visual perception. Dr. Baraldi served as an Associate Editor of the IEEE TRANS. NEURAL NETWORKS from 2001 to 2006.

PUBLIC INTEREST STATEMENT

Synonym of scene-from-image reconstruction and understanding, vision is an inherently ill-posed cognitive task; hence, it is difficult to solve and requires a priori knowledge in addition to sensory data to become better posed for numerical solution. In the inherently ill-posed cognitive domain of computer vision, this research was undertaken to validate by independent means a lightweight computer program for prior knowledge-based multi-spectral color naming, called Satellite Image Automatic Mapper (SIAM), eligible for automated near real-time transformation of large-scale Earth observation (EO) image datasets into European Space Agency (ESA) EO Level 2 information product, never accomplished to date at the ground segment. An original protocol for wall-to-wall thematic map quality assessment without sampling, where legends of the test and reference map pair differ and must be harmonized, was adopted. Conclusions are that SIAM is suitable for systematic ESA EO Level 2 product generation, regarded as necessary not sufficient pre-condition to transform EO big data into timely, comprehensive and operational EO value-adding information products and services.

reflectance space hyperpolyhedralization into static color names. Typically, a vocabulary of MS color names in a MS data (hyper)cube and a dictionary of land cover (LC) class names in the scene-domain do not coincide and must be harmonized (reconciled). The present Part 1—Theory provides the multidisciplinary background of a priori color naming. The subsequent Part 2—Validation accomplishes a GEO-CEOS stage 4 *Val* of the test SIAM-WELD annual map time-series in comparison with a reference 30 m resolution 16-class USGS National Land Cover Data 2006 map, based on an original protocol for wall-to-wall thematic map quality assessment without sampling, where the test and reference maps feature the same spatial resolution and spatial extent, but whose legends differ and must be harmonized.

Subjects: Algorithms & Complexity; Automation; Cognitive Artificial Intelligence; Expert systems; GIS, Remote Sensing & Cartography; Image Processing; Intelligent Systems; Machine Learning; Pattern Analysis; Quality Control & Reliability; Real-Time Systems; Systems Architecture

Keywords: Artificial intelligence; binary relationship; Cartesian product; cognitive science; color naming; connected-component multilevel image labeling; deductive inference; Earth observation; land cover taxonomy; high-level (attentive) and low-level (pre-attentional) vision; hybrid inference; image classification; image segmentation; inductive inference; machine learning-from-data; outcome and process quality indicators; radiometric calibration; remote sensing; surface reflectance; thematic map comparison; top-of-atmosphere reflectance; two-way contingency table; unsupervised data discretization/vector quantization; validation

1. Introduction

Jointly proposed by the intergovernmental Group on Earth Observations (GEO) and the Committee on Earth Observation Satellites (CEOS), the implementation plan for years 2005–2015 of the Global Earth Observation System of Systems (GEOSS) aimed at systematic transformation of multi-source Earth observation (EO) *big data* into timely, comprehensive and operational EO value-adding products and services (GEO, 2005), submitted to the GEO-CEOS Quality Assurance Framework for Earth Observation (QA4EO) calibration/validation (*Cal/Val*) requirements and suitable "to allow the access to the Right Information, in the Right Format, at the Right Time, to the Right People, to Make the Right Decisions" (Group on Earth Observation/Committee on Earth Observation Satellites (GEO-CEOS), 2010). In this definition of GEOSS, term *big data* identifies "a collection of data sets so large and complex that it becomes difficult to process using on-hand database management tools or traditional data processing applications. The big data challenges include capture, storage, search, sharing, transfer, analysis and visualization" (Wikipedia, 2018a), typically summarized as the five Vs of big data, specifically, volume, variety, velocity, veracity and value (IBM, 2016; Yang, Huang, Li, Liu, & Hu, 2017).

The GEOSS mission cannot be considered fulfilled by the remote sensing (RS) community to date. This is tantamount to saying the RS community is data-rich, but information-poor, a conjecture known as DRIP syndrome (Bernus & Noran, 2017). Before supporting this thesis with observations, the following definition is introduced. In this paper, an EO-IUS is defined in operating mode if and only if it scores "high" in every index of a minimally dependent and maximally informative (mDMI) set of EO outcome and process (OP) quantitative quality indicators (Q^2Is), to be community-agreed upon to be used by members of the community, in agreement with the GEO-CEOS QA4EO *Cal/Val* guidelines (GEO-CEOS, 2010). A proposed instantiation of mDMI set of EO OP-Q^2Is includes: (i) degree of automation, inversely related to human-machine interaction, (ii) effectiveness, e.g., thematic mapping accuracy, (iii) efficiency in computation time and in run-time memory occupation, e.g., inversely related to the number of system's free-parameters to be user-defined based on heuristics, (iv) robustness (vice versa, sensitivity) to changes in input data, (v) robustness to changes in input parameters to be user-defined, (vi) scalability to changes in user requirements and in sensor specifications, (vii) timeliness from data acquisition to information product generation, (viii) costs in manpower and computer power, (ix) value, e.g., semantic value of output products, economic value of output services, etc. (Baraldi, 2017, 2009; Baraldi & Boschetti, 2012a, 2012b; Baraldi, Boschetti, & Humber, 2014; Baraldi et al., 2010a, 2010b; Baraldi, Gironda, & Simonetti, 2010c; Duke, 2016). According to the Pareto formal analysis of multi-objective optimization problems, optimization of an mDMI set of OP-Q^{2}Is is an inherently-ill posed problem in the Hadamard sense (Hadamard, 1902), where many Pareto optimal solutions lying on the Pareto efficient frontier can be considered equally good (Boschetti, Flasse, & Brivio, 2004). Any EO-IUS solution lying on the Pareto efficient frontier can be considered in operating mode, therefore suitable to cope with the five Vs of spatio-temporal EO big data (Yang et al., 2017).

Stating that to date the RS community is affected by the DRIP syndrome is like saying that past and present EO image understanding systems (EO-IUSs) have been typically outpaced by the rate of data collection of EO imaging sensors, whose quality and quantity are ever-increasing at an apparently exponential rate related to the Moore law of productivity (National Aeronautics and SpaceAdministration (NASA), 2016a). In common practice, EO-IUSs are overwhelmed by sensory data they are unable to transform into EO value-adding information products and services, in compliance with the GEO-CEOS QA4EO *Cal/Val* guidelines (GEO-CEOS, 2010). If this conjecture holds true, then existing EO-IUSs cannot be considered in operating mode because unsuitable to cope with the five Vs of spatio-temporal EO big data (Yang et al., 2017). Several observations (true-facts) support this thesis. First, in 2012 the percentage of EO data ever downloaded from the European Space Agency (ESA) databases was estimated at about 10% or less (D'Elia 2012). This estimate is equal or superior (never inferior) to the percentage of ESA EO data ever used by the RS community. Since 2012, the same EO data exploitation indicator is expected to decrease, because any increase in productivity of existing EO-IUSs seems unable to match the exponential increase in the rate of collection of EO sensory data (NASA, 2016a). Second, EO-IUSs presented in the RS literature are typically assessed and compared based on the sole thematic mapping accuracy, which means their mDMI set of EO OP-Q^{2}Is remains largely unknown to date (Baraldi & Boschetti, 2012a, 2012b). As a consequence, the RS literature is unable to contradict the thesis that no EO-IUS is available in operating mode. For example, when EO data-derived thematic maps were generated by EO-IUSs based on a supervised (labeled) data learning approach, at continental or global spatial extent and with estimated accuracy not inferior to a target mapping accuracy requirement, the most limiting factors turned out to be the cost, timeliness, quality and availability of adequate supervised training data samples collected from field sites, existing maps or geospatial data archives in tabular form (Gutman et al., 2004). Third, no ESA EO data-derived Level 2 information product has ever been systematically generated at the ground segment (DLR and VEGA 2011; ESA 2015). In the ESA definition, an EO Level 2 information product is a single-date multispectral (MS) image radiometrically calibrated into surface reflectance (SURF) values corrected for atmospheric, adjacency and topographic effects, stacked with its data-derived scene classification map (SCM), whose legend includes quality layers cloud and cloud-shadow, starting from an ESA EO Level 1 product geometrically corrected and radiometrically calibrated into top-of-atmosphere reflectance (TOARF) values (European Space Agency (ESA), 2015; Deutsches Zentrum für Luft- und Raumfahrt e.V. (DLR) and VEGA Technologies, 2011; CNES, 2015).

This last observation deserves further discussion. In the words of Marr, "vision goes symbolic almost immediately without loss of information" (Marr, 1982). In agreement with this Marr's intuition, ESA defines as EO Level 2 information product an information primitive (unit of information) consisting of a stack of two coupled (inter-dependent) variables, one sub-symbolic/numeric and one symbolic, where symbolic variable means both categorical and semantics. Equivalent to two sides of the same coin, these two variables are very closely related to each other and cannot be separated, even though they seem different. The first side of the ESA EO Level 2 information primitive is a multivariate numeric variable of the highest radiometric quality, related to the concept of quantitative/unequivocal *information-as-thing* in the terminology of philosophical hermeneutics (Capurro & Hjørland, 2003), see Figure 1. The second side of the ESA EO Level 2 information unit is an EO data-derived SCM, equivalent to a categorical variable of semantic value, related to the concept of qualitative/equivocal *information-as-data-interpretation* in

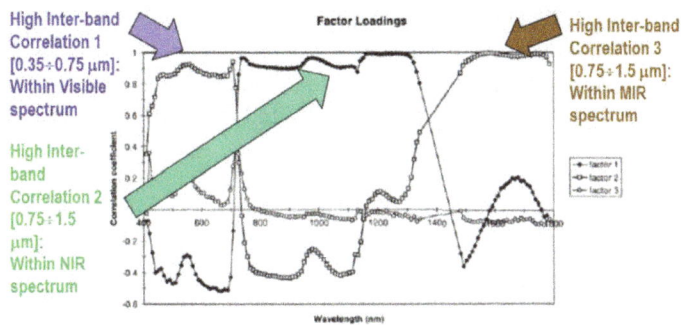

Figure 1. Courtesy of van der Meer and De John (2000). Pearson's cross-correlation (CC) coefficients for the main factors resulting from a principal component analysis and factor rotation 1, 2 and 3 for an agricultural data set based on spectral bands of the AVIRIS hyper-spectral (HS) spectro-meters. Flevoland test site, 5 July 1991. Inter-band CC values are "high" (>0.8) within the visible spectral range, the Near Infra-Red (NIR) wavelengths and the Medium IR (MIR) wavelengths. The general conclusion is that, irrespective of non-stationary local information, the global (image-wide) information content of a multi-channel image, either multi-spectral (MS) whose number N of spectral channels $\in$ {2, 9}, super-spectral (SS) with $N \in$ {10, 20}, or HS image with $N > 2 0$, can be preserved by selecting one visible band, one NIR band, one MIR band and one thermal IR (TIR) band, such as in the spectral resolution of the imaging sensor series National Oceanic and Atmospheric Administration (NOAA) Advanced Very High Resolution Radiometer (AVHRR), in operating mode from 1978 to date.

the terminology of philosophical hermeneutics (Capurro & Hjørland, 2003). In practice, ESA EO Level 2 product generation is a *chicken-and-egg* dilemma (Riano, Chuvieco, Salas, & Aguado, 2003), synonym of inherently ill-posed problem in the Hadamard sense (Hadamard, 1902). Therefore, it is very difficult to solve and requires a priori knowledge in addition to data to become better posed for numerical solution (Cherkassky & Mulier, 1998). On the one hand, no effective and efficient *Cal* of digital numbers (DNs) into SURF values corrected for atmospheric, topographic and adjacency effects is possible without an SCM, available *a priori* in addition to data to enforce a statistical stratification principle (Hunt & Tyrrell, 2012), synonym of layered (class-conditional) data analytics (Baraldi, 2017; Baraldi et al., 2010b; Baraldi & Humber, 2015; Baraldi, Humber, & Boschetti, 2013; Bishop & Colby, 2002; Bishop, Shroder, & Colby, 2003; DLR & VEGA, 2011; Dorigo, Richter, Baret, Bamler, & Wagner, 2009; Lück & van Niekerk, 2016; Riano et al., 2003; Richter & Schläpfer, 2012a, 2012b; Vermote & Saleous, 2007). On the other hand, no effective and efficient understanding (mapping) of a sub-symbolic EO image into a symbolic SCM is possible if DNs (pixels) are affected by low radiometric quality (GEO-CEOS, 2010). In an ESA EO Level 2 SCM product to be generated at the ground segment (midstream) as input to downstream applications, products and services (Mazzuccato & Robinson, 2017), the SCM legend is required to consist of a discrete, finite and hierarchical (multilevel) dictionary (Lipson, 2007; Mather, 1994; Swain & Davis, 1978) of general-purpose, user- and application-independent land cover (LC) classes, whose semantic value is "shallow" (not specialized) in hierarchy, but superior to zero semantics typical of numeric variables, in addition to quality layers cloud and cloud-shadow (ESA, 2015; DLR & VEGA, 2011; CNES, 2015). To our best knowledge, only one prototypical implementation of a sensor-specific ESA EO Level 2 product generator exists to date. Commissioned by ESA, the Sentinel 2 (atmospheric) Correction Prototype Processor (SEN2COR) is not run systematically at the ESA ground segment. Rather, it can be downloaded for free from the ESA web site to be run on user side (European Space Agency (ESA), 2015; Deutsches Zentrum für Luft- und Raumfahrt e.V. (DLR) and VEGA Technologies, 2011).

Noteworthy, a National Aeronautics and Space Administration (NASA) EO Level 2 product is defined as "a data-derived geophysical variable at the same resolution and location as Level 1 source data" (NASA, 2016b). Hence, dependence relationship "NASA EO Level 2 product → ESA EO Level 2 product" holds, where symbol "→" denotes relationship *part-of* pointing from the supplier to the client, in agreement with the standard Unified Modeling Language (UML) for graphical modeling of object-oriented software (Fowler, 2003), see Figure 2. This dependence means that although space agencies and EO data distributors claim systematic NASA EO Level 2 product generation at the ground segment, this does not imply systematic ESA EO Level 2 product generation. Rather, the vice versa holds: if ESA EO Level 2 product generation is accomplished, then NASA EO Level 2 product generation is also fulfilled.

Different from the non-standard SEN2COR SCM legend instantiation (ESA, 2015; DLR & VEGA, 2011), one example of general-purpose, user- and application-independent ESA EO Level 2 SCM legend is the standard (community-agreed) 3-level 8-class dichotomous phase (DP) taxonomy of the Food and Agriculture Organization of the United Nations (FAO)—Land

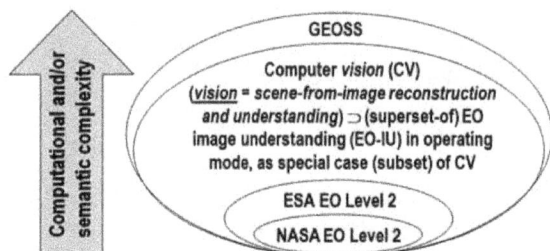

Cover Classification System (LCCS) (Di Gregorio & Jansen, 2000). The FAO LCCS-DP hierarchy is "fully nested". It comprises three dichotomous LC class-specific information layers, equivalent to a world ontology or world model (Di Gregorio & Jansen, 2000; Matsuyama & Hwang, 1990): DP Level 1—Vegetation versus non-vegetation, DP Level 2—Terrestrial versus aquatic and DP Level 3—Managed versus natural or semi-natural. The 3-level 8-class FAO LCCS-DP taxonomy is shown in Figure 3. For the sake of generality, a 3-level 8-class FAO LCCS-DP legend is added with LC class "other", synonym of "rest of the world" or "unknown", which would include quality information layers cloud and cloud-shadow. In traditional EO image classification system design and implementation requirements (Swain & Davies, 1978), the presence of output class "unknown" is considered mandatory to cope with uncertainty in *information-as-data-interpretation* tasks. Hereafter, the standard 3-level 8-class FAO LCCS-DP legend added with the mandatory output class "other", which includes quality layers cloud and cloud-shadow, is identified as "augmented" 9-class FAO LCCS-DP taxonomy. In the complete two-phase FAO LCCS hierarchy, a general-purpose 3-level 8-class FAO LCCS-DP legend is preliminary to a high-level application-dependent and user-specific FAO LCCS Modular Hierarchical Phase (MHP) taxonomy, consisting of a hierarchical (deep) battery of one-class classifiers (Di Gregorio & Jansen, 2000), see Figure 3. In recent years, the two-phase FAO LCCS taxonomy has become increasingly popular (Ahlqvist, 2008). One reason of this popularity is that the FAO LCCS hierarchy is "fully nested" while alternative LC class hierarchies, such as the Coordination of Information on the Environment (CORINE) Land Cover (CLC) taxonomy (Bossard, Feranec, & Otahel, 2000), the U.S. Geological Survey (USGS) Land Cover Land Use (LCLU) taxonomy by J. Anderson (Lillesand & Kiefer, 1979), the International Global Biosphere Programme (IGBP) DISCover Data Set Land Cover Classification System (Belward, 1996) and the EO Image Librarian LC class legend (Dumitru, Cui, Schwarz, & Datcu, 2015), start from a Level 1 taxonomy which is already multi-class. In a hierarchical EO-IUS architecture sub-mitted to a *garbage in, garbage out* (GIGO) information principle, synonym of error propaga-tion through an information processing chain, the fully-nested two-phase FAO LCCS hierarchy makes explicit the full dependence of high-level EO OP-Q^2I estimates, featured by any high-level (deep) LCCS-MHP data processing module, on previous EO OP-Q^2I values featured by lower-level LCCS modules, starting from the initial FAO LCCS-DP Level 1 vegetation/non-vegetation information layer whose relevance in thematic mapping accuracy (vice versa, in error propagation) becomes paramount for all subsequent LCCS layers. The GIGO common-sense principle applied to hierarchical semantic dependence is neither trivial nor obvious to underline (Marcus, 2018). On the one hand, it agrees with a minor portion of the RS literature where supervised data learning classification of EO image datasets at continental or global spatial extent into binary LC class vegetation/non-vegetation is considered very challenging (Gutman et al., 2004). On the other hand, it is at odd with the RS mainstream, where the semantic information gap from sub-symbolic EO data to multi-class LC taxonomies is typically filled in one step, implemented as a supervised data learning classifier (Bishop, 1995; Cherkassky & Mulier, 1998), e.g., a support vector machine, random forest or deep convolu-tional neural network (DCNN)

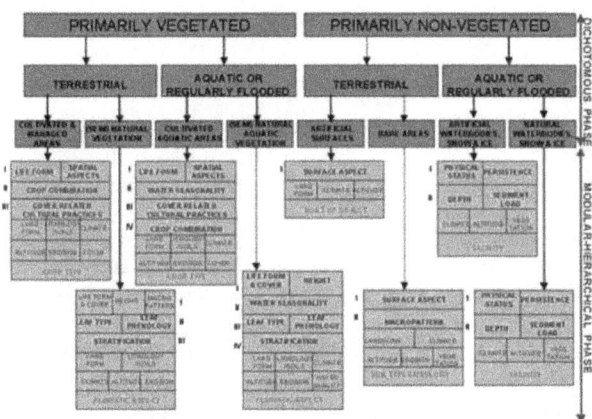

Figure 3. The fully nested 3-level 8-class FAO Land Cover Classification System (LCCS) Dichotomous Phase (DP) taxon-omy consists of a sorted set of 3 dichotomous layers: (i) vege-tation versus non-vegetation, (ii) terrestrial versus aquatic, and (iii) managed versus natural or semi-natural. They deliver as output the following 8-class LCCS-DP taxonomy. (A11) Cultivated and Managed Terrestrial (non-aquatic) Vegetated Areas. (A12) Natural and Semi-Natural Terrestrial Vegetation. (A23) Cultivated Aquatic or Regularly Flooded Vegetated Areas. (A24) Natural and Semi-Natural Aquatic or Regularly Flooded Vegetation.(B35) Artificial Surfaces and Associated Areas. (B36) Bare Areas. (B47) Artificial Waterbodies, Snow and Ice. (B48) Natural Waterbodies, Snow and Ice. The general-purpose user- and application-independent 3-level 8-class FAO LCCS-DP taxonomy is preliminary to a user- and appli-cation-specific FAO LCCS Modular Hierarchical Phase (MHP) taxonomy of one-class classifiers.

(Cimpoi, Maji, Kokkinos, & Vedaldi, 2014), which is equivalent to an unstructured black box (Marcus, 2018), inherently semiautomatic and site specific (Liang, 2004) and whose opacity contradicts the well-known engineering principles of modularity, regularity and hierarchy typical of scalable systems (Lipson, 2007).

Starting from these premises our working hypothesis was that necessary not sufficient pre-condition for a yet-unfulfilled GEOSS development (GEO, 2005) is systematic generation at the ground segment of an ESA EO Level 2 product, never accomplished to date (ESA, 2015; DLR & VEGA, 2011), whose general-purpose SCM product is constrained as follows. First, the ESA EO Level 2 SCM legend agrees with the 3-level 9-class "augmented" FAO LCCS-DP taxonomy. Second, to comply with the GEO-CEOS QA4EO *Cal/Val* requirements, the SCM product must be submitted to a GEO-CEOS stage 4 *Val*, where an mDMI set of EO OP-Q^{2}Is is evaluated by independent means at large spatial-extent and multiple time periods (GEO-CEOS, 2010). By definition, a GEO-CEOS stage 3 *Val* requires that "spatial and temporal consistency of the product with similar products are evaluated by independent means over multiple locations and time periods representing global conditions. In Stage 4 *Val*, results for Stage 3 are systematically updated when new product versions are released and as the time-series expands" (GEO-CEOS WGCV, 2015).

According to our working hypothesis, to contribute toward filling an analytic and pragmatic infor-mation gap from multi-source EO big data to ESA EO Level 2 information product as necessary not sufficient pre-condition to GEOSS development, the primary goal of this interdisciplinary study was to undertake an original (to the best of these authors' knowledge, the first) outcome and process GEO-CEOS stage 4 *Val* of an off-the-shelf lightweight computer program, the Satellite Image Automatic Mapper™ (SIAM™), presented in recent years in the RS literature where enough information was provided for the implementation to be reproduced (Baraldi, 2017; Baraldi & Boschetti, 2012a, 2012b; Baraldi et al., 2010a, 2010b, 2010c; Baraldi & Humber, 2015; Baraldi et al., 2013; Baraldi, Puzzolo, Blonda, Bruzzone, & Tarantino, 2006, 2015; Baraldi, Tiede, Sudmanns, Belgiu, & Lang, 2016). Implemented in operating mode in the C/C++ programming language, an off-the-shelf SIAM software executable runs: (i) automatically, i.e., it requires no human-machine interaction, (iii) in near real-time because it is non-iterative, more specifically it is one-pass, with a single subsystem which is two-pass (refer to the text below), and its computational complexity increases linearly with image size, and (iii)

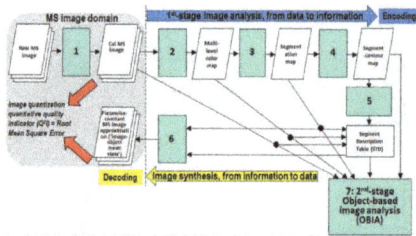

Figure 4. The SIAM lightweight computer program for prior knowledge-based MS reflectance space hyperpolyhedralization into color names, superpixel detection and vector quantiza-tion (VQ) quality assessment. It consists of six subsystems, iden-tified as 1 to 6. Phase 1-of-2 = Encodingphase/ Image analysis—Stage 1: MS data calibra-tion into top-of-atmosphere reflectance (TOARF) or surface reflectance (SURF) values. Stage 2: Prior knowledge-based SIAM decision tree for MS reflectance space partitioning (quantization, hyperpolyhedralization). Stage 3: Well-posed (deterministic) two-pass connected-component detection in the multi level color   map-domain. Connected-compo-nents in the color map-domain are connected sets of pixels fea-turing the same color label. These connected-components are also called image-objects, segments or superpixels. Stage 4: Well-posed superpixel-contour extrac-tion. Stage 5: Superpixel descrip-tion table allocation and initialization. Phase 2-of-2 = Decoding phase/Image   synthesis—Stage 6: Superpixelwise-constant input image approximation ("image-object mean view") and per-pixel VQ error estimation. (Stage 7: in cascade to the SIAM superpixel detection, a high-level object-based image analysis (OBIA) approach can be adopted).

in tile streaming mode, i.e., it requires a fixed run-time memory occupation. In addition to running on laptop and desktop computers, the SIAM lightweight computer program is eligible for use as mobile software application. By definition, a mobile software application is a lightweight computer program specifically designed to run on web services and/or mobile devices, such as tablet computers and smartphones, eventually provided with a mobile graphic user interface (GUI). An off-the-shelf SIAM software executable comprises six non-iterative subsystems for automated MS image analysis (decomposition) and synthesis (reconstruction) in linear time complexity. Its core is a one-pass prior knowledge-based decision tree (expert system) for MS reflectance space hyperpolyhedralization into static (non-adaptive-to-data) color names. Sketched in Figure 4, the SIAM software architecture is summarized as follows.

(1) MS data radiometric calibration, in agreement with the GEO-CEOS QA4EO *Cal* requirements (GEO-CEOS, 2010). The SIAM expert system instantiates a physical data model; hence, it requires as input sensory data provided with a physical meaning. Specifically, DNs must be radiome-trically *Cal* into a physical unit of radiometric measure to be community-agreed upon, such as TOARF values, SURF values or Kelvin degrees for thermal channels. Relationship TOARF $\supseteq$ SURF holds because SURF is a special case of TOARF in clear sky and flat terrain conditions (Chavez, 1988), i.e., TOARF $\approx$ SURF + atmospheric noise + topographic effects + surface adjacency effects. In a spectral decision tree for MS color space hyperpolyhedralization (partition-ing), this relationship means that MS hyperpolyhedra (envelopes, manifolds) in "noisy" TOARF values include "noiseless" hyperpolyhedra in SURF values as special case of the former according to relationship *subset-of*, while the vice versa does not hold, see Figure 5.

(2) One-pass prior knowledge-based SIAM decision tree for MS reflectance space hyperpolyhedra-lization into three static codebooks (vocabularies) of sub-symbolic/semi-symbolic color names as codewords, see Figure 5. Provided with inter-level parent–child relationships, the SIAM's three-level vocabulary of static color names features a ColorVocabularyCardinality value which decreases from fine to intermediate to coarse, refer to Table 1 and Figure 6. MS reflec-tance space hyperpolyhedra for color naming are difficult to think of and impossible to visualize when the MS data space dimensionality is superior to three. This is not the case of basic color (BC) names adopted in human languages (Berlin & Kay, 1969), whose mutually exclusive and totally exhaustive perceptual polyhedra, neither necessarily convex nor connected, are intuitive to think of and easy to visualize in a 3D monitor-typical red-green-blue (RGB) data cube, see Figure7 (Benavente, Vanrell, & Baldrich, 2008; Griffin, 2006). When each pixel of an MS image is mapped onto a color space partitioned into a set of mutually exclusive and totally exhaustive hyperpolyhedra equivalent to a vocabulary of BC names, then a 2D multilevel color map (2D

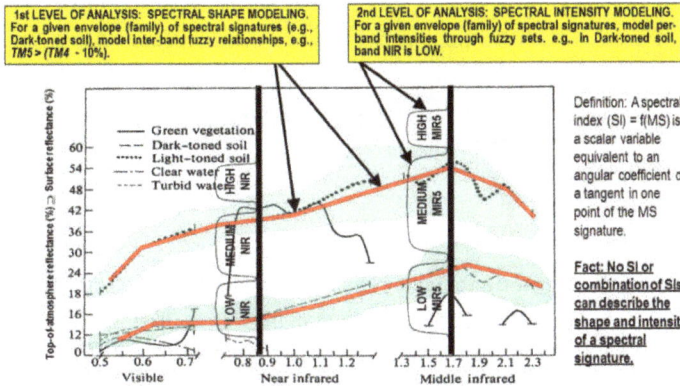

Figure 5. Examples of land cover (LC) class-specific families of spectral signatures in top-of-atmosphere reflec-tance (TOARF) values which include surface reflectance (SURF) values as a special case in clear sky and flat terrain conditions. A within-class family of spectral signatures (e.g., dark-toned soil) in TOARF values forms a buffer zone (hyperpolyhedron, envelope, manifold). The SIAM decision tree models each target family of spectral signatures in terms of multivariate shape and mul-tivariate intensity information components as a viable alter-native to multivariate analysis of spectral indexes. A typical spectral index is a scalar band ratio equivalent to an angular coefficient of a tangent in one point of the spectral signature. Infinite functions can feature the same tangent value in one point. In practice, no spectral index or combination of spec-tral indexes can reconstruct the multivariate shape and multi-variate intensity information components of a spectral signature.

gridded dataset of a multilevel variable) is generated automatically (without human-machine interaction) in near real-time (with computational complexity increasing linearly with image size), where the number k of 2D map levels (color strata, color names) belongs to range {1, ColorVocabularyCardinality}. Popular synonyms of measurement space hyperpolyhedralization (discretization, partition) are vector quantization (VQ) in inductive machine learning-from-data (Cherkassky & Mulier, 1998; Elkan, 2003; Fritzke, 1997a, 1997b; Lee, Baek, & Sung, 1997; Linde, Buzo, & Gray, 1980; Lloyd, 1982; Patanè and Russo, 2001, 2002), and deductive fuzzification of a numeric variable into fuzzy sets in fuzzy logic (Zadeh, 1965). Typical inductive learning-from-data VQ algorithms aim at minimizing a known VQ error function, e.g., a root mean square vector quantization error (RMSE), given a number of k discretization levels selected by a user based on *a priori* knowledge and/or heuristic criteria. One of the most widely used VQ heuristics in RS and computer vision (CV) applications is the k-means VQ algorithm (Elkan, 2003; Lee et al., 1997; Linde et al., 1980; Lloyd, 1982), capable of convex Voronoi tessellation of a multi-variate data space (Cherkassky & Mulier, 1998; Fritzke, 1997a). For example, in a

Table 1. The SIAM computer program is an EO system of systems scalable to any past, existing or future MS imaging sensor provided with radiometric calibration metadata parameters					
SIAM, r88v7	**Input bands**	**Prior knowledge-based color map legends: Number of output spectral categories = Vocabulary of multi-spectral (MS) color names**			
		Fine discretization levels	**Intermediate discretization levels**	**Coarse discretization levels**	**Inter-sensor discretization levels (*)**
L-SIAM	7—B, G, R, NIR, MIR1, MIR2, TIR	96	48	18	**33** (*): employed for inter-sensor post-classification change/no-change detection
S-SIAM	4—G, R, NIR, MIR1	68	40	15	
AV-SIAM	4—R, NIR, MIR1, TIR	83	43	17	
Q-SIAM	4—B, G, R, NIR	61	28	12	

It encompasses the following subsystems. (i) 7-band Landsat-like SIAM™ (L-SIAM™), with input channels Blue (B), Green (G), Red (R), Near Infra-Red (NIR), Medium IR1 (MIR1), Medium IR2 (MIR2), and Thermal IR (TIR). (ii) 4-band (channels G, R, NIR, MIR1) SPOT-like SIAM™ (S-SIAM™). (iii) 4-band (channels R, NIR, MIR1, and TIR) Advanced Very High Resolution Radiometer (AVHRR)-like SIAM™ (AV-SIAM™). (iv) 4-band (channels B, G, R, and NIR) QuickBird-like SIAM™ (Q-SIAM™)

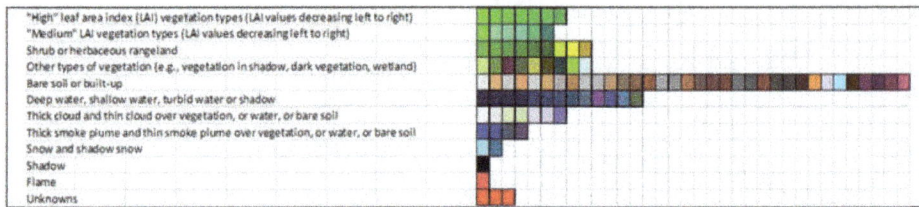

Figure 6. Prior knowledge-based color map legend adopted by the Landsat-like SIAM (L-SIAM™, release 88 version 7) implementation. For the sake of representation compactness, pseudo-colors of the 96 spectral categories are gathered along the same raw if they share the same parent spectral category in the deci-sion tree, e.g., "strong" vega-tion, equivalent to a spectral end-member. The pseudo-color of a spectral category is chosen as to mimic natural colors of pixels belonging to that spec-tral category. These 96 color names at fine color granularity are aggregated into 48 and 18 color names at intermediate and coarse color granularity respectively, according to par-ent–child relationships defined *a priori*, also refer to Table 1.

bag-of-words model applied to CV tasks, a numeric color space is typically discretized into a categorical color variable (codebook of codewords) by an inductive VQ algorithm, such as k-means; next, the categorical color variable is simplified by a 1st-order histo-gram representation, which disregards word grammar, semantics and even word-order, but keeps multiplicity; finally, the frequency of each color codeword is used as a feature for training a supervised data learning classifier (Cimpoi et al., 2014). Unlike the k-means VQ algorithm where the system's free-parameter k is user-defined based on heuristics and the VQ error is estimated from the unlabeled dataset at hand, a user can fix the target VQ error value, so that it is the free-parameter k to be dynamically learned from the finite unlabeled dataset at hand by an inductive VQ algorithm (Patané & Russo, 2001, 2002), such as ISODATA (Memarsadeghi, Mount, Netanyahu, & Le Moigne, 2007). It means there is no universal number k of static hyperpolyhedra in a vector data space

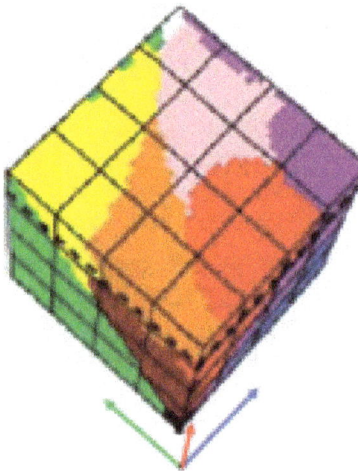

Figure 7. Courtesy of Griffin (2006). Monitor-typical RGB cube partitioned into percep-tual polyhedra corresponding to a discrete and finite diction-ary of basic color (BC) names, to be community-agreed upon in advance to be employed by members of the community. The mutually exclusive and totally exhaustive polyhedra are neither necessarily convex nor connected. In practice BC names belonging to a finite and discrete color vocabulary are equivalent to Vector Quantization (VQ) levels belonging to a VQ codebook (Cherkassky & Mulier, 1998).

suitable for satisfying any VQ problem of interest if no target VQ error is specified in advance. As a viable strategy to cope with the inherent ill-posedness of inductive VQ problems (Cherkassky & Mulier, 1998), the SIAM expert system provides its three pre-

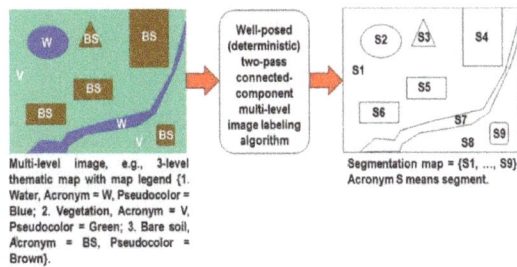

Multi-level image, e.g., 3-level thematic map with map legend {1. Water, Acronym = W, Pseudocolor = Blue; 2. Vegetation, Acronym = V, Pseudocolor = Green; 3. Bare soil, Acronym = BS, Pseudocolor = Brown}.

Segmentation map = {S1, ..., S9}. Acronym S means segment.

Figure 8. One segmentation map is deterministically gener-ated from one multilevel image, such as a thematic map, but the vice versa does not hold, i.e., many multilevel images can generate the same segmenta-tion map. In this example, nine image-objects/segments S1–S9 can be detected in the 3-level thematic map shown at left. Each segment consists of a connected set of pixels sharing the same multilevel map label. Each stratum/layer/level con-sists of one or more segments, e.g., stratum Vegetation (V) consists of two disjoint seg-ments, S1 and S8. In any mul-tilevel (categorical, nominal, qualitative) image domain, three labeled spatial primitives (spatial units) coexist and are provided with parent–child relationships: pixel with a level-label and a pixel identifier (ID, e.g., the row-column coordinate pair), segment (polygon) with a level-label and a segment ID, and stratum (multi-part poly-gon) with a level-label equiva-lent to a stratum ID. This overcomes the ill-fated dichot-omy between traditional unla-beled sub-symbolic pixels versus labeled sub-symbolic segments in the numeric (quantitative) image domain traditionally coped with by the object-based image analysis (OBIA) paradigm (Blaschke et al., 2014).

defined VQ levels with a per-pixel RMSE estimation required for VQ quality assurance, in compliance with the GEO-CEOS QA4EO *Val* guidelines, refer to point (6) below.

(3) Well-posed (deterministic) two-pass detection of connected-components in the multilevel color map-domain (Dillencourt, Samet, & Tamminen, 1992; Sonka, Hlavac, & Boyle, 1994), where the number k of map levels belongs to range {1, ColorVocabularyCardinality}, see Figure 8. These discrete and finite connected-components consist of connected sets of pixels featuring the same color label. Each connected-component is either (0D) pixel, (1D) line or (2D) polygon in the Open Geospatial Consortium (OGC) nomenclature (OGC 2015). They are typically known as *superpixels* in the CV literature (Achanta et al., 2011), homogeneous *segments* or *image-objects* in the object-based image analysis (OBIA) literature (Blaschke et al., 2014; Matsuyama & Hwang, 1990; Nagao & Matsuyama, 1980; Shackelford & Davis, 2003a, 2003b), and texture elements, i.e., *texels*, in human vision (Julesz, 1986; Julesz, Gilbert, Shepp, & Frisch, 1973). Whereas the physical model-based SIAM expert system requires no human-machine interaction to detect top-down superpixels whose shape and size can be any, superpixels detected bottom-up in statis-tical model-based CV algorithms typically require a pair of statistical model's free-para-meters to be user-defined based on heuristics, such as a first heuristic-based geometric threshold equal to the superpixel maximum area and a second heuristic-based geometric threshold forcing a superpixel to stay compact in shape (Achanta et al., 2011). In a multilevel image domain where k is the number of levels (image-wide strata), indivi-dual (0D) pixels with label 1 to k, superpixels as connected sets of pixels featuring the same label 1 to k, and strata (layers), equal to discrete and finite collections of superpixels, mutually disjoint, but belonging to the same level 1 to k, co-exist as non-alternative labeled spatial units provided with a parent–child relationship, where each superpixel is a 2-tuple (superpixel ID, level 1-of-k) and each pixel is a 2-tuple (raw-column coordinate pair, superpixel ID), see Figure 8.

(4) Well-posed 4- or 8-adjacency cross-aura representation in linear time of superpixel-contours, see Figure 9. These cross-aura contour values allow estimation of a scale-invariant planar shape index of compactness (Baraldi, 2017; Baraldi & Soares, 2017; Soares, Baraldi, & Jacobs, 2014), eligible for use by a high-level OBIA approach (Blaschke et al., 2014), see Figure 4.

(5) Superpixel/segment description table (Matsuyama & Hwang, 1990; Nagao & Matsuyama, 1980), to describe superpixels in a 1D tabular form (list) in combination with their 2D raster representation in the image-domain, referred to as "literal bit map" by Marr (1982), to take advantage of each data

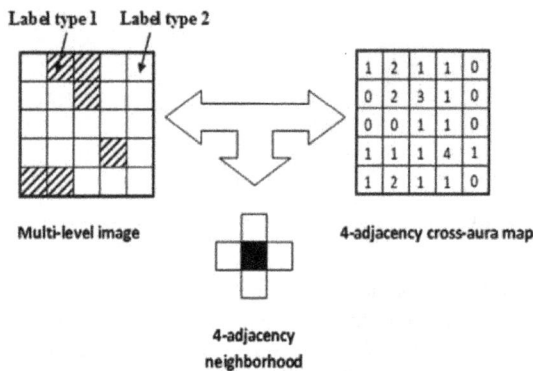

Figure 9. Example of a 4-adja-cency cross-aura map, shown at right, generated in linear time from a two-level image shown at left.

structure and overcome their shortcomings. Computationally, local spatial searches are more efficient in the 2D raster image-domain than in the 1D list representation, because "most of the spatial relationships that must be examined in early vision (encompassing the raw and full primal sketch for token detection and texture segmentation, respectively) are rather local" (Marr, 1982). Vice versa, if we had to examine global or "scattered, pepper-and-salt-like (spatial) configurations, then a (2D) bit map would probably be no more efficient than a (1D) list" (Marr, 1982).

(6) Superpixelwise-constant input image approximation (reconstruction), also known as "image-object mean view" in commercial OBIA applications (Trimble, 2015), followed by a per-pixel RMSE estimation between the original MS image and the reconstructed piecewise-constant MS image. This VQ error estimation strategy enforces a product quality assurance policy considered mandatory by the GEO-CEOS QA4EO *Val* guidelines. For example, VQ quality assurance supported by SIAM allows a user to adopt quantitative (objective) criteria in the selection of pre-defined VQ levels, equivalent to color names, to fit user- and application-specific VQ error requirement specifications.

An example of the SIAM output products automatically generated in linear time from a 13-band 10 m-resolution Sentinel-2A image radiometrically calibrated into TOARF values is shown in Figure 10.

The potential impact on the RS community of a GEO-CEOS stage 4 *Val* of an off-the-shelf SIAM lightweight computer program for automated near real-time prior knowledge-based MS reflectance space hyperpolyhedralization, superpixel detection and per-pixel VQ quality assessment is expected to be relevant, with special emphasis on existing or future *hybrid* (combined deductive and inductive) EO-IUSs. In the RS discipline, there is a long history of prior knowledge-based MS reflectance space partitioners for static color naming, alternative to SIAM's, developed but never validated by space agencies, public organizations and private companies for use in hybrid EO-IUSs in operating mode, see Figure 11. Examples of hybrid EO image pre-processing applications in the quantitative/sub-symbolic domain of *information-as-thing*, where a numeric input variable is statistically class-conditioned (masked) by a static color naming first stage to generate as output another numeric variable considered more informative than the input one, are large-scale MS image compositing (Ackerman et al. 1998; Lück & van Niekerk, 2016; Luo, Trishchenko, & Khlopenkov, 2008), MS image atmospheric correction and topographic correction (Baraldi, 2017; Baraldi et al., 2010b; Baraldi & Humber, 2015; Baraldi et al., 2013; Bishop & Colby, 2002; Bishop et al., 2003; DLR & VEGA, 2011; Dorigo et al., 2009; Lück & van Niekerk, 2016; Riano et al., 2003; Richter & Schläpfer, 2012a, 2012b; Vermote & Saleous, 2007), see Figure 12, MS image adjacency effect correction (DLR & VEGA, 2011) and radiometric quality assessment of pan-sharpened MS imagery (Baraldi, 2017; Despini, Teggi, & Baraldi, 2014). Examples of hybrid EO image classification applications in the qualitative/equivocal/categorical domain of *information-as-data-interpretation* and statistically class-conditioned by a static color naming first stage are cloud and cloud-shadow quality layer detection (Baraldi, 2015, 2017; Baraldil., DLR & VEGA, 2011; Lück & van Niekerk, 2016), single-date LC classification (DLR & VEGA, 2011; GeoTerraImage, 2015; Lück & van Niekerk, 2016; Muirhead & Malkawi, 1989; Simonetti et al. 2015a), multi-temporal post-classification LC change (LCC)/no-change detection (Baraldi, 2017; Baraldi et al., 2016; Simonetti et al., 2015a; Tiede,

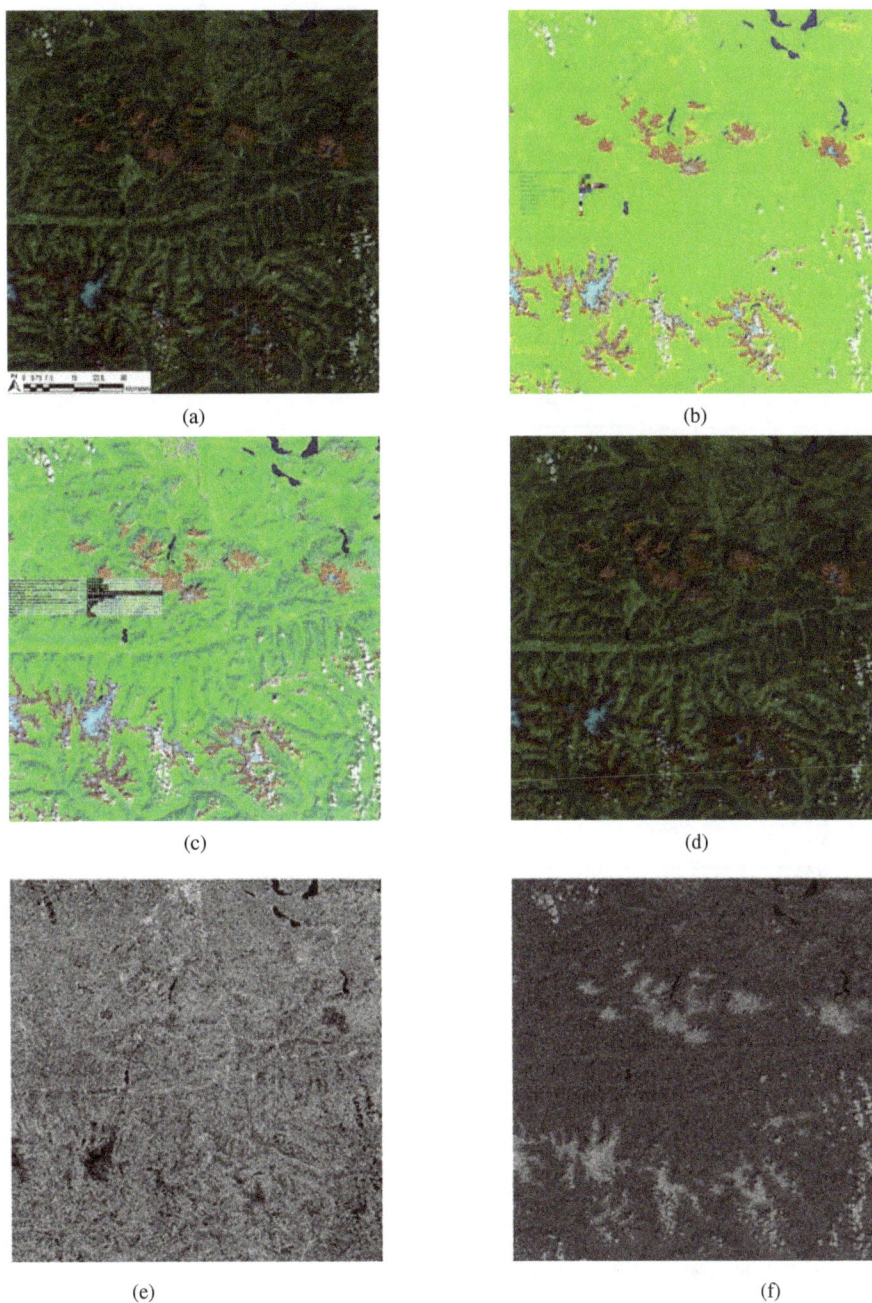

(a)

(b)

(c)

(d)

(e)

(f)

Figure 10. See note.

Baraldi, Sudmanns, Belgiu, & Lang, 2016), multi-temporal vegetation gradient detection and quantization into fuzzy sets (Arvor, Madiela, & Corpetti, 2016), multi-temporal burned area detection (Boschetti, Roy, Justice, & Humber, 2015), and prior knowledge-based LC mask refinement (cleaning) of supervised data samples employed as input to supervised data learning EO-IUSs (Baraldi et al., 2010a, 2010b). Due to their large application domain, ranging from low- (pre-attentional) to high-level (attentional) vision tasks, existing hybrid EO-IUSs in operating mode, whose statistical data models are class-conditioned by static color naming, become natural candidates for the research and development (R&D) of an EO-IUS in operating mode, capable of systematic transformation of multi-source single-date MS imagery into ESA EO Level 2 product at the ground segment.

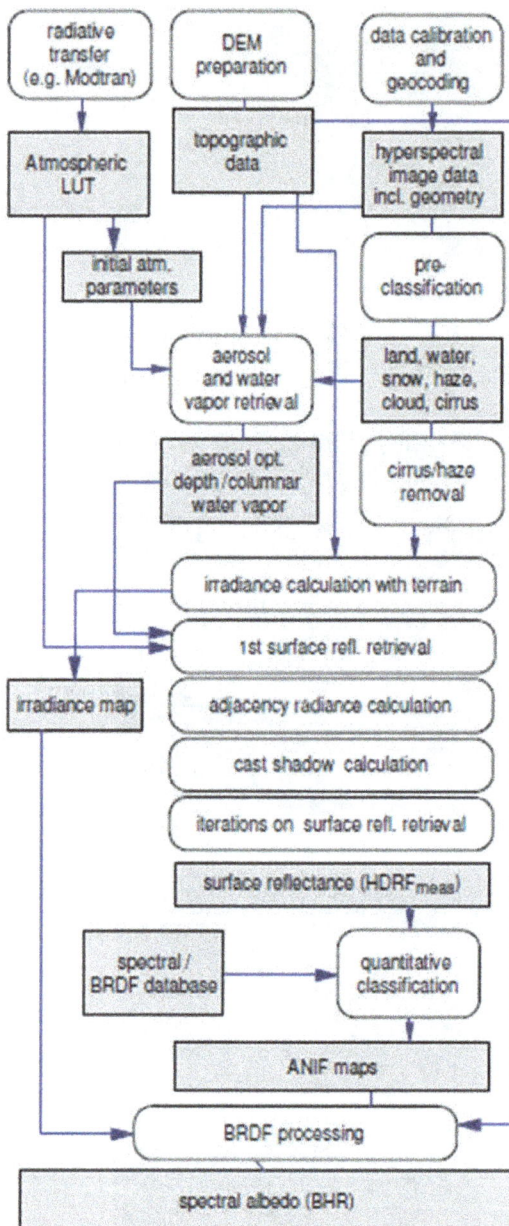

Figure 11. Same as in Schläpfer et al. (2009), courtesy of Daniel Schläpfer, ReSe Applications Schläpfer. A complete ("aug-mented") hybrid inference workflow for MS image correc-tion from atmospheric, adja-cency and topographic effects. It combines a standard Atmospheric/Topographic Correction for Satellite Imagery (ATCOR) commercial software workflow (Richter & Schläpfer, 2012a, 2012b), with a bidirec-tional reflectance distribution function (BRDF) effect correction. Processing blocks are represented as circles and out-put products as rectangles. This hybrid (combined deductive and inductive) workflow alter-nates deductive/prior knowl-edge-based with inductive/ learning-from-data inference units, starting from initial con-ditions provided by a first-stage deductive Spectral Classification of surface reflec-tance signatures (SPECL) deci-sion tree for color naming (pre-classification), implemented within the ATCOR commercial software toolbox (Richter & Schläpfer, 2012a, 2012b). Categorical variables generated by the pre-classification and classification blocks are employed to stratify (mask) unconditional numeric variable distributions, in line with the statistic stratification principle (Hunt & Tyrrell, 2012). Through statistic stratification (class-conditional data analytics), inherently ill-posed inductive learning-from-data algorithms are

provided with prior knowl-edge required in addition to data to become better posed for numerical solution, in agreement with the machine learning-from-data literature (Cherkassky & Mulier, 1998).

Figure 12. Left: Zoomed area of a Landsat 7 ETM+ image of Colorado, USA (path: 128, row: 021, acquisition date: 2000–08- 09), depicted in false colors (R: band ETM5, G: band ETM4, B: band ETM1), 30 m resolution, radiometrically calibrated into TOARF values. Right: Output product automatically gener-ated without human-machine interaction by the stratified topographic correction (STRATCOR) algorithm proposed in Baraldi et al. (2010c), whose input datasets are one Landsat image, its data-derived L-SIAM color map at coarse color gran-ularity, consisting of 18 spec-tral categories for stratification purposes (see Table 1), and a standard 30 m resolution Shuttle Radar Topography Mission (SRTM) digital elevation model (DEM).

The terminology adopted in the rest of this paper is mainly driven from the multidisciplinary domain of cognitive science, see Figure 13. Popular synonyms of deductive inference are top-down, prior knowledge-based, learning-from-rule and physical model-based inference. Synonyms of inductive inference are bottom-up, learning-from-data, learning-from-examples and statistical model-based inference (Baraldi, 2017; Baraldi & Boschetti, 2012a, 2012b; Liang, 2004). Hybrid inference systems combine statistical and physical data models to take advantage of the unique features of each and overcome their shortcomings (Baraldi, 2017; Baraldi & Boschetti, 2012a, 2012b; Cherkassky & Mulier, 1998; Liang, 2004). For example, in biological cognitive systems "there is never an absolute beginning" (Piaget, 1970), where an *a priori* genotype provides initial conditions to an inductive learning-from-examples phenotype (Parisi, 1991). Hence, any biological cognitive system is a hybrid inference system where inductive/phenotypic learning-from-examples mechanisms explore the neighborhood of deduc-tive/genotypic initial conditions in a solution space (Parisi, 1991). In line with biological cognitive systems, an artificial hybrid inference system can alternate deductive and inductive inference algo-rithms, starting from a deductive inference first stage for initialization purposes, see Figure 11. It means that no deductive inference subsystem, such as SIAM, should be considered stand-alone, but eligible for use in a hybrid inference system architecture to initialize (pre-condition, stratify) inductive learning-from-data algorithms, which are inherently ill-posed, difficult to solve and require *a priori* knowledge in addition to data to become better posed for numerical solution, as clearly acknowledged by the machine learning-from-data literature (Bishop, 1995; Cherkassky & Mulier, 1998).

To comply with the GEO-CEOS stage 4 *Cal/Val* requirements, the selected ready-for-use SIAM software executable had to be validated by independent means on a radiometrically calibrated EO image time-series at large spatial extent. This input data set was identified in the open-access USGS 30 m resolution Web Enabled Landsat Data (WELD) annual composites of the conterminous U.S. (CONUS) for the years 2006–2009, radiometrically calibrated into TOARF values (Homer, Huang, Yang, Wylie, & Coan, 2004; Roy et al., 2010; WELD 2015). The 30 m resolution 16-class U. S. National Land Cover Data (NLCD) 2006 map, delivered in 2011 by the USGS Earth Resources Observation Systems (EROS) Data Center (EDC) (EPA 2007; Vogelmann et al., 2001; Vogelmann, Sohl, Campbell, & Shaw, 1998; Wickham, Stehman, Fry, Smith, & Homer, 2010; Wickham et al., 2013; Xian & Homer, 2010), was selected as reference thematic map at continental spatial extent. The USGS 16-class NLCD 2006 map legend is summarized in Table 2. To account for typical non-stationary geospatial statistics, the USGS NLCD 2006 thematic map was partitioned into 86 Level III ecoregions of North America collected from the Environmental Protection Agency (EPA) (EPA 2013; Griffith & Omernik, 2009).

In this experimental framework, the test SIAM-WELD annual color map time-series for the years 2006–2009 and the reference USGS NLCD 2006 map share the same spatial extent and spatial resolution, but their map legends are not the same. These working hypotheses are neither trivial nor conventional in the RS literature, where thematic map quality assessment strategies typically adopt an either random or non-random sampling strategy and assume that the test and reference

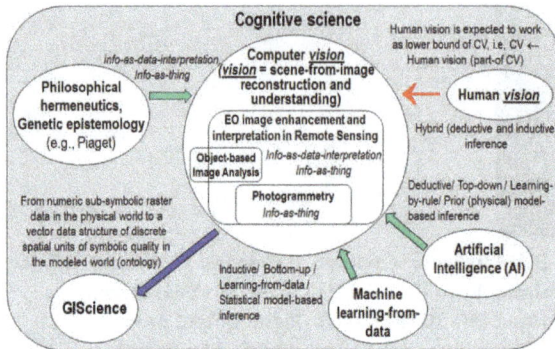

Figure 13. Like engineering, remote sensing (RS) is a metascience, whose goal is to transform knowledge of the world, provided by other scien-tific disciplines, into useful user- and context-dependent solutions in the world. Cognitive science is the inter-disciplinary scientific study of the mind and its processes. It examines what cognition (learning) is, what it does and how it works. It especially focuses on how information/knowledge is represented, acquired, processed and trans-ferred within nervous systems (distributed processing systems in humans, such as the human brain, or other animals) and machines (e.g., computers). Neurophysiology studies ner-vous systems, including the brain. Human vision is expected to work as lower bound of CV, i.e., human vision → (*part-of*) CV, such that inherently ill-posed CV is required to comply with human visual perception phenomena to become better conditioned for numerical solution.

thematic map dictionaries coincide (Stehman & Czaplewski, 1998). Starting from a stratified random sampling protocol presented in Baraldi et al. (2014), the secondary contribution of the present study was to develop a novel protocol for wall-to-wall comparison without sampling of two thematic maps featuring the same spatial extent and spatial resolution, but whose legends can differ.

For the sake of readability this paper is split into two, the present Part 1—Theory and the subsequent Part 2—Validation. An expert reader familiar with static color naming in cognitive science, spanning from linguistics to human vision and CV, can skip the present Part 1, either totally or in part. To make this paper self-contained and provided with a relevant survey value, the Part 1 is organized as follows. The multi-disciplinary background of color naming is discussed in Chapter 2. Chapter 3 reviews the long history of prior knowledge-based decision trees for MS color naming presented in the RS literature. To cope with thematic map legends that do not coincide and must be harmonized (reconciled, associated, translated) (Ahlqvist, 2005), such as dictionaries of MS color names in the image-domain and LC class names in the scene-domain, Chapter 3 proposes an original hybrid inference guideline to identify a categorical variable-pair relationship, where prior beliefs are combined with additional evidence inferred from new data. An original measure of categorical variable-pair association (harmonization) in a binary relation-ship is proposed in Chapter 4. In the subsequent Part 2, GEO-CEOS stage 4 *Val* results are collected by an original protocol for wall-to-wall thematic map quality assessment without sampling, where legends of the test SIAM-WELD annual map time-series and reference USGS NLCD 2006 map are harmonized. Conclusions are that the annual SIAM-WELD map time-series for the years 2006–2009 provides a first example of GEO-CEOS stage 4 *validated* ESA EO Level 2 SCM product, where the Level 2 SCM legend is the "augmented" 2-level 4-class FAO LCCS taxonomy at the DP Level 1 (vegetation/non-vegetation) and DP Level 2 (terrestrial/aquatic), added with extra class "rest of the world".

2. Problem background of color naming in cognitive science

Within the cognitive science domain, vision is synonym of scene-from-image reconstruction and understanding, see Figure 13. Encompassing both biological vision and CV, vision is a cognitive (*information-as-data-interpretation*) problem inherently ill-posed in the Hadamard sense (Hadamard, 1902); hence, it is very difficult to solve. Vision is non-polynomial (NP)-hard in computational complex-ity (Frintrop, 2011; Tsotsos, 1990) and requires *a priori* knowledge in addition to sensory data to become better posed for numerical solution (Cherkassky & Mulier, 1998). It is inherently ill-posed because affected by, first, data dimensionality reduction from the 4D spatio-temporal scene-domain to the (2D) image-domain and, second, by a semantic information gap from ever-varying sensations in

Table 2. Definition of the USGS NLCD 2001/2006/2011 classification taxonomy, Level II. [2] Alaska only				
USGS NLCD 2001/2006/2011 Classification Scheme (Legend), Level II				FAO LCCS-DP, level 1: A = Veg, B = Non-Veg, and level 2: 1 = Terrestrial, 2 = Aquatic
Code	ID	Name	Land cover (LC) Class Definition	ID
11	OW	Open water	OW: Areas of open water, generally with less than 25% cover of vegetation or soil	B4—Non-vegetated aquatic
12	PIS	Perennial Ice/ Snow	PIS: Areas characterized by a perennial cover of ice and/or snow, generally greater than 25% of total cover.	B4
21 22 23 24	DOS DLI DMI DHI	Developed, Open Space Developed, Low Intensity Developed, Medium Intensity Developed, High Intensity	DOS: Includes areas with a mixture of some constructed materials, but mostly vegetation in the form of lawn grasses. Impervious surfaces account for less than 20 percent of total cover. These areas most commonly include large-lot single-family housing units, parks, golf courses, and vegetation planted in developed settings for recreation, erosion control, or aesthetic purposes. DLI, DMI, DHI: refer to the "National Land Cover Database 2006 (NLCD2006)," Multi-Resolution Land Characteristics Consortium (MRLC), 2013.	B3—Non-vegetated terrestrial/A1—Vegetated terrestrial
31	BL	Barren Land (Rock/Sand/ Clay)	BL: Barren areas of bedrock, desert pavement, scarps, talus, slides, volcanic material, glacial debris, sand dunes, strip mines, gravel pits and other accumulations of earthen material. Generally, vegetation accounts for less than 15% of total cover. As a consequence of this constraint, class BL covers only 1.21% of the CONUS total surface.	B3
41 42 43	DF EF MF	Deciduous Forest Evergreen Forest Mixed Forest	DF: Areas dominated by trees generally greater than 5 m tall, and greater than 20% of total vegetation cover. More than 75 percent of the tree species shed foliage simultaneously in response to seasonal change. EF: Areas dominated by trees generally greater than 5 m tall, and greater than 20% of total vegetation cover. More than 75 percent of the tree species maintain their leaves all year. Canopy is never without green foliage. MF: Mixed Forest—Areas dominated by trees generally greater than 5 m tall, and greater than 20% of total vegetation cover. Neither deciduous nor evergreen species are greater than 75 percent of total tree cover.	A1

(Continued)

Table2. (Continued)

USGS NLCD 2001/2006/2011 Classification Scheme (Legend), Level II				FAO LCCS-DP, level 1: A = Veg, B = Non-Veg, and level 2: 1 = Terrestrial, 2 = Aquatic
Code	ID	Name	Land cover (LC) Class Definition	ID
51 52	– SS	Dwarf Scrub [2] Scrub/Shrub	SS: Areas dominated by shrubs; less than 5 m tall with shrub canopy typically greater than 20% of total vegetation. This class includes true shrubs, young trees in an early successional stage or trees stunted from environmental conditions. The aforementioned definition of class BL means that class SS may feature a vegetated cover which accounts for 15% of total cover or more.	A1/B3
71 72 73 74	GH – – –	Grassland/ Herbaceous Sedge Herbaceous [2] Lichens [2] Moss [2]	GH: Areas dominated by grammanoid or herbaceous vegetation, generally greater than 80% of total vegetation. These areas are not subject to intensive management such as tilling, but can be utilized for grazing. The aforementioned definition of class BL means that class GH may feature a vegetated cover which accounts for 15% of total cover or more.	A1/B3
81 82	PH CC	Pasture/Hay Cultivated Crops	PH: Areas of grasses, legumes, or grass-legume mixtures planted for livestock grazing or the production of seed or hay crops, typically on a perennial cycle. Pasture/hay vegetation accounts for greater than 20 percent of total vegetation. CC: Areas used for the production of annual crops, such as corn, soybeans, vegetables, tobacco, and cotton, and also perennial woody crops such as orchards and vineyards. Crop vegetation accounts for greater than 20% of total vegetation. This class also includes all land being actively tilled.	A1
90 95	WW EHW	Woody Wetlands Emergent Herbaceous Wetland	WW: Areas where forest or shrubland vegetation accounts for greater than 20 percent of vegetative cover and the soil or substrate is periodically saturated with or covered with water. EHW: Areas where perennial herbaceous vegetation accounts for greater than 80% of vegetative cover and the soil or substrate is periodically saturated with or covered with water.	A2—Vegetated aquatic

For further details, refer to the "National Land Cover Database 2006 (NLCD2006)," Multi-Resolution Land Characteristics Consortium (MRLC), 2013. The right column instantiates a possible binary relationship R: A ⇒ B ⊆ A × B from set A = NLCD legend to set B = 2-level 4-class Dichotomous Phase (DP) taxonomy of the Food and Agriculture Organization of the United Nations (FAO)—Land Cover Classification System (LCCS) (Di Gregorio & Jansen, 2000), refer to Figure 3.

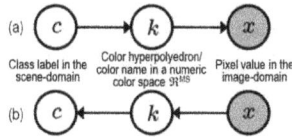

Figure 14. Graphical model of color naming, adapted from Shotton et al. (2009). Let us con-sider x as a (sub-symbolic) numeric variable, such as MS color values of a population of spatial units, with x $\in$ $\Re^{MS}$, while c represents a categorical variable of symbolic classes in the physical world, with c = 1, ..., Object Class Legend Cardinality. (a) According to Bayesian theory, posterior probability p(c|x) $\propto$ p(x|c)p(c) = p(c) $\sum\limits_{ColorName=k=1}^{ColorVocabularyCardinality}$ p(x|k)p(k|c), where color names, equivalent to color hyperpolyhedra in a numeric color space $\Re^{MS}$, provide a parti tion of the domain of change, $\Re^{MS}$, of numeric variable x. (b) For discriminative inference, the arrows in the graphical model are reversed using Bayes rule. Hence, a vocabulary of color names, physically equivalent to a parti-tion of a numeric color space $\Re^{MS}$ into color name-specific hyper-polyhedra, is conceptually equivalent to a latent/hidden/hypothetical variable linking observables (sub-symbolic sen-sory data) in the real world, spe-cifically, color values, to a categorical variable of semantic (symbolic) quality in the mental model of the physical world (world ontology, world model).

the (2D) image-domain to stable percepts in the mental model of the 4D scene-domain (Matsuyama & Hwang, 1990). On the one hand, ever-varying sensations collected from the 4D spatio-temporal physical world are synonym of observables, numeric/quantitative variables of sub-symbolic value or sensory data provided with a physical unit of measure, such as TOARF or SURF values, but featuring no semantics corresponding to abstract concepts, like perceptual categories or mental states. On the other hand, in a modeled world, also known as world ontology, mental world or "world model" (Matsuyama & Hwang, 1990), stable percepts are nominal/categorical/qualitative variables of sym-bolic value, i.e., they are categorical variables provided with semantics, such as LC class names belonging to a hierarchical FAO LCCS taxonomy of the world (Di Gregorio & Jansen, 2000), see Figure 3.

In statistics, the popular concept of latent/hidden variables was introduced to fill the information gap from input observables to target categorical variables. Latent/hidden variables are not directly measured, but inferred from observable numeric variables to link sensory data in the real world to categorical variables of semantic quality in the modeled world. "The terms hypothetical variable or hypothetical construct may be used when latent variables correspond to abstract concepts, like perceptual categories or mental states" (Baraldi, 2017; Shotton, Winn, Rother, & Criminisi, 2009; Wikipedia, 2018b). Hence, to fill the semantic gap from low-level numeric variables of sub-symbolic quality to high-level categorical variables of semantic value, hypothetical variables, such as categorical BC names (Benavente et al., 2008; Berlin & Kay, 1969; Gevers, Gijsenij, van de Weijer, & Geusebroek, 2012; Griffin, 2006), are expected to be mid-level categorical variables of "semi-symbolic" quality, i.e., hypothetical variables are nominal variables provided with a semantic value located "low" in a hierarchical ontology of the world, such as the hierarchical FAO LCCS taxonomy (Di Gregorio & Jansen, 2000), but always superior to zero, where zero is the semantic value of sub-symbolic numeric variables, see Figure 14.

In vision, spatial topological and spatial non-topological information components typically dominate color information (Baraldi, 2017; Matsuyama & Hwang, 1990). This thesis is proved by the undisputable fact that achromatic (panchromatic) human vision, familiar to everybody when wearing sunglasses, is nearly as effective as chromatic vision in scene-from-image reconstruction and understanding. Driven from perceptual evidence in human vision typically investigated by cognitive science, see Figure 13, a necessary not sufficient condition for a CV system to prove it fully exploits spatial topological and spatial non-topological information components in addition to color is to perform nearly the same when input with either panchromatic or color imagery. Stemming from *a priori* knowledge of human vision available in addition to sensory data, this necessary not sufficient condition can be adopted to make an inherently ill-posed CV system design and implementation problem better constrained for numerical solution.

Deeply investigated in CV (Frintrop, 2011; Sonka et al., 1994), content-based image retrieval (Smeulders, Worring, Santini, Gupta, & Jain, 2000) and EO image applications proposed by the RS community (Baraldi, 2017; Baraldi & Boschetti, 2012a, 2012b; Matsuyama & Hwang, 1990; Nagao & Matsuyama, 1980; Shackelford & Davis, 2003a, 2003b), well-known visual features are: (i) color values, typically discretized by humans into a finite and discrete vocabulary of BC

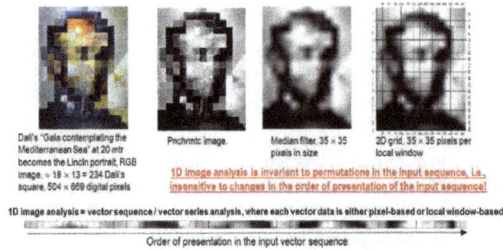

Figure 15. Example of 1D image analysis. Synonym of 1D analy-sis of a 2D gridded dataset, it is affected by spatial data dimensionality reduction. The (2D) image at left is trans-formed into the 1D vector data stream shown at bottom, where vector data are either pixel-based or spatial context-sensi-tive, e.g., local window-based. This 1D vector data stream, either pixel-based or local win-dow-based, means nothing to a human photointerpreter. When it is input to a traditional inductive data learning classi-fier, this 1D vector data stream is what the inductive classifier actually sees when watching the (2D) image at left. Undoubtedly, computers are more successful than humans in 1D image analysis, invariant to permutations in the input vector data sequence, such as in orderless pooling encoders (Cimpoi et al., 2014). Nonetheless, humans are still far more successful than com-puters in 2D image analysis, synonym of spatial topology-preserving (retinotopic) image analysis (Tsotsos, 1990), sensi-tive to permutations in the input vector data sequence, such as in order-sensitive pool-ing encoders (Cimpoi et al., 2014).

names (Benavente et al., 2008; Berlin & Kay, 1969; Gevers et al., 2012; Griffin, 2006); (ii) local shape (Baraldi, 2017; Wenwen Li et al. 2013); (iii) texture, defined as the perceptual spatial grouping of texture elements known as texels (Baraldi, 2017; Julesz, 1986; Julesz et al., 1973) or tokens (Marr, 1982); (iv) inter-object spatial topological relationships, e.g., adjacency, inclusion, etc., and (v) inter-object spatial non-topological relationships, e.g., spatial distance, angle measure, etc. In vision, color is the sole visual property available at the imaging sensor's spatial resolution, i.e., at pixel level. In other words, pixel-based information is spatial context independent. i.e., per-pixel information is exclusively related to color properties. Among the aforementioned visual variables, per-pixel color values are the sole non-spatial (spatial context-insensitive) numeric variable.

Neglecting the fact that spatial topological and spatial non-topological information components typically dominate color information in both the (2D) image-domain and the 4D spatio-temporal scene-domain involved with the cognitive task of vision (Matsuyama & Hwang, 1990), traditional EO-IUSs adopt a 1D image analysis approach, see Figure 15. In 1D image analysis, a 1D streamline of vector data, either spatial context-sensitive (e.g., window-based or image object-based like in OBIA approaches) or spatial context-insensitive (pixel-based), is processed insensitive to changes in the order of presentation of the input sequence. In practice 1D image analysis is invariant to permutations, such as in orderless pooling encoders (Cimpoi et al., 2014). When vector data are spatial context-sensitive then 1D image analysis ignores spatial topological information. When vector data are pixel-based then 1D image analysis ignores both spatial topological and spatial non-topological information components. Prior knowledge-based color naming of a spatial unit x in the image-domain, where x is either (0D) point, (1D) line or (2D) polygon defined according to the OGC nomenclature (OGC, 2015), is a special case of 1D image analysis, either pixel-based or image object-based, where spatial topological and/or spatial non-topological information are ignored, such as in SIAM's static color naming (Baraldi et al., 2006).

Alternative to 1D image analysis, 2D image analysis relies on a sparse (distributed) 2D array (2D regular grid) of local spatial filters, suitable for spatial topology-preserving (retinotopic) feature mapping (DiCarlo, 2017; Fritzke, 1997a; Martinetz, Berkovich, & Schulten, 1994; Tsotsos, 1990), sensitive to permutations in the input vector data sequence, such as in order-sensitive pooling encoders (Cimpoi et al., 2014), see Figure 16. The human brain's organizing principle is topology-preserving feature mapping (Feldman, 2013). In the biological visual system, topology-preserving feature maps are primarily spatial, where activation domains of physically adjacent processing units in the 2D array of convolutional filters are spatially adjacent regions in the 2D visual field. Provided with a superior degree of biological plausibility in modeling 2D spatial topological and spatial non-topological information components, distributed pro-cessing systems capable of 2D image analysis, such as physical model-based ("hand-crafted") 2D wavelet filter banks (Mallat, 2016) and end-to-end inductive learning-from-data DCNNs, typically

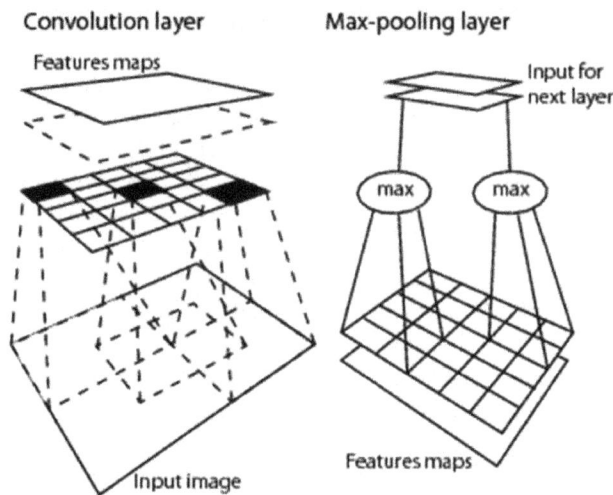

Figure 16. 2D image analysis, synonym of spatial topology-preserving (retinotopic) feature mapping in a (2D) image-domain (Tsotsos, 1990). Activation domains of physi-cally adjacent processing units in the 2D array of convolutional spatial filters are spatially adjacent regions in the 2D visual field. Provided with a superior degree of biological plausibility in modeling 2D spatial topological and spatial non-topological information, distributed processing systems capable of 2D image analysis, such as deep convolutional neural networks (DCNNs), typi-cally outperform traditional 1D image analysis approaches. Will computers become as good as humans in 2D image analysis?

outperform 1D image analysis approaches (Cimpoi et al., 2014; DiCarlo, 2017), although DCNNs are the subject of increasing criticisms by the artificial intelligence (AI) community (DiCarlo, 2017; Etzioni, 2017; Marcus, 2018). This apparently trivial consideration is at odd with a relevant portion of the RS literature, where pixel-based 1D image analysis is mainstream, followed in popularity by spatial context-sensitive 1D image analysis implemented within the OBIA paradigm (Blaschke et al., 2014). Undoubtedly, computers are more successful than humans in 1D image analysis, invariant to permutations in the input vector data sequence (Cimpoi et al., 2014). Nonetheless, humans are still far more successful than computers in 2D image analysis, synonym of spatial topology-preserving feature mapping (Tsotsos, 1990), which implies sensitivity to permutations in the input vector data sequence (Cimpoi et al., 2014).

Since traditional EO-IUSs adopt a 1D image analysis approach, where dominant spatial information is omitted either totally or in part in favor of secondary color information, it is useful to turn attention to the multidisciplinary framework of cognitive science to shed light on how humans cope with color informa-tion. According to cognitive science which includes linguistics, the study of languages, see Figure 13, humans discretize (fuzzify) ever-varying quantitative (numeric) photometric and spatio-temporal sensa-tions into stable qualitative/categorical/nominal percepts, eligible for use in symbolic human reasoning based on a convergence-of-evidence approach (Matsuyama & Hwang, 1990). In their seminal work, Berlin and Kay proved that 20 human languages, spoken across space and time in the real world, partition quantitative color sensations collected in the visible portion of the electromagnetic spectrum, see Figure 1, onto the same "universal" vocabulary of eleven BC names (Berlin & Kay, 1969): black, white, gray, red, orange, yellow, green, blue, purple, pink and brown. In a 3D monitor-typical red-green-blue (RGB) data cube, BC names are intuitive to think of and easy to visualize. They provide a mutually exclusive and totally exhaustive partition of a monitor-typical RGB data cube into RGB polyhedra neither necessarily connected nor convex, see Figure 7 (Benavente et al., 2008; Griffin, 2006). Since they are community-agreed upon to be used by members of the same community, RGB BC polyhedra are prior knowledge-based, i.e., stereotyped, non-adaptive-to-data (static), general-purpose, application- and user-independent. Multivariate measurement space partitioning into a discrete and finite set of mutually exclusive and totally exhaustive hyperpolyhedra copes with the transformation of a numeric variable into a categorical variable, see Figure 7. Numeric variable discretization is a typical problem in many scientific disciplines, such as inductive VQ in machine learning-from-data (Cherkassky & Mulier, 1998) and deductive numeric variable fuzzification into discrete fuzzy sets, e.g., low, medium and high, in fuzzy logic (Zadeh, 1965), refer to Chapter 1.

To summarize, human languages refer to human colorimetric perception in terms of a stable, prior knowledge-based vocabulary (codebook) of BC names (codewords) non-adaptive to data, physically equivalent to a discrete and finite set of mutually exclusive and totally exhaustive hyperpolyhedra, neither necessarily convex nor connected in a numeric MS color space, identified as $\Re^{MS}$, where MS >2, e.g., MS = 3 like in a monitor-typical RGB data cube, see Figure 7. These BC names are conceptually equivalent to a latent/hypothetical categorical variable of semi-symbolic quality, see Figure 14, capable of linking sub-symbolic sensory data in the real world, specifically color values in color space $\Re^{MS}$, to categorical variables of semantic (symbolic) quality in the world model, also known as world ontology or mental world, made of abstract concepts, like perceptual categories of real-world objects or mental states.

In an analytic model of vision based on a convergence-of-evidence approach, the first original contribution of the present Part 1 is to encode prior knowledge about color naming into a CV system by design, as described hereafter. Irrespective of their Pearson inter-feature cross-correlation, if any, it is easy to prove that individual sources of visual evidence, such as color, local shape, texture and inter-object spatial relationships, are statistically independent because, in general, Pearson's linear cross-correlation does not imply causation (Baraldi, 2017; Baraldi & Soares, 2017; Pearl, 2009). According to a "naive" hypothesis of conditional independence of visual features color, local shape, texture and inter-object spatial relationships, when target classes of observed objects in the real-world scene are c = 1, ..., ObjectClassLegendCardinality, for a given discrete spatial unit x in the image-domain, either 0D point, 1D line or 2D polygon (OGC, 2015), then the well-known "naïve" Bayes classification formulation (Bishop, 1995) becomes

$$p(c|\,ColorValue(x), ShapeValue(x), TextureValue(x), SpatialRelationships(x, Neigh(x)))$$

$$= p(c|F_i, \ldots, F_{I=4}) = p(c) \prod_{i=1}^{I=4} p(F_i|c) = p(c) \bullet p(ColorValue(x)|c) \bullet p(ShapeValue(x)|c)$$

$$\bullet\, p(TextureValue(x)|c) \bullet p(SpatialRelationships(x, Neigh(x))|c) \leq \min\{p(c|\,ColorValue(x)),$$

$$p(c|\,ShapeValue(x)), p(c|\,TextureValue(x)), p(c|\,SpatialRelationships(x, Neigh(x)))\}, c$$

$$= 1, \ldots, ObjectClassLegendCardinality$$

$$(1)$$

where ColorValue(x) belongs to a MS measurement space $\Re^{MS}$, i.e., ColorValue(x) $\in \Re^{MS}$, and Neigh(x) is a generic 2D spatial neighborhood of spatial unit x in the (2D) image-domain. Equation (1) shows that any convergence-of-evidence approach is more selective than each individual source of evidence, in line with a focus-of-visual attention mechanism (Frintrop, 2011). For the sake of simplicity, if priors are ignored because considered equiprobable in a maximum class-conditional likelihood inference approach alternative to a maximum a posteriori optimization criterion, then Equation (1) becomes

$$p(c|\,ColorValue(x), ShapeValue(x), TextureValue(x), SpatialRelationships(x, Neigh(x))) \propto$$

$$p(ColorValue(x)|c) \bullet p(ShapeValue(x)|c) \bullet p(TextureValue(x)|c) \bullet p(SpatialRelationships$$

$$(x, Neigh(x))|\,c) = [\sum_{ColorName=1}^{ColorVocabularyCardinality} p(ColorValue(x)|ColorName)p(ColorName|c)] \bullet$$

$$p(ShapeValue(x)|\,c)\, p(TextureValue(x)|c) \bullet p(SpatialRelationships(x, Neigh(x))|\,c)c$$

$$= 1, \ldots, ObjectClassLegendCardinality$$

$$(2)$$

where color space $\Re^{MS}$ is partitioned into hyperpolyhedra, equivalent to a discrete and finite vocabulary of static color names, with ColorName = 1, ..., ColorVocabularyCardinality. To further simplify Equation (2), its canonical interpretation based on frequentist statistics can be relaxed by fuzzy logic (Zadeh, 1965), so that the logical-AND operator is replaced by a fuzzy-AND (min) operator, inductive class-conditional probability $p(x|\quad c) \in [0, 1]$, where

$$\sum_{c=1}^{ObjectClassLegendCardinality} p(x|c) > = 0,$$ is replaced by a deductive membership (compatibility) function m

$(x|c) \in [0, 1]$, where $\sum_{c=1}^{ObjectClassLegendCardinality} m(x|c) > = 0$ according to the principles of fuzzy logic, where compatibility/membership does not mean probability, and color space hyperpolyhedra are considered mutually exclusive and totally exhaustive. If these simplifications are adopted, then Equation (2) becomes

$m(c| ColorValue(x), ShapeValue(x), TextureValue(x), SpatialRelationships(x, Neigh(x))) \propto min$

$\{\sum_{ColorName=1}^{ColorVocabularyCardinality} m(ColorValue(x)| ColorName)m(ColorName|c), m(ShapeValue(x)| c),$
$m(TextureValue(x)| c), m(SpatialRelationships(x, Neigh(x))| c)\} = min\{m(ColorName^*| c),$
$m(ShapeValue(x)| c), m(TextureValue(x)| c), m(SpatialRelationships(x, Neigh(x))| c)\}, c = 1, \ldots,$
$ObjectClassLegendCardinality, where ColorName^* \in \{1, ColorVocabularyCardinality\}, such that$
$m(ColorValue(x)|ColorName^*) = 1 and m(ColorName^*| c) \in \{0, 1\}.$

$$(3)$$

In Equation (3), the following considerations hold.

- Each numeric ColorValue(x) in color space $\Re^{MS}$ belongs to a single color name (hyperpolyhedron) ColorName* in the static color name vocabulary, i.e., $\forall$ ColorValue(x) $\in \Re^{MS}$, $\sum_{ColorName=1}^{ColorVocabularyCardinality} m(ColorValue(x)|ColorName) = m(ColorValue(x)|ColorName^*) = 1$ holds, where m(ColorValue(x)| ColorName) $\in$ {0, 1} is a binary (crisp, hard) membership function, with ColorName = 1, ..., ColorVocabularyCardinality and ColorName* $\in$ {1, ColorVocabularyCardinality}.

- Set A = VocabularyOfColorNames, with cardinality |A| = a = ColorVocabularyCardinality, and set B = LegendOfObjectClassNames, with cardinality |B| = b = ObjectClassLegendCardinality, can be considered a bivariate categorical random variable where two univariate categorical variables A and B are generated from a single population. A binary relationship from set A to set B, R: A ⇒ B, is a subset of the 2-fold Cartesian product (product set) A × B, whose size is rows × columns = a × b, hence, R: A ⇒ B ⊆ A × B. The Cartesian product of two sets A × B is a set whose elements are ordered pairs. Hence, the Cartesian product is non-commutative, A × B ≠ B × A. In agreement with common sense, see Table 3, binary relationship R: VocabularyOfColorNames ⇒ LegendOfObjectClassNames is a set of ordered pairs where each ColorName can be assigned to none, one or several classes of observed scene-objects with class index c = 1, ..., ObjectClassLegendCardinality, whereas each class of observed objects can be assigned with none, one or several color names to define the class-specific colorimetric attribute. Binary membership values m(ColorName| c) $\in$ {0, 1} and m(c| ColorName) $\in$ {0, 1}, with c = 1, ..., ObjectClassLegendCardinality and ColorName = 1, ..., ColorVocabularyCardinality, can be community-agreed upon based on various kinds of evidence, whether viewed all at once or over time, such as a combination of prior beliefs with additional evidence inferred from new data in agreement with a Bayesian updating rule, largely applied in AI, including design and implementation of expert systems. In Bayesian updating, Bayesian inference is applied iteratively (Ghahramani, 2011; Wikipedia, 2018c): after observing some evidence, the resulting posterior probability can be treated as a prior probability and a new posterior probability computed from new evidence. A binary relationship R: A ⇒ B ⊆ A × B, where sets A and B are categorical variables generated from a single population, guides the interpretation process of a two-way *contingency table*, also known as association matrix, cross tabulation, bivariate table or bivariate frequency table (BIVRFTAB) (Kuzera & Pontius, 2008; Pontius & Connors, 2006; Pontius & Millones, 2011), such that BIVRFTAB = FrequencyCount(A × B). In the conventional domain of frequentist inference with no reference to prior beliefs, a BIVRFTAB is the 2-fold Cartesian product A × B instantiated by the bivariate frequency counts of the two univariate categorical variables A and B generated from a single population. Hence, binary relationship R: A ⇒ B ⊆ A × B ≠

Table 3. Example of a binary relationship R: A ⇒ B ⊆ A × B from set A = VocabularyOfColorNames, with cardinality |A| = a = ColorVocabularyCardinality = 11, and the set B = LegendOfObject ClassNames, with cardinality |B| = b = ObjectClassLegendCardinality = 3

			Target classes of individuals (entities in a conceptual model for knowledge representation built upon an ontology language)		
			Class 1, Water body	Class 2, Tulip flower	Class 3, Italian tile roof
Color names	Black			√	
	Blue		√	√	
	Brown		√	√	√
	Grey				
	Green		√	√	
	Orange			√	
	Pink			√	
	Purple			√	
	Red			√	√
	White			√	
	Yellow			√	

The latter dictionary is a superset of the typical taxonomy of land cover (LC) classes adopted by the RS community. "Correct" entry-pairs (marked with √) must be: (i) selected by domain experts based on a hybrid combination of deductive prior beliefs with inductive evidence from data, refer to Table 5, and (ii) community-agreed upon

FrequencyCount(A × B) = BIVRFTAB, where one instantiation of the former guides the interpretation process of the latter. In greater detail, for any BIVRFTAB instance, either square or non-square, there is a binary relationship R: A ⇒ B ⊆ A × B that guides the interpretation process, where "correct" binary entry-pair cells of the 2-fold Cartesian product A × B are equal to 1 and located either off-diagonal (scattered) or on-diagonal, if a main diagonal exists when the BIVRFTAB is square. When a BIVRFTAB is estimated from a geospatial population with or without sampling, it is called *overlapping area matrix* (OAMTRX) (Baraldi et al., 2014; Baraldi, Bruzzone, & Blonda, 2005; Baraldi et al., 2006; Beauchemin & Thomson, 1997; Lunetta & Elvidge, 1999; Ortiz & Oliver, 2006; Pontius & Connors, 2006). When the binary relationship R: A ⇒ B is a bijective function (both 1–1 and onto), i.e., when the two categorical variables A and B estimated from a single population coincide, then the BIVRFTAB instantiation is square and sorted; it is typically called confusion matrix (CMTRX) or error matrix (Congalton & Green, 1999; Lunetta & Elvidge, 1999; Pontius & Millones, 2011; Stehman & Czaplewski, 1998). In a CMTRX, the main diagonal guides the interpretation process. For example, a square OAMTRX = FrequencyCount(A × B), where A = test thematic map legend, B = reference thematic map legend such that cardinality a = b, is a CMTRX if and only if A = B, i.e., if the test and reference codebooks are the same sorted set of concepts or categories. In general the class of (square and sorted) CMTRX instances is a special case of the class of OAMTRX instances, either square or non-square, i.e., OAMTRX ⊃ CMTRX. A similar consideration holds about summary Q^2Is generated from an OAMTRX or a CMTRX, i.e., Q^2I (OAMTRX) ⊃ Q^2I(CMTRX) (Baraldi et al., 2014, 2005, 2006).

Equation (3) shows that for any spatial unit x in the image-domain, when a hierarchical CV classification approach estimates posterior m(c| ColorValue(x), ShapeValue(x), TextureValue(x), SpatialRelationships(x, Neigh(x))) starting from an *a priori* knowledge-based near real-time color naming first stage, where condition m(ColorValue(x)| ColorName*) = 1 holds, if condition m(ColorName*| c) = 0 is true according to a static community-agreed binary relationship R: VocabularyOfColorNames ⇒ LegendOfObjectClassNames (and vice versa) known *a priori*, see

Table 3, then m(c| ColorValue(x), ShapeValue(x), TextureValue(x), SpatialRelationships(x, Neigh (x))) = 0 irrespective of any second-stage assessment of spatial terms ShapeValue(x), TextureValue(x) and SpatialRelationships(x, Neigh(x)), whose computational model is typically difficult to find and computationally expensive. Intuitively, Equation (3) shows that static color naming of any spatial unit x, either (0D) pixel, (1D) line or (2D) polygon, allows the color-based stratification of unconditional multivariate spatial variables into color class-conditional data distributions, in agreement with the statistic stratification principle (Hunt & Tyrrell, 2012) and the divide-and-conquer (*dividi-et-impera*) problem solving approach (Bishop, 1995; Cherkassky & Mulier, 1998; Lipson, 2007). Well known in statistics, the principle of statistic stratification guarantees that "stratification will always achieve greater precision provided that the strata have been chosen so that members of the same stratum are as similar as possible in respect of the characteristic of interest" (Hunt & Tyrrell, 2012).

Whereas 3D color polyhedra are easy to visualize and intuitive to think of in a true- or false-color RGB data cube, see Figure 7, hyperpolyhedra are difficult to think of and impossible to visualize in a MS reflectance space whose spectral dimensionality MS >3, with spectral channels ranging from visible to thermal portions of the electromagnetic spectrum, see Figure 1. Since it is non-adaptive-to-data, any static hyperpolyhedralization of a MS measurement space must be based on *a priori* physical knowledge available in addition to sensory data. Equivalent to a physical data model, static hyperpolyhedralization of a MS data space requires all spectral channels to be provided with a physical unit of radiometric measure, i.e., MS data must be radiometrically calibrated, in compliance with the GEO-CEOS QA4EO *Cal* requirements (GEO-CEOS, 2010, 2010), refer to Chapter 1.

Noteworthy, sensory data provided with a physical unit of measure can be input to both statistical/inductive and physical/deductive models, including hybrid (combined deductive and inductive) inference systems, refer to Chapter 1. On the contrary, uncalibrated dimensionless sensory data can be input to statistical data models exclusively. Although considered mandatory by the GEO-CEOS QA4EO *Cal* guidelines (GEO-CEOS, 2010) and regarded as a well-known "prerequisite for physical model-based analysis of airborne and satellite sensor measurements in the optical domain" (Schaepman-Strub, Schaepman, Painter, Dangel, & Martonchik, 2006), EO data *Cal* is ignored by relevant portions of the RS literature focusing on statistical EO data analytics, such as supervised learning-from-data function regression and classification (Bishop, 1995; Cherkassky & Mulier, 1998). One consequence is that, to date, statistical model-based EO-IUSs dominate the RS literature as well as commercial EO image processing software toolboxes, which typically consist of overly complicated collections of inherently ill-posed inductive machine learning-from-data algorithms (Bishop, 1995; Cherkassky & Mulier, 1998) to choose from based on heuristics (Baraldi, 2017; Baraldi & Boschetti, 2012a, 2012b). This is in contrast with the hybrid inference framework adopted by all biological cognitive systems (Parisi, 1991). In the words of O. Etzioni, "with all due respect to (machine learning-from-data scientists), thought is not a vector, and AI is not a problem in statistics" (Etzioni, 2017; Marcus, 2018).

3. Related works in static MS reflectance space hyperpolyhedralization
In the RS discipline, there is a long history of hybrid EO-IUSs in operating mode, suitable for either low-level EO image enhancement (pre-processing) or high-level EO image understanding (classification), where an *a priori* knowledge-based decision tree for static MS reflectance space hyperpolyhedralization is plugged into the hybrid CV system architecture without *Val* by independent means, in disagreement with the GEO-CEOS QA4EO *Val* requirements (GEO-CEOS, 2010; Group on Earth Observation/Committee on Earth Observation Satellites (GEO-CEOS WGCV, 2015), refer to Chapter 1.

In recent years, the SIAM stratification of single-date MS imagery into MS color names was applied to MS image topographic correction, which is a traditional *chicken-and-egg* dilemma (Bishop & Colby, 2002; Bishop et al., 2003; Riano et al., 2003), synonym of inherently ill-posed problem in the Hadamard sense (Hadamard, 1902). When an inherently ill-posed MS image topographic correction was better conditioned for numeric solution by a prior knowledge-based SIAM color naming (masking) first stage, it required no human-machine interaction to run (Baraldi et al., 2010c), in compliance with process requirements of systematic ESA EO Level 2 product generation (ESA, 2015; DLR & VEGA, 2011; CNES, 2015), see Figure 12.

In the Atmospheric/Topographic Correction for Satellite Imagery (ATCOR) commercial software product, several deductive spectral pixel-based decision trees are implemented for use in different stages of an EO data enhancement pipeline (Baraldi et al., 2013; Baraldi & Humber, 2015; Dorigo et al., 2009; Richter & Schläpfer, 2012a, 2012b; Schläpfer, Richter & Hueni, 2009), see Figure 11. One of the ATCOR's prior knowledge-based per-pixel decision trees delivers as output a haze/cloud/water (and snow) classification mask file ("image_out_hcw.bsq"). In addition, ATCOR includes a so-called prior knowledge-based decision tree for Spectral Classification of surface reflectance signatures (SPECL) (Baraldi & Humber, 2015; Baraldi et al., 2013; Dorigo et al., 2009), see Table 4. Unfortunately, SPECL has never been tested by its authors in the RS literature, although it has been validated by independent means (Baraldi & Humber, 2015; Baraldi et al., 2013).

Supported by NASA, atmospheric effect removal by the Landsat Ecosystem Disturbance Adaptive Processing System (LEDAPS) project relies on exclusion masks for water, cloud, shadow and snow surface types detected by a simple set of prior knowledge-based spectral decision rules applied per pixel. Quantitative analyses of LEDAPS products led by its authors revealed that these exclusion masks are prone to errors, to be corrected in future LEDAPS releases (Vermote & Saleous, 2007). Unfortunately, to date, in a recent comparison of cloud and cloud-shadow detectors, those implemented in LEDAPS scored low among alternative solutions (Foga et al., 2017).

In the 1980s, to provide an automatic alternative to a visual and subjective assessment of the cloud cover on Advanced Very High Resolution Radiometer (AVHRR) quicklook images in the ESA Earthnet archive, Muirhead and Malkawi developed a simple algorithm to classify daylight AVHRR images on a pixel-by-pixel basis into land, cloud, sea, snow or ice and sunglint, such that the classified quicklook image was presented in appropriate pseudo-colors, e.g., green: land, blue: sea, white: cloud, etc. (Muirhead & Malkawi, 1989).

Developed independently by NASA (Ackerman et al. 1998) and the Canadian Center for Remote Sensing (CCRS) (Luo et al., 2008), pixel-based static decision trees contribute, to date, to the systematic generation of clear-sky Moderate Resolution Imaging Spectroradiometer (MODIS) image composites in operating mode, see Figure 17.

To pursue high-level LC/LCC detection through time, extensions to the time domain of a single-date *a priori* spectral rule base for VQ of an MS reflectance space have become available to the general public in 2015 through the Google Earth Engine (GEE) platform (Simonetti et al. 2015b) or in the form of a commercial LC/LCC map product at national scale (GeoTerraImage, 2015). These are both post-classification approaches for LC/LCC detection based on a time-series of single-date per-pixel prior knowledge-based MS decision-tree classification maps. In practice, MS image time-series analysis in the domain of numeric sub-symbolic variables is replaced by MS color map time-series analysis in the domain of categorical semi-symbolic variables. These two post-classification approaches share the same operational limitations, specifically, they are Landsat sensor series-specific, pixel-based, where spatial topological and spatial non-topological information components are totally ignored, and their post-classification overall accuracy (OA) $\in [0, 1]$ is not superior to the product of the single-date classification accuracies in the time-series (Lunetta & Elvidge, 1999). For example,

Bi-temporal post-classification LC change/no-change(LCC)detection overall accuracy(OA),
$\quad$ OA-LCC$_{1,2} \in [0, 1]$, OA-LCC$_{1,2} = $ f(OA of the LC maps at time T1 and time T2, identified as OA-LC$_1$
$\quad$ and OA-LC$_2$ respectively) $\leq$ OA-LC$_1 \times$ OA-LC$_2$, where OA-LC$_1 \in [0, 1]$ and OA-LC$_2 \in [0, 1]$

$$(4)$$

Index	Spectral Categories	Spectral Rule (based on reflectance measured at Landsat TM central wave bands: b1 is located at 0.48 μm, b2 at 0.56 μm, b3 at 0.66 μm, b4 at 0.83 μm, b5 at 1.6 μm, b7 at 2.2 μm)	Pseudo-color
1	Snow/ice	b4/b3 ≤ 1.3 AND b3 ≥ 0.2 AND b5 ≤ 0.12	
2	Cloud	b4 ≥ 0.25 AND 0.85 ≤ b1/b4 ≤ 1.15 AND b4/b5 ≥ 0.9 AND b5 ≥ 0.2	
3	Bright bare soil/sand/cloud	b4 ≥ 0.15 AND 1.3 ≤ b4/b3 ≤ 3.0	
4	Dark bare soil	b4 ≥ 0.15 AND 1.3 ≤ b4/b3 ≤ 3.0 AND b2 ≤ 0.10	
5	Average vegetation	b4/b3 ≥ 3.0 AND (b2/b3 ≥ 0.8 OR b3 ≤ 0.15) AND 0.28 ≤ b4 ≤ 0.45	
6	Bright vegetation	b4/b3 ≥ 3.0 AND (b2/b3 ≥ 0.8 OR b3 ≤ 0.15) AND b4 ≥ 0.45	
7	Dark vegetation	b4/b3 ≥ 3.0 AND (b2/b3 ≥ 0.8 OR b3 ≤ 0.15) AND b3 ≤ 0.08 AND b4 ≤ 0.28	
8	Yellow vegetation	b4/b3 ≥ 2.0 AND b2 ≥_b3 AND b3 ≥ 8.0 AND b4/b5 ≥ 1.5 [a]	
9	Mix of vegetation/soil	2.0 ≤ b4/b3 ≤ 3.0 AND 0.05 ≤ b3 ≤ 0.15 AND b4 ≥ 0.15	
10	Asphalt/dark sand	b4/b3 ≤ 1.6 AND 0.05 ≤ b3 ≤ 0.20 AND 0.05 ≤ b4 ≤ 0.20[a] AND 0.05 ≤ b5 ≤ 0.25 AND b5/b4 ≥ 0.7[a]	
11	Sand/bare soil/cloud	b4/b3 ≤ 2.0 AND b4 ≥ 0.15 AND b5 ≥ 0.15[a]	
12	Bright sand/bare soil/cloud	b4/b3 ≤ 2.0 AND b4 ≥ 0.15 AND (b4 ≥ 0.25b OR b5 ≥ 0.30[b])	
13	Dry vegetation/soil	(1.7 ≤ b4/b3 ≤ 2.0 AND b4 ≥ 0.25[c]) OR (1.4 ≤ b4/b3 ≤ 2.0 AND b7/b5 ≤ 0.83[c])	
14	Sparse veg./soil	(1.4 ≤ b4/b3 ≤ 1.7 AND b4 ≥ 0.25[c]) OR (1.4 ≤ b4/b3 ≤ 2.0 AND b7/b5 ≤ 0.83 AND b5/b4 ≥ 1.2[c])	
15	Turbid water	b4 ≤ 0.11 AND b5 ≤ 0.05[a]	
16	Clear water	b4 ≤ 0.02 AND b5 ≤ 0.02[a]	
17	Clear water over sand	b3 ≥ 0.02 AND b3 ≥ b4 + 0.005 AND b5 ≤ 0.02[a]	
18	Shadow		
19	Not classified (outliers)		

Table 4. Rule set (structural knowledge) and order of presentation of the rule set (procedural knowledge) adopted by the prior knowledge-based MS reflectance space quantizer called Spectral Classification of surface reflectance signatures (SPECL), implemented within the ATCOR commercial software toolbox (Dorigo et al., 2009; Richter & Schläpfer, 2012a, 2012b)

[a] These expressions are optional and only used if b5 is present. [b] Decision rule depends on presence of b5. [c] Decision rule depends on presence of b7.

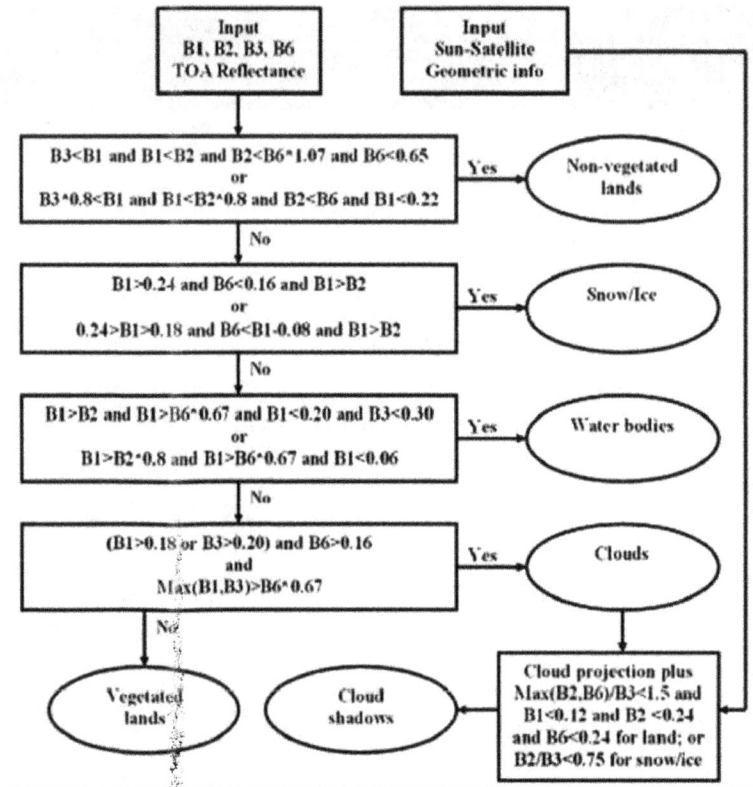

Figure 17. Courtesy of Luo et al.(2008). Canada Centre for Remote Sensing (CCRS)'s flow chart of a physical model-based per-pixel MODIS image classi-fier integrated into a clear-sky multi-temporal MODIS image compositing system in operat-ing mode. Acronym B stands for MODIS Band.

In Equation (4), if OA-LC$_1$ = 0.90 and OA-LC$_2$ = 0.90, then OA-LCC$_{1,2}$ ≤ 0.81. Hence, post-classification analysis is recommended for its simplicity if and only if single-date OA values are "high" through the time-series (Baraldi, 2017; Baraldi et al., 2016; Tiede et al., 2016). In other words, a necessary not sufficient pre-condition for multi-temporal image analysis to score "high" in accuracy according to a conceptually simple and computationally efficient post-classification LC change/no-change detection approach is that single-date image classi-fication accuracies score individually "high" through the time-series. The two aforementioned post-classification approaches were both inspired by a year 2006 SIAM instantiation of a static decision tree for Landsat reflectance space hyperpolyhedralization, presented in pseudo-code in the RS literature (Baraldi et al., 2006) and further developed into the SIAM application software available to date (Baraldi, 2017, 2011; Baraldi & Boschetti, 2012a, 2012b; Baraldi et al., 2014, 2010a, 2010b; Baraldi & Humber, 2015; Baraldi et al., 2013; Baraldi, Wassenaar, & Kay, 2010d).

In Boschetti et al. (2015), a year 2013 SIAM instantiation was successfully employed to accom-plish post-classification burned area detection in MS image time-series.

Among the aforementioned static decision trees for MS color naming, only SIAM claims scal-ability to several families of EO imaging sensors featuring different spectral resolutions, see Table 1.

It is obvious but not trivial to emphasize to the RS community that, in human vision and CV, an *a priori* vocabulary of general-purpose data- and application-independent BC names is equivalent to a static sub-symbolic or semi-symbolic categorical variable non-coincident with

GEO-CEOS stage 4 validation of the Satellite Image Automatic Mapper...

161

a symbolic categorical variable whose levels are user- and application-specific classes of objects observed in the 4D spatio-temporal scene-domain, refer to Table 3 and Equation (3). The very same consideration holds for any discrete and finite set of spectral endmembers in mixed pixel analysis, which "cannot always be inverted to unique LC class names" (Adams et al., 1995). It means that spectral endmembers in hyperspectral (HS) image analysis are conceptually equivalent to a static, user- and application-independent vocabulary of BC names, corresponding to a mutually exclusive and totally exhaustive set of neither necessarily convex nor connected hyperpolyhedra in a hyperspectral color space. Whereas the SIAM expert system has been successfully applied to HS imagery for fully automated color naming and superpixel detection (Baraldi, 2017), in the RS literature spectral endmembers detection in HS imagery is traditionally dealt with by inductive learning-from-data algorithms (Ghamisi et al., 2017), which are typically site specific and semiautomatic (Liang, 2004).

Quite surprisingly, the non-coincident assumption between an *a priori* vocabulary of sub-symbolic color names A in the (2D) image-domain and an application-specific legend of symbolic classes B of real-world objects in the 4D scene-domain, where A ≠ B always holds true, refer to Table 3 and Equation (3), appears somehow difficult to acknowledge by relevant portions of the RS community. For example, in the DigitalGlobe Geospatial Big Data platform (GBDX), a patented prior knowledge-based decision tree for pixel-based very high resolution WorldView-2 and WorldView-3 image mapping onto static MS reflectance space hyperpolyhedra (GBDX Registered Name: protogenV2LULC, Provider: GBDX) was proposed to RS end-users as an "Automated Land Cover Classification" (DigitalGlobe, 2016). This program name can be considered somehow misleading because it refers to no EO image mapper of DNs into LC class names, but to a static sub-symbolic color space partitioner, where DNs are mapped onto color name-specific hyperpolyhedra, see Equation (3). Due to the confusion between color names in the (2D) image-domain with target LC classes in the 4D scene-domain, see Table 3, the "Automated Land Cover Classification" computer program is affected by several "known issues" (DigitalGlobe, 2016): "Vegetation: Thin cloud (cloud edges) might be misinterpreted as vegetation; Water: False positives maybe present due to certain types of concrete roofs or shadows; Soils: Ceramic roofing material and some types of asphalt may be misinterpreted as soil," etc.

In Salmon et al. (2013), a year 2006 SIAM's *a priori* dictionary of static sub-symbolic MS color names was downscaled in cardinality and sorted in the order of presentation to form a bijective function with a legend of symbolic classes of target objects in the scene-domain. In practice these authors forced a non-square BIVRFTAB to become a (square and sorted) CMTRX, where the main diagonal guides the interpretation process (Congalton & Green, 1999), to make it more intuitive and familiar to RS practitioners. In general, no binary relationship R: A ⇒ B between an *a priori* vocabulary A of static sub-symbolic color names and a user- and application-dependent dictionary B of symbolic classes of observed objects in the scene-domain is a bijective function, refer to Table 3 and Equation (3). As a consequence of its unrealistic hypothesis in color information/knowledge representation, the 1D image classification approach proposed in Salmon et al. (2013) scored low in accuracy. Unfortunately, to explain their poor MS image classification outcome these authors concluded that, in their experiments, a year 2006 SIAM's static dictionary of color names was useless to identify target LC classes. The lesson to be gained by these authors' experience is that well-established RS practices, such as 1D image analysis based on supervised data learning algorithms and thematic map quality assessment by means of a square and sorted CMTRX where test and reference thematic legends are the same, can become malpractices when an *a priori* dictionary of static color names is employed for MS image classification purposes in agreement with Equation (3) and common sense, see Table 3. This lesson learned is supported by the fact that one of the same co-authors of paper (Salmon et al., 2013) reached opposite conclusions when a year 2013 SIAM application software, the same investigated by the present paper, was employed successfully in detecting burned areas from MS image time-series according to a convergence of color names with spatio-temporal visual properties in agreement with Equation (3) (Boschetti et al., 2015).

4. Original hybrid eight-step guideline for identification of a categorical variable-pair binary relationship

Summarized in Chapter 1, our experimental project required to compare an annual time-series of test SIAM-WELD maps of sub-symbolic color names, see Figure 6, with a reference USGS NLCD 2006 map whose legend of symbolic LC classes is summarized in Table 2. Since these test and reference map legends do not coincide, they must be reconciled/harmonized through a binary relationship R: VocabularyOfColorNames ⇒ LegendOfObjectClassNames (and vice versa), refer to Equation (3).

The harmonization of ontologies and the comparison of thematic maps with different legends are the subject of research of a minor body of literature, e.g., refer to works in ontology-driven geographic information systems (ODGIS) (Fonseca, Egenhofer, Agouris, & Camara, 2002; Guarino, 1995; Sowa, 2000). Ahlqvist writes that "to negotiate and compare information stemming from different classification systems (Bishr, 1998; Mizen, Dolbear, & Hart, 2005)... a translation can be achieved by *matching the concepts in one system with concepts in another*, either directly or through an intermediate classification (Feng & Flewelling, 2004; Kavouras & Kokla, 2002)" (Ahlqvist, 2005). Stehman describes four common types of thematic map-pair comparisons (Stehman, 1999). In the first type, different thematic maps, either crisp or fuzzy, of the same region of interest and employing the same sorted set (legend) of LC classes are compared (Kuzera & Pontius, 2008). In the second type, which includes the first type as a special case, thematic maps, either crisp of fuzzy, of the same region of interest, but featuring map legends that differ in their basic terms with regard to semantics and/or cardinality and/or order of presentation are compared. The third and fourth types of thematic maps comparison regard maps of different surface areas featuring, respectively, the same dictionary or different dictionaries of basic terms. Whereas a large portion of the RS community appears concerned with the aforementioned first type of map comparisons exclusively, the protocol proposed in (Baraldi et al., 2014) focuses on the second type, which includes the first type as a special case. In Couclelis (2010), the author observed that inter-dictionary concept matching ("conceptual matching") (Ahlqvist, 2005) is an inherently equivocal *information-as-data-interpretation* process (Capurro & Hjørland, 2003), see Table 3. In common practice, two independent human domain-experts (cognitive agents, knowledge engineers) are likely to identify different binary associations between two codebooks of codewords (Laurini & Thompson, 1992). The conclusion is that no "universal best match" of two different codebooks can exist, but identification of the most appropriate binary relationship between two different nomenclatures becomes a subjective matter of negotiation to become community-agreed upon (Baraldi, 2017; Capurro & Hjørland, 2003; Couclelis, 2010).

To streamline the inherently subjective selection of "correct" entry-pairs in a binary relationship R: A ⇒ B ⊆ A × B between two univariate categorical variables A and B estimated from a single population, an original hybrid eight-step guideline was designed for best practice, where deductive/top-down prior beliefs and inductive/bottom-up learning-from-data inference are combined. This hybrid protocol is sketched hereafter as the second original and pragmatic contribution of the present Part 1 to fill the gap from EO sensory data, mapped onto BC names, to ESA EO Level 2 product, whose SCM features semantics. As an example, let us consider a binary relationship R: A ⇒ B = VocabularyOfColorNames ⇒ LegendOfObjectClassNames ⊆ A × B where rows are a test set of three semi-symbolic color names, say, A = {MS green-as-"*Vegetation*", MS white-as-"*Cloud*", "*Unknowns*"}, where |A| = a = ColorVocabularyCardinality = TC = 3 is the row (test) cardinality, and where columns are a reference set of three symbolic LC classes, say, B = { "*Evergreen Forest*", "*Deciduous Forest*", "*Others*"}, where |B| = b = ObjectClassLegendCardinality = RC = 3 is the column (reference) cardinality.

(1) Display multivariate frequency distributions of the two univariate categorical variables estimated from a single population in the BIVRFTAB = FrequencyCount(A × B) whose size is TC × RC.

(2) Estimate probabilities in the BIVRFTAB cells.

(3) Compute class-conditional probability $p(r \mid t)$ of reference class $r = 1, ..., RC$, given test class $t = 1, ..., TC$.

(4) Reset to zero all $p(r \mid t)$ below $TH1 \in [0, 1]$ (e.g., $TH1 = 9\%$), otherwise set that cell to 1. Let us identify this contingency table instantiation as $DataDrivenConditionalProb(r|t)(x, y)$, $x = 1, ...,$ RC, $y = 1, ..., TC$.

(5) Compute class-conditional probability $p(t \mid r)$ of test class $t = 1, ..., TC$, given reference class $r = 1, ..., RC$.

(6) Reset to zero all $p(t \mid r)$ below $TH2 \in [0, 1]$ (e.g., $TH2 = 6\% \leq TH1$), otherwise set that cell to 1. Let us identify this contingency table instantiation as $DataDrivenConditionalProb(t|r)(x, y)$, $x = 1, ..., RC$, $y = 1, ..., TC$.

(7) Compute $DataDrivenTemporaryCells(x, y) = \max\{DataDrivenConditionalProb(t|r)(x, y),$ $DataDrivenConditionalProb(r|t)(x, y)\}$, $x = 1, ..., RC$, $y = 1, ..., TC$. At this point, based exclusively on bottom-up evidence stemming from frequency data, in the 2-fold Cartesian product $A \times B$ each cell is equal to 0 or 1. Then that cell is termed either "temporary non-correct" or "temporary correct".

(8) Top-down scrutiny by a human domain-expert of each cell in the BIVRFTAB, which is either "temporary correct" or "temporary non-correct" at this point, to select those cells to be finally considered as "correct entry-pairs". Actions undertook by this top-down scrutiny are twofold.

  • Switch any data-derived "temporary correct" cell to a "final non-correct" cell if it is provided with a strong prior belief of conceptual mismatch. For example, based on experimental evidence a test spectral category MS white-as-"*Cloud*" can match a reference LC class "*Evergreen Forest*": this data-derived entry-pair match must be considered non-correct in the final R: $A \Rightarrow B$ following semantic scrutiny by a human expert.

  • Switch any data-derived "temporary non-correct" cell to a "final correct" cell if it is provided with a strong prior belief of conceptual match. For example, the test spectral category MS green-as-"*Vegetation*" is considered a superset of the reference LC class "*Deciduous Forest*" irrespective of whether there are frequency data in support of this conceptual relationship.

Table 5 shows an example of how this protocol can be employed in practice. In Table 5 the last step 8 identifies an inherently equivocal *information-as-data-interpretation* process, where a human decision maker has a pro-active role in providing frequency data with semantics (symbolic meanings) (Capurro & Hjørland, 2003). It is highly recommended that any inherently subjective *information-as-data-interpretation* activity occurs as late as possible in the information processing workflow, to avoid propagation of "errors" due to personal preferences not yet community-agreed upon. Noteworthy, in the proposed eight-step guideline there are two "hidden" system's free-parameters to be user-defined based on heuristics, equivalent to a trial-and-error strategy: variables $TH1$ and $TH2$ are two numeric thresholds in range $[0, 1]$ for binary (hard, crisp) decision making, whose normalized range of change and intuitive meaning in terms of probability should make their selection easy and, to a certain extent, application- and user-independent.

5. Original measure of association (harmonization) in a categorical variable-pair binary relationship eligible for guiding the interpretation process of a two-way contingency table

Traditional scalar indicators of bivariate categorical variable association estimated from a BIVRFTAB = FrequencyCount($A \times B$), either square or non-square, include the Pearson's chi-square index of statistical independence and the normalized Pearson's chi-square index, also known as Cramer's coefficient V (Sheskin, 2000). These frequentist statistics of independence do not apply to a binary relationship R: $A \Rightarrow B \subseteq A \times B$, such as that shown in Table 3, where there is no frequency count, i.e., binary relationship R: $A \Rightarrow B \subseteq A \times B \neq$ FrequencyCount($A \times B$) = BIVRFTAB, refer to Chapter 4. Hereafter, a scalar indicator of association (harmonization, reconciliation) between two

Table 5. 8-step guideline for best practice in the identification of a dictionary-pair binary relationship based on a hybrid combination of (top-down) prior beliefs, if any, with (bottom-up) frequentist inference

STEP 1. Dictionary-pair relationship, multivariate occurrence distributions

		Reference Classification (RC)				
		EvergreenF	DeciduousF	Others		
Test Classification (TC)	Vegetation	10	30	60	100	
	Cloud	2	0	10	12	
	Unknowns	0	5	100	105	Tot.
		12	35	170		217

STEP 2. Dictionary-pair relationship, multivariate probability distributions

		Reference Classification (RC)				
		EvergreenF	DeciduousF	Others		
Test Classification (TC)	Vegetation	0.046082949	0.138248848	0.276498	0.460829	
	Cloud	0.00921659	0	0.046083	0.0553	
	Unknowns	0	0.023041475	0.460829	0.483871	Tot.
		0.055299539	0.161290323	0.78341		1

STEP 3. Dictionary-pair relationship, cond. prob. (RC|TC)

		Reference Classification (RC)				
		EvergreenF	DeciduousF	Others		
Test Classification (TC)	Vegetation	0.1	0.3	0.6	1	
	Cloud	0.166666667	0	0.833333	1	
	Unknowns	0	0.047619048	0.952381	1	

STEP 4. Crisp membership function (RC|TC) > TH1 = 0.09.

		Reference Classification (RC)			
		EvergreenF	DeciduousF	Others	
Test Classification (TC)	Vegetation	1	1	1	
	Cloud	1	0	1	
	Unknowns	0	0	1	

STEP 5. Dictionary-pair relationship, cond. prob. (TC|RC)

		Reference Classification (RC)			
		EvergreenF	DeciduousF	Others	
Test Classification (TC)	Vegetation	0.833333333	0.857142857	0.352941	
	Cloud	0.166666667	0	0.058824	
	Unknowns	0	0.142857143	0.588235	
		1	1	1	

(Continued)

Table5. (Continued)

STEP 6. Crisp membership function (TC|RC) > TH2 = 0.06 ≤ TH1 = 0.09.

		Reference Classification (RC)				
		EvergreenF	DeciduousF	Others		
Test Classification (TC)	Vegetation	1	1	1		
	Cloud	1	0	0		
	Unknowns	0	1	1		

STEP 7. OR{Crisp membership function (TC|RC), Crisp membership function(RC|TC)}

		Reference Classification (RC)				
		EvergreenF	DeciduousF	Others		
Test Classification (TC)	Vegetation	1	1	1		
	Cloud	1	0	1		
	Unknowns	0	1	1		

STEP 8. Top-down (driven-by-prior knowledge) scrutiny of bottom-up (data-driven)

"temporary correct" or "temporary non-correct" cells

		Reference Classification (RC)				
		EvergreenF	DeciduousF	Others		
Test Classification (TC)	Vegetation	1	1	1		
	Cloud	0	0	1		
	Unknowns	0	0	1		

univariate categorical variables (codebooks of codewords), A and B, collected from a single population, called Categorical Variable Pair Association Index (CVPAI) in range [0, 1], is estimated from a binary relationship R: A ⇒ B, such that CVPAI(R: A ⇒ B) ∈ [0, 1].

Proposed in Baraldi et al. (2014), a CVPAI version 1, CVPAI1(R: A ⇒ B) ∈ [0, 1], is maximized (tends to 1, meaning maximum harmonization) if the binary relationship R: A ⇒ B from set A = test categorical variable, e.g., vocabulary of color names, to set B = reference categorical variable, e.g., dictionary of LC class names, is a bijective function, i.e., the binary relationship R: A ⇒ B is a function, therefore to each instance in test set A (of color names) corresponds a single instance in reference set B (of LC class names), and this function is both injective (one-to-one, for any instance in reference set B of LC class names there is no more than one instance in test set A of color names) and surjective (onto, for any instance in reference set B of LC class names there is at least one instance in test set A of color names), see Figure 18.

Hereafter, original formulations of CVPAI version 2, CVPAI2(R: A ⇒ B) ∈ [0, 1], and CVPAI version 3, CVPAI3(R: A ⇒ B) ∈ [0, 1], complementary not alternative to the CVPAI1 formulation presented in (Baraldi et al., 2014), are proposed as the third original and analytic contribution of the present Part 1 of this paper. Unlike the CVPAI1 expression, a novel CVPAI2 formulation was constrained as follows, see Figure 18. (i) The "most discriminative" test-to-reference inter-set binary relation R: A ⇒ B is a function, i.e., each test color name in test set A matches with only one reference LC class name in reference set B. (ii) The "most discriminative" reference-to-test class relation is either a surjective function, i.e., each reference LC class in set B matches with at least one test color name in set B, or a bijective function, both surjective and injective as a special case of the former, i.e., each reference LC class in set B matches with only one test color name in set A, see Figure 18. In short, CVPAI2(R: A ⇒ B) is

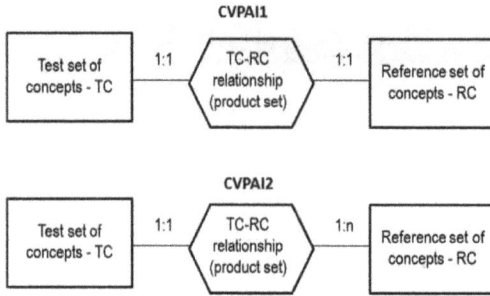

Figure 18. Entity-relationship conceptual model representa-tion of the binary relationship R: A ⇒ B from set A = test cate-gorical variable to set B = reference categorical vari-able, provided with the min: max cardinality required by the Categorical Variable-Pair Association Index (CVPAI) for-mulation 1 (CVPAI1) and for-mulation 2 (CVPAI2) to score maximum in range [0, 1]. Inequality CVPAI1 ≤ CVPAI2 holds, i.e., the latter is a relaxed version of the former. In parti-cular, CVPAI1 is maximum (equal 1) when the binary rela-tionship R: A ⇒ B from set A = test categorical variable to set B = reference categorical variable is a bijective function, both injective (one-to-one) and surjective (onto). CVPAI2 is maximum when the binary relationship R: A ⇒ B is either a surjective function or a bijec-tive function.

maximum (tends to 1, meaning maximum harmonization) when the binary relationship R: A ⇒ B is either a surjective function or a bijective function. Since the CVPAI2 formulation relaxes the CVPAI1 formulation, see Figure 18, it is always true that CVPAI2(R: A ⇒ B) ≥ CVPAI1(R: A ⇒ B) ∈ [0, 1]. The two formulations CVPAI1 and CVPAI2 are complementary not alternative because they have been designed to be maximized by different distributions of "correct" entry-pair cells in a binary relationship R: A ⇒ B ⊆ A × B, see Figure 18. Both formulations CVPAI1(R: A ⇒ B) and CVPAI2(R: A ⇒ B) are independent of frequency counts generated by a bivariate categorical variable distribution to be displayed in a BIVRFTAB = FrequencyCount(A × B) ≠ R: A ⇒ B ⊆ A × B. Depending on the problem at hand, where two specific instantiations of test set A and reference set B are estimated from the same population, either CVPAI1 or CVPAI2 can be considered better suited for estimating the degree of harmonization between sets A and B, see Figure 18. If test set A is a vocabulary of BC names and reference test B is a legend of LC class names, then the novel CVPAI2 formulation is recommended for use. An alternative formulation to CVPAI2(R: A ⇒ B) is CVPAI3(R: A ⇒ B). They are both maximized by the same distribution of "correct" entry-pair cells in a binary relationship R: A ⇒ B ⊆ A × B.

The analytic formulation of CVPAI2(R: A ⇒ B), see Figure 18, is proposed as follows. In a binary relationship R: A ⇒ B ⊆ A × B, set A is a test codebook of cardinality |A| = TC as rows and set B is a reference codebook of cardinality |B| = RC as columns, so that the size of the 2-fold Cartesian product A × B is TC×RC. The total number of "correct" entry-pair cells in R: A ⇒ B is identified as CE, where $0 \leq CE \leq TC \times RC$. In addition, symbol " = = " is adopted to mean "equal to". The CVPAI2 formulation is constrained as follows.

(a) $CE = \sum_{t=1}^{TC} \sum_{r=1}^{RC} CE_{t,r}$, with $CE_{t,r} \in \{0, 1\} =$
{non − correct entry − pair(t,r) = 0, correct entry − pair(t,r) = 1},
$CE \in \{0, RC \times TC\}$.

(b) If (CE = = 0) then CVPAI2 = 0 must hold. It means that, when no "correct'" cell exists, then the degree of conceptual match between the two categorical variables is zero.

(c) If (CE = = TC×RC) then CVPAI2 → 0 must hold. It means that when all table cells are considered "correct", then no entry-pair is discriminative (informative), i.e., nothing makes the difference between the two categorical variables.

(d) If

$$\left\{ \begin{array}{l} \left[\left(\sum_{t=1}^{TC} CE_{t,r} = CE_{+,r} \right) > 0, r = 1, \ldots, RC \right] AND \\ \left[\left(\sum_{r=1}^{RC} CE_{t,r} = CE_{t,+} \right) == 1, t = 1, \ldots, TC \right] \end{array} \right\},$$

where $CE_{+,r}$ is the total sum of correct entry-cell pairs along reference column r, with $r = 1, ..., RC$, $CE_{t,+}$ is the total sum of correct entry-cell pairs along test row t, with $t = 1, ..., TC$, then $CVPAI2$ must be maximum, i.e., $CVPAI2 = 1$. It means that if for each test class $t = 1, ..., TC$ there is one single match and for each reference class $r = 1, ..., RC$ there is at least one match, then $CVPAI2$ must be maximum, such that $CVPAI2 = 1$, see Figure 18.

(e) If [*not* condition(b) *AND not* condition(c) *AND not* condition(d)] then $CVPAI2 \in (0,1)$.

In a square binary relationship R: A $\Rightarrow$ B where $TC == RC$, to maximize the CVPAI2 (to become equal to 1), submitted to condition (d), the binary relationship must be a 1–1 function (an injective function, 1–1 forward and 1–1 backward). To satisfy the set of aforementioned constraints (a) to (e), the following set of original equations is proposed.

$$CVPAI2 \in [0, 1], CVPAI2 = \frac{1}{RC + TC} \left(\sum_{r=1}^{RC} f_{RC}(CE_{+,r}) + \sum_{t=1}^{TC} f_{TC}(CE_{t,+}) \right), \tag{5}$$

with

$$f_{RC}(i) = \begin{cases} 0 \, if \, i = 0, \\ 1 \, if \, i > 0, \end{cases} \quad i \in \{0, TC\} \subset I_0^+, \text{ where } i = CE_{+,r}, r \in \{1, RC\}, \tag{6}$$

$$f_{TC}(j) = \begin{cases} 0 \, if \, j = 0, \\ GaussianMembership(j, Center = 1, StnDev = RC/3) = \\ e^{-\frac{1}{2}\frac{(j-1)^2}{\left(\frac{RC}{3}\right)^2}} \in [0, 1], \quad if \, j > 0, \text{ with } j \in \{0, RC\} \subset I_0^+, \text{ where } j = CE_{t,+}, t \in \{1, TC\}. \end{cases} \tag{7}$$

Although it is maximized by the same distribution of "correct" entry-pair cells in a binary relationship R: A $\Rightarrow$ B $\subseteq$ A × B, a novel CVPAI3 expression is a more severe formulation of CVPAI2, i.e., $1 \geq CVPAI2 \geq CVPAI3 \in [0, 1]$. The proposed CVPAI3 formulation alternative to CVPAI2 is the following.

$$CVPAI3 \in [0, 1], CVPAI3 = \min \left\{ \frac{\sum_{r=1}^{RC} f_{RC}(CE_{+,r})}{RC}, \frac{\sum_{t=1}^{TC} f_{TC}(CE_{t,+})}{TC} \right\}. \tag{8}$$

In Equation (7), GaussianMembership(j, Center = 1, StnDev = RC/3) → 0 when it covers approximately 99.73% of the area underneath the Gaussian curve at distance j ≈ ± 3 · StnDev = ± 3 · RC/3 = ± RC from its Center = 1. If j = 1, then GaussianMembership(j, Center = 1, StnDev = RC/3) = 1. It is trivial to prove that Equations (5–7) satisfy the aforementioned requirements (a) to (d). By means of a numeric example, it can be shown that requirement (e) is satisfied too. For example, estimated from the binary relationship instantiated at step 8 of Table 5, CVPAI2 = (1/6)* (1 + 1 + 1 + 1 + 1 + exp(−0.5*(3−1)^2/(3/3)^2)) = (1/6) * (5 + 0.1353) = 0.8558. Intuitively closer to 1 than 0, this CVPAI2 value shows that the harmonization between the two test and reference nominal variables is (fuzzy) "high" (>0.8) in Table 5.

To appreciate the conceptual difference between the CVPAI1 and CVPAI2 formulations maximized by different distribution of "correct" entry-pairs in a binary relationship R: A $\Rightarrow$ B $\subseteq$ A × B, see Figure 18, let us compare a test vocabulary A of color names, such as SIAM's, see Figure 6, with a reference dictionary B of LC class names, such as the USGS NLCD's, see Table 2. In terms of capability of color names to discriminate LC class names, the ideal test-to-reference binary relationship is a function where one color name matches with only one reference LC class. On the other hand, the color attribute of a real-world LC class can be typically linked to one or more discrete color names, see Table 3. In this realistic example the range of change of an estimated CVPAI2 value would be (0, 1], up to its maximum value equal to 1, while the range of change of the CVPAI1 formulation proposed in (Baraldi et al., 2014) would be (0, 1), below its maximum value equal to 1, see Figure 18.

Another example where the difference between the CVPAI1 and CVPAI2 formulations is highlighted is when the test dictionary A is a specialized version of the reference dictionary B, according to a parent–child relationship. For example, a test taxonomy of LC classes is A = LegendOfObjectClassNames_A = {LC class *"Dark-tone bare soil"*, LC class *"Light-tone bare soil"*, LC class *"Deciduous Forest"*, LC class *"Evergreen Forest"*} and a reference LC class taxonomy is B = LegendOfObjectClassNames_B = {LC class *"Bare soil"*, LC class *"Forest"*}. Based on our prior knowledge-based understanding of these two semantic dictionaries A and B, a reasonable binary relationship can be considered R: A ⇒ B = {(LC class *"Dark-tone bare soil"*, LC class *"Bare soil"*); (LC class *"Light-tone bare soil"*, LC class *"Bare soil"*); (LC class *"Deciduous Forest"*, LC class *"Forest"*); (LC class *"Evergreen Forest"*, LC class *"Forest"*)}. In this case, the CVPAI1 formulation scores below its maximum, i.e., CVPAI1 ∈ (0, 1), while the expected CVPAI2 value would score maximum, i.e., CVPAI2 = 1, meaning that the two vocabularies are harmonized because one is the specialization of the other, featuring a parent–child relationship.

These two examples illustrate the intuitive meaning and practical use of the normalized quantitative indicator CVPAI2 ∈ [0, 1] in an EO-IUS implementation based on a convergence-of-evidence approach, in agreement with Equation (3). When the semantic information gap from sub-symbolic sensory data to a symbolic set B = LegendOfObjectClassNames is filled by an EO-IUS starting from a static color naming first stage, provided with a semi-symbolic set A = VocabularyOfColorNames, if the binary relationship R: A ⇒ B ⊆ A × B features a degree of association CVPAI2 ∈ [0, 1], then (1—CVPAI2) ∈ [0, 1] is the semantic information gap from sub-symbolic sensory data to the symbolic LegendOfObjectClassNames left to be filled by further stages in the hierarchical EO-IUS pipeline, where spatial information is masked by first-stage color names. If CVPAI2 = 1, then secondary color information discretized by set A = VocabularyOfColorNames suffices to detect target set B = LegendOfObjectClassNames with no further need to investigate primary spatial information in a hierarchical convergence-of-evidence image classification approach, refer to Equation (3).

6. Conclusions

To pursue the GEO-CEOS visionary goal of a GEOSS implementation plan for years 2005–2015 not-yet accomplished by the RS community, this interdisciplinary work aimed at filling an analytic and pragmatic information gap from EO image big data to systematic ESA EO Level 2 product generation at the ground segment, never achieved to date by any EO data provider and postulated as necessary not sufficient pre-condition to GEOSS development. For the sake of readability this paper is split into two, the present Part 1 - Theory and the following Part 2 - Validation.

The original contribution of the present Part 1 is fourfold. A first lesson was learned from published works on prior knowledge-based MS reflectance space hyperpolyhedralization into static (non-adaptive-to-data) color names, according to the principle of color naming discovered by linguistics and investigated by CV in the realm of cognitive science, see Figure 13. In color naming, a static vocabulary of sub-symbolic color names is equivalent to a set of mutually exclusive and totally exhaustive (hyper)polyhedra, neither necessarily convex nor connected, in a color data (hyper)cube. It was observed that well-established RS practices, such as 1D image analysis based on supervised data learning algorithms, where dominant spatial information is neglected in favor of secondary color information, and thematic map quality assessment where test and reference map legends are required to coincide, can become malpractices when an *a priori* dictionary of static color names is employed for MS image classification based on a convergence-of-evidence approach, such as in Bayesian naïve classification, see Equation (3). When test and reference thematic map legends A and B are the same, the binary relationship R: A ⇒ B ⊆ A × B becomes a bijective function (both 1–1 and onto) and the main diagonal of the 2-fold Cartesian product A × B guides the interpretation process of a bivariate frequency table, BIVRFTAB = FrequencyCount(A × B), equal to a square and sorted confusion matrix, CMTRX. This constraint makes a CMTRX, whose input categorical variables A and B coincide, intuitive to understand and more familiar to RS practitioners. Noteworthy, inequality R: A ⇒ B ⊆ A × B ≠ FrequencyCount(A × B) = BIVRFTAB always holds true, where one instance of the binary relationship R guides the interpretation process of the two-way contingency table BIVRFTB. Quite surprisingly, the non-coincident assumption between an *a priori* vocabulary A of static sub-symbolic color names in the measurement color space and a user- and application-dependent legend B of symbolic classes of real-world objects in the scene-domain, where inequality A ≠ B always holds, appears somehow difficult to acknowledge by relevant portions of the RS community, in contrast with common sense, see Table 3.

Second, Equation (3) was proposed as an analytic expression of a biologically plausible hybrid (combine deductive and inductive) CV system suitable for convergence of color and spatial evidence, in agreement with a Bayesian approach to vision proposed by Marr (1982), with the principle of statistic stratification well known in statistics and with the divide-and-conquer (*divide-et-impera*) problem solving criterion widely adopted in structured engineering. In compliance with common sense, see Table 3, Equation (3) shows that a static color naming first stage can be employed for stratification purposes of further spatial-context sensitive image classification stages. In the static color naming first stage, a binary relationship R: A ⇒ B ⊆ A × B from a vocabulary A of general-purpose static color names in the MS color space to a taxonomy B of LC class names in the 4D spatio-temporal scene-domain, such as the standard FAO LCCS taxonomy shown in Figure 3, can be established by human experts based on top-down prior beliefs, if any, in combination with bottom-up evidence inferred from new data, as described in Table 5. Once established and community-agreed upon, a binary relationship R: A ⇒ B ⊆ A × B from a vocabulary A of static color names in a MS color space to a standard legend of LC classes in the 4D spatio-temporal scene-domain becomes equivalent to an *a priori* knowledge base in a Bayesian updating framework, where Bayesian inference is applied iteratively: after observing some evidence, the resulting posterior probability can be treated as prior probability. For example, once community-agreed upon, Table 3 becomes equivalent to an *a priori* knowledge base available in addition to sensory data, and a new posterior probability can be computed from new data, e.g., to pursue image classification in agreement with Equation (3).

Third, for best practice a hybrid eight-step protocol, sketched in Table 5, was proposed to infer a binary relationship, R: A ⇒ B ⊆ A × B, from categorical variable A to categorical variable B estimated from the same population, where codebooks A and B can differ in cardinality, semantics or in the order of presentation of codewords. This protocol streamlines a hybrid combination of deductive prior beliefs by human domain experts with inductive evidence from data. It is of practical use because identification of a binary relationship R: A ⇒ B is mandatory to guide the interpretation process of a bivariate frequency table, BIVRFTAB = FrequencyCount(A × B), where A ≠ B in general. Only if A = B then BIVRFTAB becomes equal to the well-known square and sorted CMTRX, where the main diagonal guides the interpretation process.

Fourth, in compliance with the GEO-CEOS QA4EO *Val* guidelines, two original and alternative formulations, CVPAI2(R: A ⇒ B ⊆ A × B) ∈ [0, 1] and CVPAI3(R: A ⇒ B ⊆ A × B) ∈ [0, 1], were proposed as categorical variable-pair degree of association (harmonization) in a binary relationship, R: A ⇒ B, from categorical variable A to categorical variable B estimated from the same population, where A ≠ B in general. When CVPAI2 or CVPAI3 is maximum, equal to 1, then the two categorical variables A and B are considered fully harmonized.

To comply with the GEO-CEOS QA4EO *Cal/Val* requirements, the subsequent Part 2 of this paper presents and discusses a GEO-CEOS stage 4 *Val* of the annual SIAM-WELD map time-series for the years 2006 to 2009 in comparison with the reference USGS NLCD 2006 map, based on an original protocol for wall-to-wall inter-map comparison without sampling where the test and reference maps feature the same spatial resolution and spatial extent, but whose legends are not the same and must be harmonized.

Abbreviations

AI:	Artificial Intelligence
ATCOR:	Atmospheric/Topographic Correction commercial softare product
AVHRR:	Advanced Very High Resolution Radiometer

BC:	Basic Color
BIVRFTAB:	Bivariate Frequency Table
Cal:	Calibration
Cal/Val:	Calibration and Validation
CCRS:	Canada Centre for Remote Sensing
CEOS:	Committee on Earth Observation Satellites
CLC:	CORINE Land Cover (taxonomy)
CMTRX:	Confusion Matrix
CNES:	Centre national d'études spatiales
CONUS:	Conterminous U.S.
CORINE:	Coordination of Information on the Environment
CV:	Computer Vision
CVPAI:	Categorical Variable Pair Association Index (in range [0, 1])
DCNN:	Deep Convolutional Neural Network
DLR:	Deutsches Zentrum für Luft- und Raumfahrt (German Aerospace Center)
DN:	Digital Number
DP:	Dichotomous Phase (in the FAO LCCS taxonomy)
DRIP:	Data-Rich, Information-Poor (scenario)
EDC:	EROS Data Center
EO-IU:	EO Image Understanding
EO-IUS:	EO-IU System
EPA:	Environmental Protection Agency
EROS:	Earth Resources Observation Systems
ESA:	European Space Agency
FAO:	Food and Agriculture Organization
GEO:	Intergovernmental Group on Earth Observations
GEOSS:	Global EO System of Systems
GIGO:	Garbage In, Garbage Out principle of error propagation
GIS:	Geographic Information System
GIScience:	Geographic Information Science
GUI:	Graphic User Interface
HS:	Hyper-Spectral
IGBP:	International Global Biosphere Programme
IUS:	Image Understanding System
LC:	Land Cover
LCC:	Land Cover Change
LCCS:	Land Cover Classification System (taxonomy)
LCLU:	Land Cover Land Use
LEDAPS:	Landsat Ecosystem Disturbance Adaptive Processing System
mDMI:	Minimally Dependent and Maximally Informative (set of quality indicators)
MHP:	Modular Hierarchical Phase (in the FAO LCCS taxonomy)

MIR:	Medium InfraRed
MODIS:	Moderate Resolution Imaging Spectroradiometer
MS:	Multi-Spectral
NASA:	National Aeronautics and Space Administration
NIR:	Near InfraRed
NLCD:	National Land Cover Data
NOAA:	National Oceanic and Atmospheric Administration
OA:	Overall Accuracy
OAMTRX:	Overlapping Area Matrix
OBIA:	Object-Based Image Analysis
ODGIS:	Ontology-Driven GIS
OGC:	Open Geospatial Consortium
OP:	Outcome (product) and Process
OP-Q^2I:	Outcome and Process Quantitative Quality Index
QA:	Quality Assurance
QA4EO:	Quality Accuracy Framework for Earth Observation
Q^2I:	Quantitative Quality Indicator
R&D:	Research and Development
RGB:	monitor-typical Red-Green-Blue data cube
RMSE:	Root Mean Square Error
RS:	Remote Sensing
SCM:	Scene Classification Map
SEN2COR:	Sentinel 2 (atmospheric) Correction Prototype Processor
SIAM™ :	Satellite Image Automatic Mapper™
STRATCOR:	Stratified Topographic Correction
SS:	Super-Spectral
SURF:	Surface Reflectance
TIR:	Thermal InfraRed
TOA:	Top-Of-Atmosphere
TOARF:	TOA Reflectance
TM (superscript):	(non-registered) Trademark
UML:	Unified Modeling Language
USGS:	US Geological Survey
Val:	Validation
VQ:	Vector Quantization
WELD:	Web-Enabled Landsat Data set
WGCV:	Working Group on Calibration and Validation

Acknowledgments

Prof. Ralph Maughan, Idaho State University, is kindly acknowledged for his contribution as active conservationist and for his willingness to share his invaluable photo archive with the scientific community as well as the general public. Andrea Baraldi thanks Prof. Raphael Capurro, Hochschule der Medien, Germany, and Prof. Christopher Justice, Chair of the Department of Geographical Sciences, University of Maryland, College Park, MD, for their support. Above all, the authors acknowledge the fundamental contribution of Prof. Luigi Boschetti, currently at the Department of Forest, Rangeland and Fire Sciences, University of Idaho, Moscow, Idaho, who conducted by independent means all experiments whose results are proposed in this validation paper. The authors also wish to thank the Editor-in-Chief, Associate Editor and reviewers for their competence, patience and willingness to help.

Funding

To accomplish this work Andrea Baraldi was supported in part by the National Aeronautics and Space Administration (NASA) under Grant No. NNX07AV19G issued through the Earth Science Division of the Science Mission Directorate. Andrea Baraldi, Dirk Tiede and Stefan Lang were supported in part by the Austrian Science Fund (FWF) through the Doctoral College GIScience (DK W1237-N23).

Competing interests

As a source of potential competing interests of professional or financial nature, coauthor Andrea Baraldi reports he is the sole developer and intellectual property right (IPR) owner of the Satellite Image Automatic Mapper[TM] (non-registered trademark) computer program validated by an independent third-party in this research and licensed by the one-man-company Baraldi Consultancy in Remote Sensing (siam.andreabaraldi.com) to academia, public institutions and private companies, eventually free-of-charge. Throughout his scientific career Andrea Baraldi has put in place approved plans for managing any potential conflict arising from his IPR ownership of the SIAM[TM] computer software.

Author details

Andrea Baraldi[1,2,3,4]
E-mail: andrea6311@gmail.com
ORCID ID: http://orcid.org/0000-0001-5196-9944
Michael Laurence Humber[2]
E-mail: mhumber@umd.edu
Dirk Tiede[3]
E-mail: dirk.tiede@sbg.ac.at
ORCID ID: http://orcid.org/0000-0002-5473-3344
Stefan Lang[3]
E-mail: stefan.lang@sbg.ac.at
[1] Department of Agricultural Sciences, University of Naples Federico II, Portici, Italy.
[2] Department of Geographical Sciences, University of Maryland, College Park, MD, USA.
[3] Department of Geoinformatics – Z_GIS, University of Salzburg, Salzburg, Austria.
[4] Italian Space Agency (ASI), Rome, Italy.

Cover image

Source: Satellite Image Automatic Mapper (SIAM) multi-level map generated automatically, without human-machine interaction, and in near real-time, specifically, in linear time complexity with image size, from the 30 m resolution annual Web-Enabled Landsat Data (WELD) image composite for the year 2006 of the conterminous U.S. (CONUS), radiometrically calibrated into top-of-atmosphere reflectance (TOARF) values. The multi-level map's legend, shown at bottom left, is the SIAM intermediate discretization level, consisting of 48 basic color (BC) names reassembled into 19 spectral macro-categories by an independent human expert, according to a proposed hybrid (combined deductive and inductive) eight-step protocol for identification of a categorical variable-pair binary relationship, from a vocabulary of BC names to a dictionary of land cover (LC) class

names. No discrete and finite vocabulary of color names, such as SIAM's, equivalent to a set of mutually exclusive and totally exhaustive (hyper)polyhedra, neither necessarily convex nor connected, in a MS reflectance (hyper)space, should ever be confused with a symbolic (semantic) taxonomy of LC class names in the 4D geospace-time scene-domain. Black lines across the SIAM-WELD 2006 color map represent the boundaries of the 86 Environmental Protection Agency (EPA) Level III ecoregions of the CONUS, suitable for regional-scale statistical stratification required to intercept geospatial non-stationary statistics, typically lost when a global spatial average, e.g., at continental spatial extent, is superimposed on the local computational processes.

Notes

Figure 2. Graphical representation of a dependence relationship *part-of*, denoted with symbol "→" pointing from the supplier to the client in agreement with the standard Unified Modeling Language (UML) for graphical modeling of object-oriented software (Fowler, 2003), between computer vision (CV), whose special case is EO image understanding (EO-IU) in operating mode, where relationship *subset-of*, denoted with symbol "⊃" meaning specialization with inheritance from the superset to the subset, holds true, and a Global Earth Observation System of Systems (GEOSS) (GEO, 2005), such that "NASA EO Level 2 product → ESA EO Level 2 product ⊂ EO-IU in operating mode ⊂ CV → GEOSS". Synonym of 4D spatio-temporal scene from (2D) image reconstruction and understanding, vision is acknowledged to be a cognitive problem very difficult to solve because: (i) non-polynomial (NP)-hard in computational complexity (Frintrop, 2011; Tsotsos, 1990), (ii) inherently ill-posed in the Hadamard sense, as it is affected by: (I) a 4D-to-2D data dimensionality reduction from the scene- to the image-domain, e.g., responsible of occlusion phenomena, and (II) a semantic information gap from ever-varying sub-symbolic sensory data (sensations) in the image-domain to stable symbolic percepts in the modeled world (mental world, world ontology, world model) (Fonseca et al., 2002; Laurini & Thompson, 1992; Matsuyama & Hwang, 1990; Sonka et al., 1994; Sowa, 2000). A NASA Earth observation (EO) Level 2 product, defined as "a data-derived geophysical variable at the same resolution and location as Level 1 source data" (NASA 2016b), is *part-of* the ESA EO Level 2 product, defined as follows (ESA, 2015; DLR & VEGA, 2011): (a) a single-date multi-spectral (MS) image whose digital numbers (DNs) are radiometrically calibrated into surface reflectance (SURF) values corrected for atmospheric, adjacency and topographic effects, stacked with (b) its data-derived general-purpose, user- and application-independent scene classification map (SCM), whose thematic map legend includes quality layers cloud and cloud-shadow (CNES, 2015). Working hypothesis "NASA EO Level 2 product → ESA EO Level 2 ⊂ EO-IU in operating mode → GEOSS" postulates that no GEOSS can exist if the necessary not sufficient pre-condition of systematic ESA EO Level 2 product generation is accomplished in advance as the mandatory first step in a hierarchical EO-IU workflow for scene-from-image reconstruction and understanding in operating mode.

Figure 10. (a) Sentinel-2A (S2A) MSI Level-1C image radiometrically calibrated (*Cal*) into top-of-atmosphere reflectance (TOARF) values by the ESA data provider, depicting an Earth surface located south of the city of Salzburg, Austria. The city area is visible around the middle of the image upper boundary (Lat-long coordinates: 47°48'25.0"N 13°02'43.6"E). Acquired on 2015-09-11. Spatial resolution: 10 m. Image size: 110 × 110 km. TOARF values in range [0, 1] are byte-coded in range {0, 255}. The MS image is shown in RGB false colors, where monitor channel R = Medium InfraRed (MIR) = S2 Band 11, channel G = Near IR (NIR) = S2 Band 8, channel B = (visible) Blue = S2 Band 2. No histogram stretching is applied for visualization purposes. (b) L-SIAM color map at coarse color granularity, consisting of 18 spectral categories depicted in pseudo colors shown in the map legend. Coarse-granularity color categories are generated by merging color hyperpolyhedra at fine color granularity, according to pre-defined parent‾ child relationships, refer to Table 1. (c) L-SIAM color map at fine color granularity, consisting of 96 spectral categories depicted in pseudo colors shown in the map legend, refer to Table 1. (d) Superpixelwise-constant approximation of the input image ("image-object mean view") generated from the L-SIAM's 96 color map at fine granularity. Depicted in false colors: R = MIR = S2 Band 11, G = NIR = S2 Band 8, B = (visible) Blue = S2 Band 2. Spatial resolution: 10 m. No histogram stretching is applied for visualization purposes. (e) 8-adjacency cross-aura contour map in range {0, 8} automatically generated from the L-SIAM's 96 color map at fine granularity. It shows contours of connected sets of pixels featuring the same color label. These connected-components are also called image-objects, segments or superpixels. (f) Per-pixel scalar difference between the input MS image shown in (a) and the superpixelwise-constant MS image reconstruction shown in (d). This scalar difference is computed as the per-pixel Root Mean Square Error (RMSE) in range {0, 255}, which is the same domain of change of input byte-coded pixel values. The RMSE is a well-known vector quantization (VQ) error estimate to be minimized. Image-wide RMSE statistics: Min = 0, Max = 130, Mean = 2.60, Stdev = 3.45. Histogram stretching is applied for visualization purposes. The highest RMSE values are located in pixels belonging to segments labeled as snow and cloud, which tend to be larger in size and whose class-specific within-segment variance tends to be "high".

References

Achanta, R., Shaji, A., Smith, K., Lucchi, A., Fua, P., & Susstrunk, S. (2011). SLIC superpixels compared to state-of-the-art superpixel methods. *IEEE Transactions Pattern Analysis Machine Intelligent, 6* (1), 1–8.

Ackerman, S. A., Strabala, K. I., Menzel, W. P., Frey, R. A., Moeller, C. C., & Gumley, L. E. (1998). Discriminating clear sky from clouds with MODIS. *Journal of Geophysical Research, 103*(32), 141–157. doi:10.1029/1998JD200032

Adams, J. B., Donald, E. S., Kapos, V., Almeida Filho, R., Roberts, D. A., Smith, M. O., & Gillespie, A. R. (1995). Classification of multispectral images based on fractions of endmembers: Application to land-cover change in the Brazilian Amazon. *Remote Sensing Environment, 52*, 137–154. doi:10.1016/0034-4257 (94)00098-8

Ahlqvist, O. (2005). Using uncertain conceptual spaces to translate between land cover categories. *Intrenational Journal of Geographical Information Science, 19*, 831–857.

Ahlqvist, O. (2008). In search of classification that supports the dynamics of science: The FAO Land Cover Classification System and proposed modifications. *Environment and Planning B: Planning and Design, 35*, 169–186. doi:10.1068/b3344

Arvor, D., Madiela, B. D., & Corpetti, T. (2016). Semantic pre-classification of vegetation gradient based on linearly unmixed Landsat time series. In *Geoscience and Remote Sensing Symposium (IGARSS), IEEE International* (pp. 4422–4425).

Baraldi, A. (2009). Impact of radiometric calibration and specifications of spaceborne optical imaging sensors on the development of operational automatic remote sensing image understanding systems. *IEEE Journal of Selected Topics in Applied Earth Observations and Remote Sensing, 2*(2), 104–134. doi:10.1109/JSTARS.2009.2023801

Baraldi, A. (2011). Fuzzification of a crisp near-real-time operational automatic spectral-rule-based decision-tree preliminary classifier of multisource multispectral remotely sensed images. *IEEE Transactions on Geoscience and Remote Sensing, 49*, 2113–2134. doi:10.1109/TGRS.2010.2091137

Baraldi, A. (2015). "Automatic Spatial Context-Sensitive Cloud/Cloud-Shadow Detection in Multi-Source Multi-Spectral Earth Observation Images – AutoCloud+," Invitation to tender ESA/AO/1-8373/15/I-NB – "VAE: Next Generation EO-based Information Services", DOI: 10.13140/RG.2.2.34162.71363. arXiv: 1701.04256. Retrieved Jan., 8, 2017 from https://arxiv.org/ftp/arxiv/papers/1701/1701.04256.pdf

Baraldi, A., 2017. "Pre-processing, classification and semantic querying of large-scale Earth observation spaceborne/airborne/terrestrial image databases: Process and product innovations". Ph.D. dissertation Agricultural and Food Sciences, University of Naples "Federico II", Department of Agricultural Sciences, Italy. Ph.D. defense: 16 May 2017. Retrieved January 30, 2018, from https://www.researchgate.net/publication/317333100_Pre-processing_classification_and_semantic_querying_of_large-scale_Earth_observation_spaceborneairborneterrestrial_image_databases_Process_and_product_innovations doi:10.13140/RG.2.2.25510.52808

Baraldi, A., & Boschetti, L. (2012a). Operational automatic remote sensing image understanding systems: Beyond Geographic Object-Based and Object-Oriented Image Analysis (GEOBIA/GEOOIA) - Part 1: Introduction. *Remote Sensing, 4*, 2694–2735. doi:10.3390/rs4092694

Baraldi, A., & Boschetti, L. (2012b). Operational automatic remote sensing image understanding systems: Beyond Geographic Object-Based and Object-Oriented Image Analysis (GEOBIA/GEOOIA) - Part 2: Novel system architecture, information/knowledge representation, algorithm design and implementation. *Remote Sensing, 4*, 2768–2817.

Baraldi, A., Boschetti, L., & Humber, M. (2014). Probability sampling protocol for thematic and spatial quality assessment of classification maps generated from spaceborne/airborne very high resolution images. *IEEE Transactions on Geoscience and Remote Sensing, 52*(1), 701–760. doi:10.1109/TGRS.2013.2243739

Baraldi, A., Bruzzone, L., & Blonda, P. (2005). Quality assessment of classification and cluster maps without ground truth knowledge. *IEEE Transactions on Geoscience and Remote Sensing, 43*, 857–873. doi:10.1109/TGRS.2004.843074

Baraldi, A., Durieux, L., Simonetti, D., Conchedda, G., Holecz, F., & Blonda, P. (2010a). Automatic spectral rule-based preliminary classification of radiometrically calibrated SPOT-4/-5/IRS, AVHRR/ MSG,AATSR, IKONOS/QuickBird/OrbView/GeoEye and DMC/SPOT-1/-2 imagery- Part I: System design and implementation. *IEEE Transactions on Geoscience and Remote Sensing, 48*, 1299–1325. doi:10.1109/TGRS.2009.2032457

Baraldi, A., Durieux, L., Simonetti, D., Conchedda, G., Holecz, F., & Blonda, P. (2010b). "Automatic spectral rule-based preliminary classification of radiometrically calibrated SPOT-4/-5/IRS, AVHRR/ MSG,AATSR, IKONOS/QuickBird/OrbView/GeoEye and DMC/SPOT-1/-2 imagery- Part II: Classification accuracy assessment," IEEE Trans. *IEEE Transactions on Geoscience and Remote Sensing, 48*, 1326–1354. doi:10.1109/TGRS.2009.2032064

Baraldi, A., Gironda, M., & Simonetti, D. (2010c). Operational two-stage stratified topographic correction of spaceborne multispectral imagery employing an automatic spectral-rule-based decision-tree preliminary classifier. *IEEE Transactions Geoscience Remote Sensing, 48*(1), 112–146. doi:10.1109/TGRS.2009.2028017

Baraldi, A. (2015). "Automatic Spatial Context-Sensitive Cloud/Cloud-Shadow Detection in Multi-Source Multi-Spectral Earth Observation Images – AutoCloud+," Invitation to tender ESA/AO/1-8373/15/I-NB – "VAE: Next Generation EO-based Information Services", doi: 10.13140/RG.2.2.34162.71363. arXiv: 1701.04256. Retrieved Jan., 8, 2017 from https://arxiv.org/ftp/arxiv/papers/1701/1701.04256.pdf

Baraldi, A. (2015). "Automatic Spatial Context-Sensitive Cloud/Cloud-Shadow Detection in Multi-Source Multi-Spectral Earth Observation Images – AutoCloud+," Invitation to tender ESA/AO/1-8373/15/I-NB – "VAE: Next Generation EO-based Information Services", doi: 10.13140/RG.2.2.34162.71363. arXiv: 1701.04256. Retrieved Jan., 8, 2017 from https://arxiv.org/ftp/arxiv/papers/1701/1701.04256.pdf

Baraldi, A., & Humber, M. (2015). Quality assessment of pre-classification maps generated from spaceborne/airborne multi-spectral images by the Satellite Image Automatic Mapper™ and Atmospheric/Topographic Correction™-Spectral Classification software products: Part 1 – Theory. *IEEE Journal of Selected Topics Applications Earth Observation Remote Sensing, 8*(3), 1307–1329.

Baraldi, A., Humber, M., & Boschetti, L. (2013). Quality assessment of pre-classification maps generated from spaceborne/airborne multi-spectral images by the Satellite Image Automatic Mapper™ and Atmospheric/Topographic Correction™-Spectral Classification software products: Part 2 – Experimental results. *Remote Sensing, 5*, 5209–5264. doi:10.3390/rs5105209

Baraldi, A., Puzzolo, V., Blonda, P., Bruzzone, L., & Tarantino, C. (2006). Automatic spectral rule-based preliminary mapping of calibrated Landsat TM and ETM+ images. *IEEE Transactions on Geoscience and Remote Sensing, 44*, 2563–2586. doi:10.1109/TGRS.2006.874140

Baraldi, A., & Soares, J. V. B. (2017). Multi-objective software suite of two-dimensional shape descriptors for object-based image analysis, subjects: computer vision and pattern recognition (cs.CV). arXiv:1701.01941. Retrieved January, 8, 2017 from https://arxiv.org/ftp/arxiv/papers/1701/1701.01941.pdf

Baraldi, A., Tiede, D., Sudmanns, M., Belgiu, M., & Lang, S. (2016, September 14–16). Automated near real-time Earth observation Level 2 product generation for semantic querying. *GEOBIA*. Enschede, The Netherlands: University of Twente Faculty of Geo-Information and Earth Observation (ITC).

Baraldi, A., Wassenaar, T., & Kay, S. (2010d). Operational performance of an automatic preliminary spectral rule-based decision-tree classifier of spaceborne very high resolution optical images. *IEEE Transactions on Geoscience and Remote Sensing, 48*, 3482–3502. doi:10.1109/TGRS.2010.2046741

Beauchemin, M., & Thomson, K. (1997). The evaluation of segmentation results and the overlapping area matrix. *International Journal of Remote Sensing, 18*, 3895–3899. doi:10.1080/014311697216720

Belward, A. (Ed.). (1996). *The IGBP-DIS global 1Km land cover data set "DISCover": Proposal and implementation plans*. IGBP-DIS working paper 13. Ispra, Varese, Italy: International Geosphere Biosphere Programme. European Commission Joint Research Center.

Benavente, R., Vanrell, M., & Baldrich, R. (2008). Parametric fuzzy sets for automatic color naming. *Journal of the Optical Society of America A, 25*, 2582–2593. doi:10.1364/JOSAA.25.002582

Berlin, B., & Kay, P. (1969). *Basic color terms: Their universality and evolution*. Berkeley: University of California.

Bernus, P., & Noran, O. (2017). Data rich – but information poor. In L. Camarinha-Matos, H. Afsarmanesh, & R. Fornasiero (Eds), *Collaboration in a Data-Rich World: PRO-VE 2017* (Vol. 506, pp. 206–214). IFIP Advances in Information and Communication Technology.

Bishop, C. M. (1995). *Neural networks for pattern recognition*. Oxford, UK: Clarendon.

Bishop, M. P., & Colby, J. D. (2002). Anisotropic reflectance correction of SPOT-3 HRV imagery. *International Journal of Remote Sensing, 23*(10), 2125–2131. doi:10.1080/01431160110097231

Bishop, M. P., Shroder, J. F., & Colby, J. D. (2003). Remote sensing and geomorphometry for studying relief production in high mountains. *Geomorphology, 55*(1-4), 345–361. doi:10.1016/S0169-555X(03)00149-1

Bishr, Y. (1998). Overcoming the semantic and other barriers to GIS interoperability. *International Journal of Geographical Information Science, 12*, 299–314. doi:10.1080/136588198241806

Blaschke, T., Hay, G. J., Kelly, M., Lang, S., Hofmann, P., Addink, E., ... Tiede, D. (2014). Geographic object-based image analysis - towards a new paradigm. *ISPRS Journal of Photogrammetry and Remote Sensing, 87*, 180–191. doi:10.1016/j.isprsjprs.2013.09.014

Boschetti, L., Flasse, S. P., & Brivio, P. A. (2004). Analysis of the conflict between omission and commission in low spatial resolution dichotomic thematic products: The Pareto boundary. *Remote Sensing of Environment, 91*, 280–292. doi:10.1016/j.rse.2004.02.015

Boschetti, L., Roy, D. P., Justice, C. O., & Humber, M. L. (2015). MODIS–Landsat fusion for large area 30 m burned area mapping. *Remote Sensing of Environment, 161*, 27–42. doi:10.1016/j.rse.2015.01.022

Bossard, M., Feranec, J., & Otahel, J. (2000). *CORINE land cover technical guide – Addendum 2000* (Technical report No 40). European Environment Agency.

Capurro, R., & Hjørland, B. (2003). The concept of information. *Annual Review of Information Science and*

Technology, 37, 343–411. doi:10.1002/aris.1440370109

Chavez, P. (1988). An improved dark-object subtraction technique for atmospheric scattering correction of multispectral data. Remote Sensing of Environment, 24, 459–479. doi:10.1016/0034-4257(88)90019-3

Cherkassky, V., & Mulier, F. (1998). Learning from data: Concepts, theory, and methods. New York, NY: Wiley.

Cimpoi, M., Maji, S., Kokkinos, I., & Vedaldi, A. (2014). Deep filter banks for texture recognition, description, and segmentation. CoRR. abs/1411.6836.

CNES. (2015). Venμs satellite sensor level 2 product. Retrieved January, 5, 2016 from https://venus.cnes.fr/en/VENUS/prod_l2.htm

Congalton, R. G., & Green, K. (1999). Assessing the accuracy of remotely sensed data. Boca Raton, FL: Lewis Publishers.

Couclelis, H. (2010). Ontologies of geographic information. International Journal of Geographical Information Science, 24(12), 1785–1809. doi:10.1080/13658816.2010.484392

D'Elia, S. (2012). Personal communication. European Space Agency.

Despini, F., Teggi, S., & Baraldi, A. (2014, September 22). Methods and metrics for the assessment of pan-sharpening algorithms. In L. Bruzzone, J. A. Benediktsson, & F. Bovolo, (Eds.), SPIE Proceedings, Vol. 9244: Image and Signal Processing for Remote Sensing XX. Amsterdam, Netherlands.

Deutsches Zentrum für Luft- und Raumfahrt e.V. (DLR) and VEGA Technologies. (2011). Sentinel-2 MSI – level 2A products algorithm theoretical basis document (Document S2PAD-ATBD-0001). European Space Agency.

DiCarlo, J. (2017). The science of natural intelligence: Reverse engineering primate visual perception. Keynote, CVPR17 Conference. Retrieved January, 30, 2018 from https://www.youtube.com/watch?v=ilbbVkIhMgo

Di Gregorio, A., & Jansen, L. (2000). Land Cover Classification System (LCCS): Classification concepts and user manual. Rome, Italy: FAO, FAO Corporate Document Repository. Retrieved February, 10, 2015 from, http://www.fao.org/DOCREP/003/X0596E/X0596e00.htm

DigitalGlobe (2016). Automated Land Cover Classification. Retrieved November, 13, 2016 from http://gbdxdocs.digitalglobe.com/docs/automated-land-cover-classification

Dillencourt, M. B., Samet, H., & Tamminen, M. (1992). A general approach to connected component labeling for arbitrary image representations. Journal of the ACM, 39, 253–280. doi:10.1145/128749.128750

Dorigo, W., Richter, R., Baret, F., Bamler, R., & Wagner, W. (2009). Enhanced automated canopy characterization from hyperspectral data by a novel two step radiative transfer model inversion approach. Remote Sensing, 1, 1139–1170. doi:10.3390/rs1041139

Duke Center for Instructional Technology. (2016). Measurement: Process and outcome indicators. Retrieved June, 20, 2016 from http://patientsafetyed.duhs.duke.edu/module_a/measurement/measurement.html

Dumitru, C. O., Cui, S., Schwarz, G., & Datcu, M. (2015). Information content of very-high-resolution SAR images: Semantics, geospatial context, and ontologies. IEEE Journal of Selected Topics in Applied Earth Observations and Remote Sensing, 8(4), 1635–1650. doi:10.1109/JSTARS.2014.2363595

Elkan, C. (2003). Using the triangle inequality to accelerate k-means. International Conference Machine Learning.

Environmental Protection Agency (EPA). (2007). "Definitions" in Multi-Resolution Land Characteristics Consortium (MRLC). Retrieved November, 13, 2013 from http://www.epa.gov/mrlc/definitions.html#2001

Environmental Protection Agency (EPA). (2013). Western ecology division. Retrieved November, 13, 2013 from http://www.epa.gov/wed/pages/ecoregions.htm

Etzioni, O. (2017). "What shortcomings do you see with deep learning?" Quora. Retrieved January, 8, 2018 from https://www.quora.com/What-shortcomings-do-you-see-with-deep-learning

European Space Agency (ESA). (2015). Sentinel-2 user handbook, standard document, issue 1 rev 2.

Feldman, J. (2013). The neural binding problem(s). Cognition Neurodynamic, 7, 1–11. doi:10.1007/s11571-012-9219-8

Feng, -C.-C., & Flewelling, D. M. (2004). Assessment of semantic similarity between land use/land cover classification systems. Computers, Environment and Urban Systems, 28, 229–246. doi:10.1016/S0198-9715(03)00020-6

Foga, S., Scaramuzza, P., Guo, S., Zhu, Z., Dilley, R., Jr, Beckmann, T., ... Laue, B. (2017). Cloud detection algorithm comparison and validation for operational Landsat data products. Remote Sensing of Environment, 194, 379–390. doi:10.1016/j.rse.2017.03.026

Fonseca, F., Egenhofer, M., Agouris, P., & Camara, G. (2002). Using ontologies for integrated geographic information systems. Transactions in GIS, 6, 231–257. doi:10.1111/1467-9671.00109

Fowler, M. (2003). UML Distilled (3rd ed.). Boston, MS: Addison-Wesley.

Frintrop, S. (2011). Computational visual attention. In A. A. Salah & T. Gevers (Eds.), Computer analysis of human behavior, advances in pattern recognition. Springer.

Fritzke, B. (1997a). Some competitive learning methods. Retrieved March, 17, 2015 from http://www.demogng.de/JavaPaper/t.html

Fritzke, B. (1997b). The LBG-U method for vector quantization - An improvement over LBG inspired from neural networks. Neural Processing Letters, 5(1), 35–45. doi:10.1023/A:1009653226428

GeoTerraImage. (2015). Provincial and national land cover 30m. Retrieved September, 22, 2015 from http://www.geoterraimage.com/productslandcover.php

Gevers, T., Gijsenij, A., van de Weijer, J., & Geusebroek, J. M. (2012). Color in computer vision. Hoboken, NJ: Wiley.

Ghahramani, Z. (2011). Bayesian nonparametrics and the probabilistic approach to modelling. Philosophical Transactions R Social A, 1–27.

Ghamisi, P., Yokoya, N., Li, J., Liao, W., Liu, S., Plaza, J., ... Plaza, A. (2017). Advances in hyperspectral image and signal processing. IEEE Geoscience Remote Sensing Magazine, 12(37), 78.

Griffin, L. D. (2006). Optimality of the basic colour categories for classification. Journal of the Royal Society Interface, 3, 71–85. doi:10.1098/rsif.2005.0076

Griffith, G. E., & Omernik, J. M. (2009). Ecoregions of the United States-Level III (EPA). In C. J. Cleveland (Ed.), Encyclopedia of earth. Washington, DC: Environmental Information Coalition, National Council for Science and the Environment.

Group on Earth Observation (GEO). (2005). The Global Earth Observation System of Systems (GEOSS) 10-Year Implementation Plan, adopted 16 February 2005. Retrieved January, 10, 2012 from http://www.earthobservations.org/docs/10-Year%20Implementation%20Plan.pdf

Group on Earth Observation/Committee on Earth Observation Satellites (GEO-CEOS). (2010). A quality assurance framework for earth observation, version 4.0. Retrieved November, 15, 2012 from http://qa4eo.org/docs/QA4EO_Principles_v4.0.pdf

Group on Earth Observation/Committee on Earth Observation Satellites (GEO-CEOS) - Working Group on Calibration and Validation (WGCV). (2015). Land Product Validation (LPV). Retrieved March, 20, 2015 from http://lpvs.gsfc.nasa.gov/

Guarino, N. (1995). Formal ontology, conceptual analysis and knowledge representation. *International Journal of Human-Computer Studies, 43*, 625–640. doi:10.1006/ijhc.1995.1066

Gutman, G., Janetos, A. C., Justice, C. O., Moran, E. F., Mustard, J. F., Rindfuss, R. R., ... Cochrane, M. A. (Eds.). (2004). *Land change science*. Dordrecht, The Netherlands: Kluwer.

Hadamard, J. (1902). Sur les problemes aux derivees partielles et leur signification physique. *Princet University Bulletin, 13*, 49–52.

Homer, C., Huang, C. Q., Yang, L. M., Wylie, B., & Coan, M. (2004). Development of a 2001 National Land-Cover Database for the United States. *Photogrammetric Engineering & Remote Sensing, 70*, 829–840. doi:10.14358/PERS.70.7.829

Hunt, N., & Tyrrell, S. (2012). Stratified sampling. Coventry University. Retrieved February, 7, 2012 from http://www.coventry.ac.uk/ec/~nhunt/meths/strati.html

IBM. (2016). The four V's of big data, IBM big data & analytics hub. Retrieved January, 30, 2018 from http://www.ibmbigdatahub.com/infographic/four-vs-big-data

Julesz, B. (1986). Texton gradients: The texton theory revisited. In *Biomedical and life sciences collection* (Vol. 54, pp.4–5). Berlin/Heidelberg: Springer.

Julesz, B., Gilbert, E. N., Shepp, L. A., & Frisch, H. L. (1973). Inability of humans to discriminate between visual textures that agree in second-order statistics – Revisited. *Perception, 2*, 391–405. doi:10.1068/p020391

Kavouras, M., & Kokla, M. (2002). A method for the formalization and integration of geographical categorizations. *International Journal of Geographical Information Science, 16*, 439–453. doi:10.1080/13658810210129120

Kuzera, K., & Pontius, R. G., Jr. (2008). Importance of matrix construction for multiple-resolution categorical map comparison. *GIScience & Remote Sensing, 45*, 249–274. doi:10.2747/1548-1603.45.3.249

Laurini, R., & Thompson, D. (1992). *Fundamentals of spatial information systems*. London, UK: Academic Press.

Lee, D., Baek, S., & Sung, K. (1997). Modified k-means algorithm for vector quantizer design. *IEEE Signal Processing Letters, 4*, 2–4. doi:10.1109/97.551685

Liang, S. (2004). *Quantitative remote sensing of land surfaces*. Hoboken, NJ: John Wiley and Sons.

Lillesand, T., & Kiefer, R. (1979). *Remote sensing and image interpretation*. New York, NY: John Wiley and Sons.

Linde, Y., Buzo, A., & Gray, R. M. (1980). An algorithm for vector quantizer design. *IEEE Transactions Communicable, 28*, 84–95. doi:10.1109/TCOM.1980.1094577

Lipson, H. (2007). Principles of modularity, regularity, and hierarchy for scalable systems. *Journal of Biological Physics and Chemistry, 7*, 125–128. doi:10.4024/40701.jbpc.07.04

Lloyd, S. P. (1982). Least squares quantization in PCM. *IEEE Transactions on Information Theory, 28*(2), 129–137. doi:10.1109/TIT.1982.1056489

Lück, W., & van Niekerk, A. (2016). Evaluation of a rule-based compositing technique for Landsat-5 TM and Landsat-7 ETM+ images. *International Journal of Applied Earth Observation and Geoinformation, 47*, 1–14. doi:10.1016/j.jag.2015.11.019

Lunetta, R., & Elvidge, D. (1999). *Remote sensing change detection: Environmental monitoring methods and applications*. London, UK: Taylor & Francis.

Luo, Y., Trishchenko, A. P., & Khlopenkov, K. V. (2008). Developing clear-sky, cloud and cloud shadow mask for producing clear-sky composites at 250-meter spatial resolution for the seven MODIS land bands over Canada and North America. *Remote Sensing of Environment, 112*, 4167–4185. doi:10.1016/j.rse.2008.06.010

Mallat, S. (2016). Understanding deep convolutional networks. *Philosophical Transactions R Social A, 374*, 1–16. doi:10.1098/rsta.2015.0203

Marcus, G. (2018). Deep learning: A critical appraisal. arXiv: 1801.00631. Retrieved January, 16, 2018 from https://arxiv.org/ftp/arxiv/papers/1801/1801.00631.pdf

Marr, D. (1982). *Vision*. New York, NY: Freeman and C.

Martinetz, T., Berkovich, G., & Schulten, K. (1994). Topology representing networks. *Neural Networks, 7*(3), 507–522. doi:10.1016/0893-6080(94)90109-0

Mather, P. (1994). *Computer processing of remotely-sensed images - an introduction*. Hoboken, NJ: Wiley.

Matsuyama, T., & Hwang, V. S. (1990). *SIGMA – A knowledge-based aerial image understanding system*. New York, NY: Plenum Press.

Mazzuccato, M., & Robinson, D. (2017). *Market creation and the European space agency* (European Space Agency (ESA) Report).

Memarsadeghi, N., Mount, D., Netanyahu, N., & Le Moigne, J. (2007). A fast implementation of the ISODATA clustering algorithm. *International Journal of Computational Geometry & Applications, 17*(1), 71–103. doi:10.1142/S0218195907002252

Mizen, H., Dolbear, C., & Hart, G. (2005, November 29–30). Ontology ontogeny: Understanding how an ontology is created and developed. In M. Rodriguez, I. Cruz, S. Levashkin, & M. J. Egenhofer (Eds.) Proceedings GeoSpatial Semantics: First International Conference, GeoS 2005 (pp. 15–29). Mexico City, Mexico, Springer Berlin Heidelberg.

Muirhead, K., & Malkawi, O. (1989, Sepptember 5–8). Automatic classification of AVHRR images. In *Proceedings Fourth AVHRR Data Users Meeting* (pp. 31–34). Rottenburg, Germany.

Nagao, M., & Matsuyama, T. (1980). *A structural analysis of complex aerial photographs*. New York, NY: Plenum.

National Aeronautics and Space Administration (NASA). (2016a). Getting petabytes to people: How the EOSDIS facilitates earth observing data discovery and use. [Online]. Retrieved December, 29, 2016 from https://earthdata.nasa.gov/getting-petabytes-to-people-how-the-eosdis-facilitates-earth-observing-data-discovery-and-use

National Aeronautics and Space Administration (NASA). (2016b). Data processing levels. [Online]. Retrieved December, 20, 2016 from https://science.nasa.gov/earth-science/earth-science-data/data-processing-levels-for-eosdis-data-products

Open Geospatial Consortium (OGC) Inc. (2015). OpenGIS® implementation standard for geographic information - simple feature access - Part 1: Common architecture. Retrieved March, 8, 2015 from http://www.opengeospatial.org/standards/is

Ortiz, A., & Oliver, G. (2006). On the use of the overlapping area matrix for image segmentation evaluation: A survey and new performance measures. *Pattern Recognition Letters, 27,* 1916–1926. doi:10.1016/j.patrec.2006.05.002

Parisi, D. (1991). La scienza cognitive tra intelligenza artificiale e vita artificiale. In *Neurosceinze e Scienze dell'Artificiale: Dal Neurone all'Intelligenza.* Bologna, Italy: Patron Editore.

Patané, G., & Russo, M. (2001). The enhanced-LBG algorithm. *Neural Networks, 14*(9), 1219–1237. doi:10.1016/S0893-6080(01)00104-6

Patané, G., & Russo, M. (2002). Fully automatic clustering system. *IEEE Transactions on Neural Networks, 13*(6), 1285–1298. doi:10.1109/TNN.2002.804226

Pearl, J. (2009). *Causality: Models, reasoning and inference.* New York, NY: Cambridge University Press.

Piaget, J. (1970). *Genetic epistemology.* New York, NY: Columbia University Press.

Pontius, R. G., Jr., & Connors, J. (2006, July 5–7). "Expanding the conceptual, mathematical and practical methods for map comparison. In M. Caetano & M. Painho (Eds). Proceedings of the 7th International Symposium on Spatial Accuracy Assessment in Natural Resources and Environmental Sciences (pp. 64–79). Lisbon. Instituto Geográfico Português.

Pontius, R. G., Jr., & Millones, M. (2011). Death to Kappa: Birth of quantity disagreement and allocation disagreement for accuracy assessment. *International Journal of Remote Sensing, 32*(15), 4407–4429. doi:10.1080/01431161.2011.552923

Riano, D., Chuvieco, E., Salas, J., & Aguado, I. (2003). Assessment of different topographic corrections in landsat-TM data for mapping vegetation types (2003). *IEEE Transactions on Geoscience and Remote Sensing, 41*(5), 1056–1061. doi:10.1109/TGRS.2003.811693

Richter, R., & Schläpfer, D. (2012a). Atmospheric/topographic correction for satellite imagery – ATCOR-2/3 user guide, version 8.2 BETA. Retrieved April, 12, 2013 from http://www.dlr.de/eoc/Portaldata/60/Resources/dokumente/5_tech_mod/atcor3_manual_2012.pdf

Richter, R., & Schläpfer, D. (2012b). Atmospheric/Topographic correction for airborne imagery – ATCOR-4 User Guide, Version 6.2 BETA, 2012. Retrieved April, 12, 2013 from http://www.dlr.de/eoc/Portaldata/60/Resources/dokumente/5_tech_mod/atcor4_manual_2012.pdf

Roy, D. P., Ju, J., Kline, K., Scaramuzza, P. L., Kovalskyy, V., Hansen, M., … Zhang, C. (2010). Web-enabled Landsat Data (WELD): Landsat ETM+ composited mosaics of the conterminous United States. *Remote Sensing of Environment, 114,* 35–49. doi:10.1016/j.rse.2009.08.011

Salmon, B., Wessels, K., van den Bergh, F., Steenkamp, K., Kleynhans, W., Swanepoel, D., … Kovalskyy, V. (2013, July 21–26). Evaluation of rule-based classifier for Landsat-based automated land cover mapping in South Africa. IEEE International Geoscience Remote Sensing Symposium (IGARSS) (pp. 4301–4304).

Schaepman-Strub, G., Schaepman, M. E., Painter, T. H., Dangel, S., & Martonchik, J. V. (2006). Reflectance quantities in optical remote sensing - Definitions and case studies. *Remote Sensing of Environment, 103,* 27–42. doi:10.1016/j.rse.2006.03.002

Schläpfer, D., Richter, R., & Hueni, A. (2009). Recent developments in operational atmospheric and radiometric correction of hyperspectral imagery, in

Proc. 6th EARSeL SIG IS Workshop, 16-19 March 2009. Retrieved July, 14, 2012 from http://www.earsel6th.tau.ac.il/~earsel6/CD/PDF/earsel-PROCEEDINGS/3054%20Schl%20pfer.pdf

Shackelford, K., & Davis, C. H. (2003a). A combined fuzzy pixel-based and object-based approach for classification of high-resolution multispectral data over urban areas. *IEEE Transactions on Geoscience and Remote Sensing, 41*(10), 2354–2364. doi:10.1109/TGRS.2003.815972

Shackelford, K., & Davis, C. H. (2003b). A hierarchical fuzzy classification approach for high-resolution multispectral data over urban areas. *IEEE Transactions on Geoscience and Remote Sensing, 41*(9), 1920–1932. doi:10.1109/TGRS.2003.814627

Sheskin, D. (2000). *Handbook of parametric and nonparametric statistical procedures.* Boca Raton, FL: Chapman & Hall/CRC.

Shotton, J., Winn, J., Rother, C., & Criminisi, A. (2009). Texton-boost for image understanding: Multi-class object recognition and segmentation by jointly modeling texture, layout, and context. *International Journal of Computer Vision, 81*(1), 2–23. doi:10.1007/s11263-007-0109-1

Simonetti, D., Marelli, A., & Eva, H. (2015b). *Impact toolbox.* (JRC Technical EUR 27358 EN).

Simonetti, D., Simonetti, E., Szantoi, Z., Lupi, A., & Eva, H. D. (2015a). First results from the phenology based synthesis classifier using Landsat-8 imagery. *IEEE Geoscience and Remote Sensing Letters, 12*(7), 1496–1500. doi:10.1109/LGRS.2015.2409982

Smeulders, A., Worring, M., Santini, S., Gupta, A., & Jain, R. (2000). Content-based image retrieval at the end of the early years. *IEEE Transactions Pattern Analysis Machine Intelligent, 22*(12), 1349–1380. doi:10.1109/34.895972

Soares, J., Baraldi, A., & Jacobs, D. (2014). *Segment-based simple-connectivity measure design and implementation* (Tech. Rep). College Park, MD: University of Maryland.

Sonka, M., Hlavac, V., & Boyle, R. (1994). *Image processing, analysis and machine vision.* London, UK: Chapman & Hall.

Sowa, J. F. (2000). *Knowledge representation: Logical, Philosophical, and Computational Foundations.* Pacific Grove, CA: Brooks/Cole.

Stehman, S. V. (1999). Comparing thematic maps based on map value. *International Journal of Remote Sensing, 20,* 2347–2366. doi:10.1080/014311299212065

Stehman, S. V., & Czaplewski, R. L. (1998). Design and analysis for thematic map accuracy assessment: Fundamental principles. *Remote Sensing of Environment, 64,* 331–344. doi:10.1016/S0034-4257(98)00010-8

Swain, P. H., & Davis, S. M. (1978). *Remote sensing: The quantitative approach.* New York, NY: McGraw-Hill.

Tiede, D., Baraldi, A., Sudmanns, M., Belgiu, M., & Lang, S. (2016, March 15–17). ImageQuerying (IQ) – Earth observation image content extraction & querying across time and space. Submitted (Oral presentation and poster session), ESA 2016 Conf. on Big Data From Space, BIDS '16. Santa Cruz de Tenerife, Spain.

Trimble. (2015). *eCognition® Developer 9.0 Reference Book.*

Tsotsos, J. K. (1990). Analyzing vision at the complexity level. *Behavioral and Brain Sciences, 13,* 423–445. doi:10.1017/S0140525X00079577

Vermote, E., & Saleous, N. (2007). *LEDAPS surface reflectance product description - Version 2.0.* University of

Maryland at College Park/Dept Geography and NASA/ GSFC Code 614.5.

Vogelmann, J. E., Howard, S. M., Yang, L., Larson, C. R., Wylie, B. K., & van Driel, N. (2001). Completion of the 1990s National Land Cover Data set for the conterminous United States from Landsat Thematic Mapper data and ancillary data sources. *Photo Engineering Remote Sensing, 67*, 650–662.

Vogelmann, J. E., Sohl, T. L., Campbell, P. V., & Shaw, D. M. (1998). Regional land cover characterization using Landsat Thematic Mapper data and ancillary data sources. *Environmental Monitoring and Assessment, 51*, 415–428. doi:10.1023/A:1005996900217

Web-Enabled Landsat Data (WELD) Tile FTP. 2015. Retrieved December, 12, 2016 from https://weld.cr.usgs.gov/

Wenwen, L. M., Goodchild, F., & Church, R. L. (2013). An efficient measure of compactness for 2D shapes and its application in regionalization problems. *International Journal of Geographical Info Science*, 1–24.

Wickham, J. D., Stehman, S. V., Fry, J. A., Smith, J. H., & Homer, C. G. (2010). Thematic accuracy of the NLCD 2001 land cover for the conterminous United States. *Remote Sensing of Environment, 114*, 1286–1296. doi:10.1016/j.rse.2010.01.018

Wickham, J. D., Stehman, S. V., Gass, L., Dewitz, J., Fry, J. A., & Wade, T. G. (2013). Accuracy assessment of NLCD 2006 land cover and impervious surface. *Remote Sensing of Environment, 130*, 294–304. doi:10.1016/j.rse.2012.12.001

Wikipedia. (2018a). Big data. Retrieved January, 30, 2018 from en.wikipedia.org/wiki/Big_data

Wikipedia. (2018b). Latent variable. Retrieved January, 30, 2018 from https://en.wikipedia.org/wiki/Latent_variable

Wikipedia. (2018c). Bayesian inference. Retrieved January, 30, 2018 from https://en.wikipedia.org/wiki/Bayesian_inference

Xian, G., & Homer, C. (2010). Updating the 2001 National Land Cover Database impervious surface products to 2006 using Landsat imagery change detection methods. *Remote Sensing of Environment, 114*, 1676–1686. doi:10.1016/j.rse.2010.02.018

Yang, C., Huang, Q., Li, Z., Liu, K., & Hu, F. (2017). Big Data and cloud computing: Innovation opportunities and challenges. *International Journal of Digital Earth, 10*(1), 13–53. doi:10.1080/17538947.2016.1239771

Zadeh, L. A. (1965). Fuzzy sets. *Information and Control, 8*, 338–353. doi:10.1016/S0019-9958(65)90241-X

GEO-CEOS stage 4 validation of the Satellite Image Automatic Mapper lightweight computer program for ESA Earth observation level 2 product generation - Part 2: Validation

Andrea Baraldi[1,2,3,4]*, Michael Laurence Humber[2], Dirk Tiede[3] and Stefan Lang[3]

*Corresponding author: Andrea Baraldi, Department of Agricultural Sciences, University of Naples Federico II, Portici NA 80055, Italy
E-mail: andrea6311@gmail.com
Reviewing Editor: Louis-Noel Moresi, University of Melbourne, Australia

Abstract: ESA defines as Earth Observation (EO) Level 2 information product a multi-spectral (MS) image corrected for atmospheric, adjacency, and topographic effects, stacked with its data-derived scene classification map (SCM), whose legend includes quality layers cloud and cloud-shadow. No ESA EO Level 2 product has ever been systematically generated at the ground segment. To fill the information gap from EO big data to ESA EO Level 2 product in compliance with the GEO-CEOS stage 4 validation (*Val*) guidelines, an off-the-shelf Satellite Image Automatic Mapper (SIAM) lightweight computer program was selected to be validated by independent means on an annual 30 m resolution Web-Enabled Landsat Data (WELD) image composite time-series of the conterminous U.S. (CONUS) for the years 2006 to 2009.

ABOUT THE AUTHORS

Andrea Baraldi (*Laurea* in Electronic Engineering, Univ. Bologna, 1989; Master in Software Engineering, Univ. Padua, 1994; PhD in Agricultural Sciences, Univ. Naples Federico II, 2017) has held research positions at the Italian Space Agency (2018-to date); Dept. Geoinformatics, Univ. Salzburg, Austria (2014–2017); Dept. Geographical Sciences, Univ. Maryland, College Park, MD (2010–2013); European Commission Joint Research Centre (2000–2002; 2005–2009); International Computer Science Institute, Berkeley, CA (1997–1999); European Space Agency Research Institute (1991–1993); Italian National Research Council (1989, 1994–1996, 2003–2004). In 2009, he founded Baraldi Consultancy in Remote Sensing. He was appointed with a Senior Scientist Fellowship at the German Aerospace Center, Oberpfaffenhofen, Germany (February 2014) and was a visiting scientist at the Ben Gurion Univ. of the Negev, Sde Boker, Israel (February 2015). His main interests center on the research and technological development of automatic near-real-time spaceborne/airborne image preprocessing and understanding systems in operating mode, consistent with human visual perception. Dr. Baraldi served as an Associate Editor of the IEEE TRANS. NEURAL NETWORKS from 2001 to 2006.

PUBLIC INTEREST STATEMENT

Synonym of scene-from-image reconstruction and understanding, vision is an inherently ill-posed cognitive task; hence, it is difficult to solve and requires a priori knowledge in addition to sensory data to become better posed for numerical solution. In the inherently ill-posed cognitive domain of computer vision, this research was undertaken to validate by independent means a lightweight computer program for prior knowledge-based multi-spectral color naming, called Satellite Image Automatic Mapper (SIAM), eligible for automated near-real-time transformation of large-scale Earth observation (EO) image datasets into European Space Agency (ESA) EO Level 2 information product, never accomplished to date at the ground segment. An original protocol for wall-to-wall thematic map quality assessment without sampling, where legends of the test and reference map pair differ and must be harmonized, was adopted. Conclusions are that SIAM is suitable for systematic ESA EO Level 2 product generation, regarded as necessary not sufficient pre-condition to transform EO big data into timely, comprehensive, and operational EO value-adding information products and services.

The SIAM core is a prior knowledge-based decision tree for MS reflectance space hyperpolyhedralization into static (non-adaptive to data) color names. For the sake of readability, this paper was split into two. The present Part 2—Validation— accomplishes a GEO-CEOS stage 4 *Val* of the test SIAM-WELD annual map time-series in comparison with a reference 30 m resolution 16-class USGS National Land Cover Data (NLCD) 2006 map. These test and reference map pairs feature the same spatial resolution and spatial extent, but their legends differ and must be harmonized, in agreement with the previous Part 1 - Theory. Conclusions are that SIAM systematically delivers an ESA EO Level 2 SCM product instantiation whose legend complies with the standard 2-level 4-class FAO Land Cover Classification System (LCCS) Dichotomous Phase (DP) taxonomy.

Subjects: Algorithms & Complexity; Automation; Cognitive Artificial Intelligence; Expert systems; GIS, Remote Sensing & Cartography; Image Processing; Intelligent Systems; Machine Learning; Pattern Analysis; Quality Control & Reliability; Real-Time Systems; Systems Architecture

Keywords: Artificial intelligence; binary relationship; Cartesian product; cognitive science; color naming; connected-component multi-level image labeling; deductive inference; earth observation; land cover taxonomy; high-level (attentive) and low-level (pre-attentional) vision; hybrid inference; image classification; image segmentation; inductive inference; machine learning-from-data; outcome and process quality indicators; radiometric calibration; remote sensing; surface reflectance; thematic map comparison; top-of-atmosphere reflectance; two-way contingency table; unsupervised data discretization/vector quantization; validation

1. Introduction

Jointly proposed by the intergovernmental Group on Earth Observations (GEO) and the Committee on Earth Observation Satellites (CEOS), the implementation plan for years 2005–2015 of the Global Earth Observation System of Systems (GEOSS) aimed at systematic transformation of multi-source Earth observation (EO) *big data* (IBM, 2016; Yang, Huang, Li, Liu, & Hu, 2017) into timely, comprehensive, and operational EO value-adding products and services (GEO, 2005), submitted to the GEO-CEOS Quality Assurance Framework for Earth Observation (QA4EO) calibration/validation (*Cal/ Val*) requirements (Group on Earth Observation/Committee on Earth Observation Satellites (GEO/ CEOS), 2010). The visionary goal of GEOSS cannot be considered fulfilled by the remote sensing (RS) community to date. In the terminology of philosophical hermeneutics, the problem is not a lack of sensory data, but our lack of knowledge in transforming big sensory data into quantitative/ unequivocal *information-as-thing* and qualitative/equivocal *information-as-data-interpretation* (Capurro & Hjørland, 2003). Such a lack of knowledge causes the so-called data-rich, information-poor (DRIP) syndrome (Bernus & Noran, 2017), supported by undisputable observations (true-facts). For example, past and present EO image understanding systems (EO-IUSs) have been typically outpaced by the rate of collection of EO sensory data, whose quality and quantity are ever-increasing at an apparently exponential rate related to the Moore law of productivity (National Aeronautics and Space Administration (NASA), 2016).

To contribute toward the visionary goal of GEOSS, this interdisciplinary work aimed at filling an analytic and pragmatic information gap from EO big sensory data to systematic European Space Agency (ESA) EO Level 2 information product generation (CNES, 2015; Deutsches Zentrum fürLuft- und Raumfahrt e.V. (DLR) and VEGA Technologies, 2011; European Space Agency (ESA), 2015), never accomplished at the ground segment by any EO data provider to date (DLR & VEGA, 2011; ESA, 2015). ESA defines as EO Level 2 information product: (i) a single-date multi-spectral (MS) image, radiometrically calibrated into surface reflectance (SURF) values corrected for atmospheric, adjacency, and topographic effects, in compliance with the GEO-CEOS QA4EO *Cal* requirements (GEO-CEOS, 2010), stacked with (ii) its data-derived Scene Classification Map (SCM), whose legend includes quality layers cloud and cloud-shadow (ESA, 2015; DLR & VEGA, 2011; CNES, 2015). In practice, ESA EO Level 2 product generation is a *chicken-and-egg* dilemma, synonym of inherently ill-posed problem in the Hadamard sense (Hadamard, 1902); therefore, it is very difficult to solve. On the one hand, no effective and efficient understanding (mapping) of a sub-symbolic EO image into a symbolic SCM is possible if DNs (pixels) are affected by low radiometric quality (GEO-CEOS, 2010). On the other hand, no effective and efficient *Cal* of digital numbers (DNs) into SURF values corrected for atmospheric, topographic and adjacency effects is possible without an SCM, available a priori in addition to sensory data to enforce a statistic stratification (layered) principle, synonym of class-conditional data analytics (Baraldi, 2017; Baraldi et al., 2010b; Baraldi & Humber, 2015; Baraldi, Humber, & Boschetti, 2013; Bishop & Colby, 2002; Bishop, Shroder, & Colby, 2003; DLR & VEGA, 2011; Dorigo, Richter, Baret, Bamler, & Wagner, 2009; Lück & Van Niekerk, 2016; Riano, Chuvieco, Salas, & Aguado, 2003, Richter & Schläpfer, 2012a, 2012b; Vermote & Saleous, 2007). Well known in statistics, the principle of statistic stratification guarantees that "stratification will always achieve greater precision provided that the strata have been chosen so that members of the same stratum are as similar as possible in respect of the characteristic of interest" (Hunt & Tyrrell, 2012).

For the sake of readability, this paper is split into two. The preliminary Part 1 - Theory postulated as working hypothesis a necessary not sufficient pre-condition for a yet-unfulfilled GEOSS development (GEO-CEOS, 2005). The proposed working hypothesis is "ESA EO Level 2 product ⊂ EO image understanding (EO-IU) in operating mode ⊂ computer vision (CV) → GEOSS," where relationship *subset-of*, denoted with symbol "⊃," means specialization with inheritance from the superset to the subset, while dependence relationship *part-of* is denoted with symbol "→," pointing from the supplier to the client in agreement with the standard Unified Modeling Language (UML) for graphical modeling of object-oriented software (Fowler, 2016). This working hypothesis postulates that no GEOSS can exist if the necessary not sufficient pre-condition of systematic ESA EO Level 2 product generation is not accomplished in advance at the ground segment. Hence, systematic ESA EO Level 2 product generation is considered a mandatory early stage in a hierarchical EO-IUS workflow, capable of scene-from-image reconstruction and understanding in operating mode to cope with the five Vs of big data, specifically, volume, variety, velocity, veracity, and value (IBM, 2016; Yang et al., 2017).

In the words of Marr, "vision goes symbolic almost immediately, right at the level of zero-crossing (first-stage primal sketch), without loss of information" (Marr, 1982). In agreement with Marr's intuition, our instantiation of an ESA EO Level 2 product generation is constrained as follows. (I) A single-date MS image is radiometrically corrected for atmospheric, adjacency, and topographic effects, automatically (without human-machine interaction) and in near real time. (II) It is stacked with its data-derived SCM, generated automatically and in near real time. The SCM legend must be general purpose and user and application independent. Unlike the non-standard SCM legend adopted by the Sentinel 2 imaging sensor-specific (atmospheric) Correction Prototype Processor (SEN2COR) developed by ESA to be run on user side (ESA, 2015; DLR & VEGA, 2011), the proposed SCM legend is selected equal to an "augmented" 3-level 9-class Dichotomous Phase (DP) taxonomy of the Food and Agriculture Organization of the United Nations (FAO)—Land Cover Classification System (LCCS) (Di Gregorio & Jansen, 2000). Such an "augmented" land cover (LC) class taxonomy in the 4D spatio-temporal scene-domain encompasses the standard fully nested 3-level 8-class FAO LCCS-DP legend in addition to a thematic layer "other" or "rest of the world," which includes quality layers cloud and cloud-shadow; see Figure 1. (III) A GEO-CEOS stage 4 *Val* of the ESA EO Level 2 outcome and process is considered mandatory to comply with the GEO-CEOS QA4EO *Cal/Val* requirements (GEO-CEOS, 2010). By definition, a GEO-CEOS Stage 3 *Val* requires that "spatial and temporal consistency of the product with similar products are evaluated by independent means over multiple locations and time periods representing global conditions. In Stage 4

Val, results for Stage 3 are systematically updated when new product versions are released and as the time-series expands" (GEO-CEOS - Working Group on Calibration and Validation (WGCV), 2015).

To contribute toward filling an analytic and pragmatic information gap from multi-source EO big imagery to systematic generation of ESA EO Level 2 product as constrained above, the primary goal of this interdisciplinary study was to undertake an original (to the best of these authors' knowledge, the first) outcome and process GEO-CEOS stage 4 *Val* (GEO-CEOS WGCV, 2015) of an off-the-shelf Satellite Image Automatic Mapper™ (SIAM™) lightweight computer program for top-down (deductive) MS reflectance space hyperpolyhedralization into MS color names, superpixel detection, and vector quantization (VQ) quality assessment. Implemented in operating mode in the C/C++ programming language, the SIAM software toolbox is "lightweight" because it runs automatically (without human-machine interaction), in near real time (it is non-iterative and its computational complexity increases linearly with image size) and in tile streaming mode (it requires a fixed runtime memory occupation) (Baraldi, 2015, 2017; Baraldi, Puzzolo, Blonda, Bruzzone, & Tarantino, 2006; Baraldi et al., 2010a, 2010b; Baraldi, Gironda, & Simonetti, 2010c; Baraldi, 2011; Baraldi & Boschetti, 2012a, 2012b; Baraldi et al., 2013; Baraldi, Tiede, Sudmanns, Belgiu, & Lang, 2016, 2017; Baraldi & Humber, 2015). In addition to running on either laptop or desktop computers, the SIAM lightweight computer program is eligible for use in mobile application software or web services. Eventually provided with a mobile user interface, a mobile application software is a lightweight computer program specifically designed to run directly on mobile devices, such as tablet computers and smartphones. The core of the non-iterative SIAM software pipeline is a one-pass prior knowledge-based decision tree (expert system) for MS reflectance space hyperpolyhedralization (quantization, partitioning) into static (non-adaptive-to-data) color names, see Figure 2 and refer to Chapter 2 and Chapter 3 in the Part 1. Presented in the RS literature where enough information was provided for the implementation to be reproduced (Baraldi et al., 2006), the SIAM expert system for MS color naming is followed by a well-posed two-pass superpixel detector in the multi-level color map-domain (Dillencourt, Samet, & Tamminen, 1992; Sonka, Hlavac, & Boyle, 1994) and a per-pixel VQ error assessment for VQ quality assurance, in agreement with the GEO-CEOS QA4EO *Val* guidelines, refer to Figure 4 in the Part 1 of this paper.

There is a long history of prior knowledge-based MS reflectance space partitioners for static color naming developed, but never validated by space agencies, public organizations, and private companies for use in hybrid (combined deductive and inductive) EO-IUSs in operating mode, refer to Chapter 3 in the Part 1 of this paper. EO value-adding products and services delivered by existing hybrid EO-IUSs whose input is a MS image class-conditioned (masked) by static color names encompass a large variety of low-level EO image enhancement tasks, ranging from MS image compositing to atmospheric and topographic correction of top-of-atmosphere reflectance (TOARF) into SURF values (Ackerman et al., 1998; Baraldi et al., 2010c; Baraldi & Humber, 2015; Baraldi et al., 2013; Despini, Teggi, & Baraldi, 2014; DLR and VEGA, 2011; Dorigo et al., 2009; Lück & Van Niekerk, 2016; Luo, Trishchenko, & Khlopenkov, 2008; Richter & Schläpfer, 2012a, 2012b; Vermote & Saleous, 2007) and high-level EO image understanding applications, including EO image time-series classification, ranging from cloud/cloud-shadow detection to burned area recognition (Arvor, Madiela, & Corpetti, 2016; Baraldi, 2015; Baraldi et al., 2010a, 2010b; Boschetti, Roy, Justice, & Humber, 2015; DLR & VEGA, 2011; Lück & Van Niekerk, 2016; Muirhead & Malkawi, 1989; Simonetti, Simonetti, Szantoi, Lupi, & Eva, 2015; GeoTerraImage, 2015). To the best of these authors' knowledge, none of these prior knowledge-based MS reflectance space partitioners presented in the RS literature has ever been submitted to a GEO-CEOS stage 4 *Val* process (GEO-CEOS WGCV, 2015), in compliance with the GEO-CEOS QA4EO *Val* requirements (GEO-CEOS, 2010). To fill this analytic and pragmatic lack, the proposed GEO-CEOS stage 4 *Val* outcome and process of an off-the-shelf SIAM lightweight computer program for prior knowledge-based MS reflectance space hyperpolyhedralization would be the first of its kind. Hence, the potential impact of the present study on the research and development (R&D), testing and validation of present or future hybrid EO-IUSs in operating mode, where static color naming is employed for MS image stratification purposes according to a convergence-of-evidence approach in agreement with a Bayesian naïve classification formulation (Baraldi, 2017; Matsuyama & Hwang, 1990), refer to Equation (3) in the Part 1 of this paper and see Figure 3, is expected to be relevant.

To comply with the GEO-CEOS stage 4 *Val* requirements (GEO-CEOS WGCV, 2015), outcome and process of an off-the-shelf SIAM computer program had to be validated by independent means on a radiometrically calibrated EO image time-series at large spatial extent. The open-access U.S. Geological Survey (USGS) 30 m resolution annual Web Enabled Landsat Data (WELD) image composite of the conterminous U.S. (CONUS) for the years 2006 to 2009, radiometrically calibrated into TOARF values (Homer, Huang, Yang, Wylie, & Coan, 2004; Roy et al., 2010; WELD, 2016), was identified as a viable input dataset. The 30 m resolution 16-class U.S. National Land Cover Data (NLCD) 2006 map, delivered in 2011 by the USGS Earth Resources Observation Systems (EROS) Data Center (EDC) (Environmental Protection Agency (EPA), 2007; Vogelmann et al., 2001; Vogelmann, Sohl, Campbell, & Shaw, 1998; Wickham, Stehman, Fry, Smith, & Homer, 2010; Wickham et al., 2013; Xian & Homer, 2010), was selected as the reference thematic map at the CONUS spatial extent. The 16-class NLCD map legend is described in Table 1. To account for typical non-stationary geospatial statistics, the USGS NLCD 2006 thematic map was partitioned into 86 Level III ecoregions of North America collected from the Environmental Protection Agency (EPA) (Environmental Protection Agency (EPA), 2013; Griffith & Omernik, 2009).

In the proposed experimental framework, the test SIAM-WELD map time-series and the reference USGS NLCD 2006 map share the same spatial extent and spatial resolution, but their map legends are not the same, differing in both semantics and cardinality. These working hypotheses are neither trivial nor conventional in the RS literature, where thematic map quality assessments typically adopt a sampling strategy, either probabilistic (random) or non-random (Baraldi et al., 2013), and assume that the test and reference thematic map dictionaries coincide (Stehman & Czaplewski, 1998). Starting from a stratified random sampling protocol presented in literature (Baraldi et al., 2013), the present Part 2 - Validation proposes an original protocol for wall-to-wall comparison without sampling of two thematic maps featuring the same spatial extent and spatial resolution, but whose legends can differ. This novel protocol incorporates two original contributions of the Part 1 where, first, a hybrid (combined deductive and inductive) eight-step guideline was proposed to streamline a human decision maker in the identification of a binary relationship, $R: A \Rightarrow B \subseteq A \times B$, from categorical variable A to categorical variable B estimated from the same population, where $A \neq B$ in general and $A \times B$ is the 2-fold Cartesian product (product set). This is an inherently ill-posed (equivocal, subjective) *information-as-data-interpretation* process (Capurro & Hjørland, 2003) belonging to the multi-disciplinary domain of cognitive science, refer to Figure 13 in the Part 1. The proposed hybrid eight-step guideline is of practical use because identification of a binary relationship, $R: A \Rightarrow B$, is mandatory to guide the interpretation process of a bivariate frequency table, $BIVRFTAB = FrequencyCount(A \times B) \neq R: A \Rightarrow B \subseteq A \times B$, where $A \neq B$ in general. Only if $A = B$ then BIVRFTAB becomes equal to the well-known square and sorted confusion matrix (CMTRX), where the main diagonal guides the interpretation process. Second, version 2 of a categorical variable-pair index of association (harmonization, matching) in a binary relationship, $R: A \Rightarrow B$, where $A \neq B$ in general, $CVPAI2(R: A \Rightarrow B) \in [0, 1]$, was proposed to cope with the entity-relationship conceptual model shown in Figure 18 of the Part 1.

The rest of the present Part 2 is organized as follows. Chapter 2 describes materials including the SIAM computer program, the time-series of annual WELD image composites, the reference USGS NLCD 2006 map and the EPA Level III ecoregion map of North America. Methods, specifically, an original protocol to compare without sampling the test SIAM-WELD map and the reference USGS NLCD 2006 map of the CONUS, whose map legends do not coincide and must be harmonized (reconciled, associated, translated) (Ahlqvist, 2005), is proposed in Chapter 3. Experimental results are presented in Chapter 4 and discussed in Chapter 5. Conclusions are reported in Chapter 6.

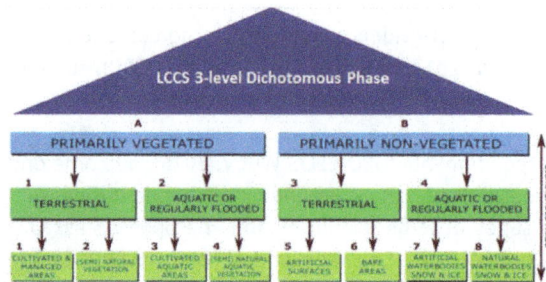

Figure 1. The fully nested 3-level 8-class FAO Land Cover Classification System (LCCS) Dichotomous Phase (DP) layers are: (i) vegetation versus non- vegetation, (ii) terrestrial versus aquatic, and (iii) managed versus natural or semi-natural. They deliver as output the fol-lowing 3-level 8-class FAO LCCS-DP taxonomy. (A11) Cultivated and Managed Terrestrial (non-aquatic) Vegetated Areas. (A12) Natural and Semi-Natural Terrestrial Vegetation. (A23) Cultivated Aquatic or Regularly Flooded Vegetated Areas. (A24) Natural and Semi-Natural Aquatic or Regularly Flooded Vegetation.(B35) Artificial Surfaces and Associated Areas. (B36) Bare Areas. (B47) Artificial Waterbodies, Snow and Ice. (B48) Natural Waterbodies, Snow and Ice. The general-pur-pose, user- and application-independent 3-level 8-class FAO LCCS-DP taxonomy is pre-liminary to a user- and appli-cation-specific FAO LCCS Modular Hierarchical Phase (MHP) taxonomy of one-class classifiers (Di Gregorio & Jansen, 2000), refer to Figure 3 in the Part 1 of this paper.

2. Materials

Presented in the RS literature, four alternative implementations of a prior knowledge-based decision tree for static MS reflectance space hyperpolyhedralization into static color names were compared for model selection. (i) The year 2006 SIAM decision tree presented in Baraldi et al. (2006). (ii) The static decision tree for Spectral Classification of surface reflectance signatures (SPECL) proposed by Dorigo et al. (2009), see Table 4 in the Part 1 of this paper, and implemented by the Atmospheric/Topographic Correction for Satellite Imagery (ATCOR) commercial software product (Richter & Schläpfer, 2012a, 2012b). (iii) The static decision tree for Single-Date Classification (SDC), proposed by Simonetti et al. (2015). (iv) The Canada Centre for Remote Sensing (CCRS) spectral decision tree is shown in Figure 17 of the Part 1 (Luo et al., 2008). Whereas the SDC, SPECL, and CCRS decision trees declare their applicability to Landsat images exclusively, SIAM claims its scalability to MS imaging sensors featuring different spectral resolution specifications; see Table 1 in the Part 1. Moreover, the SIAM decision tree outperforms its counter-parts in terms of spectral quantization capability, parameterized by the total number of detected color names, equal to 96 for the 7-band Landsat-like SIAM (L-SIAM) subsystem, see Table 1 in the Part 1, versus 13, 19, and 7 color names detected in Landsat images by the SDC, SPECL (see Tables 4 in the Part 1) and CCRS (see Figure 17 in the Part 1) decision trees, respectively. To explain their broad differences in terms of number of detected color names and scalability to MS imaging sensors whose spectral and spatial resolution specifications can vary, the four static spectral decision trees of interest were compared at the level of understanding of spectral information/knowledge representation (Marr, 1982), irrespective of the implementation of the decision rule set (structural knowledge in the decision tree) and of the order of presentation of decision rules (procedural knowledge in the decision tree).

To investigate the scalability of an a priori knowledge-based spectral decision tree to varying MS imaging sensor specifications, we started observing that, given a partition of a MS color space, $\Re^{MS}$, into a discrete and finite vocabulary (codebook) of hyperpolyhedra equivalent to color names (codewords), identified as {1, ColorVocabularyCardinality}, for any spatial unit x, either (0D) pixel, (1D) line or (2D) polygon defined according to the Open Geospatial Consortium (OGC) nomencla-ture (OGC, 2015) and featuring a numeric ColorValue(x) $\in \Re^{MS}$, the photometric attribute of spatial unit x can be assigned with a categorical ColorName* $\in$ {1, ColorVocabularyCardinality}, such that membership m(ColorValue(x)| ColorName*) = 1, see Equation (3) in the Part 1 of this paper. In practice, when spatial unit x is (0D) pixel, then any prior knowledge-based spectral decision tree for color naming can work at the sensor spatial resolution whatever it is, that is, it can work pixel-based irrespective of the spatial resolution of the imaging sensor.

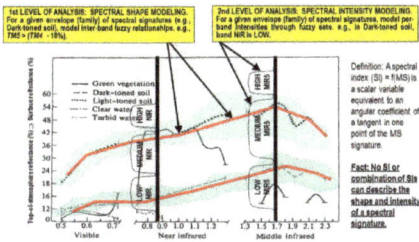

Figure 2. (same as Figure 5 in the Part 1 of this paper, duplicated for the sake of readability of the present Part2). Examples of land cover (LC) class-specific families of spectral signatures in top-of-atmosphere reflectance (TOARF) values that include surface reflectance (SURF) values as a special case in clear sky and flat terrain conditions. A within-class family of spectral signatures (e.g., dark-toned soil) in TOARF values forms a buffer zone (hyperpolyhedron, envelope, manifold). The SIAM decision tree models each target family of spectral signatures in terms of multivariate shape and multi-variate intensity information components as a viable alterna-tive to multivariate analysis of spectral indexes. A typical spec-tral index is a scalar band ratio equivalent to an angular coeffi-cient of a tangent in one point of the spectral signature. Infinite functions can feature the same tangent value in one point. In practice, no spectral index or combination of spectral indexes can reconstruct the multivariate shape and multivariate intensity information components of a spectral signature.

Because they are independent of the spatial resolution of the imaging sensor, static decision trees for color naming depend on spectral resolution specifications exclusively. Inter-sensor differ-ences in spectral resolution can vary from minor differences in a band-specific sensitivity curve to the major lack of a whole spectral channel. To gain robustness to changes in spectral resolution specifications, the necessary not sufficient pre-condition for spectral rules is to infer "strong" (robust and reliable) conjectures based on the redundant convergence of multiple independent sources of spectral evidence, each of which is individually "weak." This rationale is alternative to, for example, pruning of redundant processing elements in distributed processing systems such as multi-layer perceptrons (Bishop, 1995; Cherkassky & Mulier, 1998). If this diagnosis holds true, that is, redundancy of spectral evidence is a value-added of spectral rules to scale to varying spectral resolutions, then information redundancy of a spectral if-then rule is expected to increase mono-tonically with the collection of independent premises.

In a MS reflectance (hyper)cube, any target family of LC class-specific spectral signatures is a multivariate (hyper)polyhedron (envelope, distribution, manifold). Unfortunately, MS reflectance space hyperpolyhedra for color naming are difficult to think of and impossible to visualize when the MS data space dimensionality is superior to three, see Figure 7 in the Part 1 of this paper. Like a vector quantity has two characteristics, a magnitude and a direction, any LC class-specific MS manifold is characterized by a multivariate shape and a multivariate intensity information component; see Figure 2. Hence, spectral information redundancy required to gain robustness to changes in spectral resolution specifications can regard the modelling of both the MS shape and MS intensity information components of a target MS hyperpolyhedron. Among the spectral decision trees being compared, only the SIAM decision tree adopts two different sets of spectral rules to model the MS shape and the MS intensity as two independent spectral information components of a target manifold of MS signatures. On the contrary, in the SDC, SPECL and CCRS decision trees MS shape and MS intensity properties are modeled jointly, which negatively affects principles of modularity, regularity, and hierarchy required by scalable systems (Lipson, 2007). For example, a typical SDC spectral rule applied to a Landsat pixel vector, radiometrically calibrated into a TOARF value in range [0, 1] in each Landsat band 1 to 6, is

If NDVI < 0.5 and NIR (= Landsat band 4) ≥ 0.15 then do something else do otherwise.

In this spectral decision rule, the normalized difference vegetation index, NDVI = (NIR—Red)/(NIR + Red), where NIR = Landsat band 4 and Red = Landsat band 3, is a well-known spectral index, whose unbounded version is the band ratio NIR/Red. Noteworthy, band ratios are scalar spectral indexes widely employed in the SPECL decision tree, see Table 4 in the Part 1, and in the CCRS decision tree

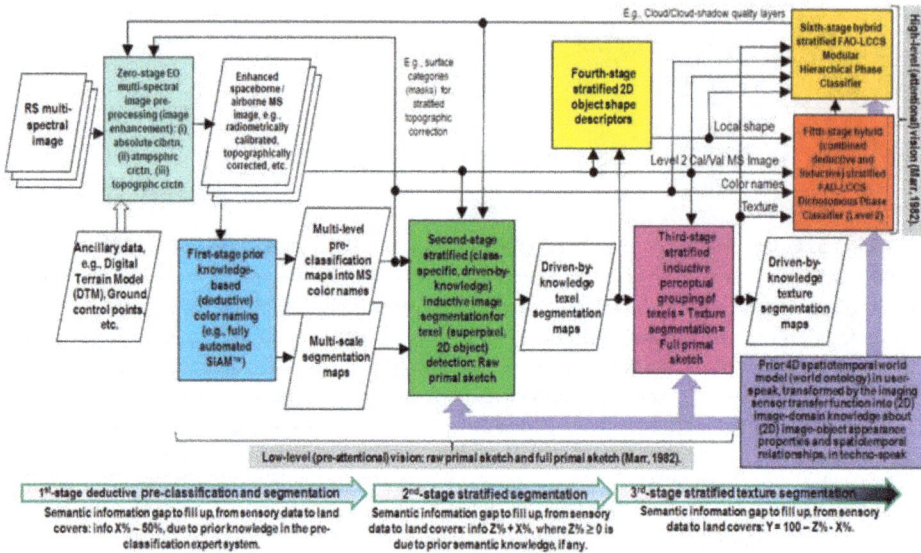

Figure 3. See note.

shown in Figures 17 of the Part 1 (Luo et al., 2008). Any scalar spectral index, either normalized band difference or band ratio, is conceptually equivalent to the slope of a tangent to the spectral signature in one point. This spectral slope is a MS shape descriptor independent of the MS intensity, that is, infinite functions with different intensity values can feature the same tangent value in one point. Although appealing due to its conceptual and numerical simplicity (Liang, 2004), any scalar spectral index is unable per se to represent either the multivariate shape information or the multivariate intensity information component of an MS signature (Baraldi, 2017). Intuitively a scalar spectral index causes a dramatic N-to-1 loss in spectral resolution by reducing an N-channel MS signature to a univariate (scalar) value, corresponding in the (2D) image-plane to a panchromatic (one-channel) image. No photointerpreter whose objective is a one-class LC classification, for example, vegetation detection, would typically consider a panchromatic image made of a univariate spectral index, for example, NDVI, either as informative as an MS image or informative enough to be mapped onto a binary map, for example, vegetation/non-vegetation, where a crisp thresholding criterion is expected to be successful enough to accomplish binary target/no-target discrimination with high accuracy at large spatial extent, different from toy problems. In general, no univariate or multivariate spectral index is representative of the multivariate shape and multivariate intensity information components of an MS manifold; see Figure 2. This obvious, but not trivial observation explains why, in spectral pattern recognition applications, lossy scalar spectral indexes are ever-increasing in number and variety in the endless search for yet-another scalar spectral index, supposedly more informative (Baraldi et al., 2010a, 2010b; Liang, 2004). In the SDC rule reported above, the first spectral term, NDVI < 0.5, constrains per se the multivariate shape of the target MS hyperpolyhedron independent of multivariate intensity; it is employed in logical combination with a second spectral term, where a MS intensity value is restrained as NIR ≥ 0.15, which constrains per se the multivariate intensity of the target MS hyperpolyhedron independent of multivariate shape. The conclusion is that, unlike the SIAM decision rule set, neither the SDC nor the SPECL nor the CCRS decision tree decomposes a target MS hyperpolyhedron, equivalent to a color name, into its multivariate shape and multivariate intensity information components to make each information component easier to be investigated by multivariate data analysis according to principles of modularity, regularity and hierarchy typical of scalable systems (Lipson, 2007). In each of its two independent sets of spectral rules for MS shape and MS intensity modelling SIAM pursues redundancy of spectral terms as a necessary condition to accomplish scalability to changes in the sensor spectral resolution specifications. Possible combinations of these two independent sets of spectral rules make the SIAM decision tree implementations, starting from that proposed in pseudo-code in Baraldi et al. (2006), capable of representing the multivariate shape and multivariate intensity information components of a target

MS hyperpolyhedron, neither necessarily convex nor connected, as a converging combination of independent functions whose individual terms are input with 1- to N-variate variables, with N equal to the total number of spectral channels. Multivariate data statistics are known to be more informative than a sequence of univariate data statistics. For example, maximum likelihood data classification, accounting for multivariate data correlation and variance (covariance), is typically more accurate than parallelepiped data classification whose rectangular decision regions, equivalent to a concatenation of univariate data constraints, poorly fit multivariate data in the presence of bivariate cross-correlation (Lillesand & Kiefer, 1979). In the RS common practice, thanks to its spectral redundancy of multivariate data statistics, the "master" 7-band Landsat-like SIAM (L-SIAM) decision tree can be down-scaled to cope with "slave" MS imaging sensors whose spectral resolution is inferior to but overlaps with Landsat's; see Table 1 in the Part 1 of this paper (Baraldi et al., 2010a, 2010b).

Moving from this decision-tree models comparison, these authors concluded that the SIAM's peculiar design in modeling MS hyperpolyhedra and the SIAM implementation complexity/redundancy, superior to that of its alternative decision trees in terms of number of rules and number of terms (premises) per rule, appeared sufficient to justify the SIAM claims of, first, a finer spectral quantization capability and, second, a superior spectral scalability to changes in sensor specifications in comparison with its alternative SDC, SPECL, and CCRS decision tree implementations. Based on these conclusions, an off-the-shelf SIAM application software was selected and considered worth of a GEO-CEOS stage 4 *Val* in compliance with the GEO-CEOS QA4EO *Val* requirements; refer to previous Chapter 1.

To pursue a GEO-CEOS stage 4 *Val* of the SIAM computer program, the 30 m resolution USGS NLCD 2006 map was selected as reference LC map at the CONUS spatial extent. When this experimental work was conducted the USGS NLCD 2006 map was the most recent release of the U.S. NLCD map series developed by the USGS EDC (Vogelmann et al., 1998, 2001; Wickham et al., 2010, 2013; Xian & Homer, 2010; EPA, 2007; Homer et al., 2004); see Figure 4. By now, the U.S. NLCD map series comprises the USGS NLCD 1992, 2001, 2001 Version 2.0, 2006 (released in 2011) and 2011 (released in 2015) editions. The timeliness from EO image collection to NLCD product delivery, which includes information layers such as tree cover fraction and impervious fraction, has steadily decreased from the about 5 years of the initial NLCD product. Made available for public

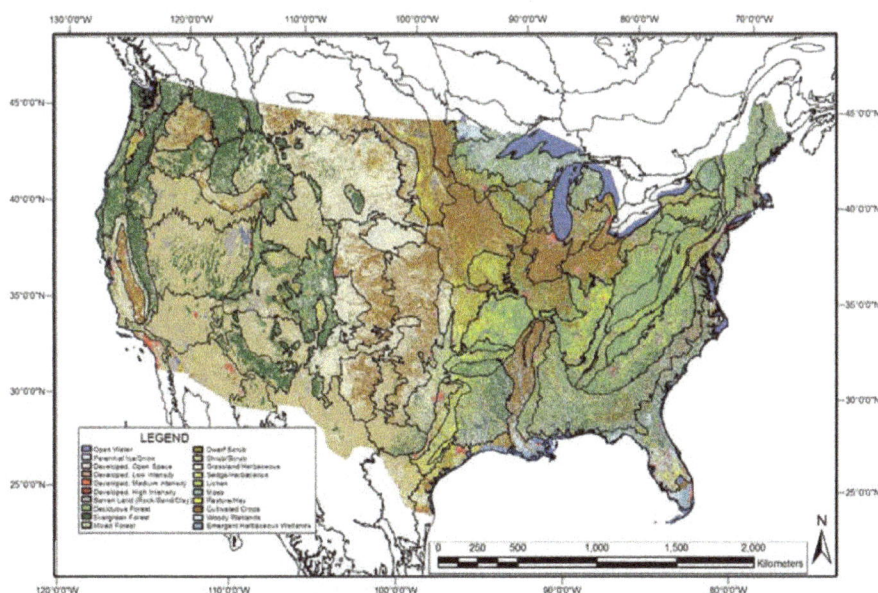

Figure 4. Reference USGS NLCD 2006 map of the CONUS. It is shown in the same scale and projection of the WELD 2006 composite depicted in Figure 6. Black lines across the USGS NLCD 2006 map represent the boundaries of the 86 EPA Level III ecoregions of the CONUS. The USGS NLCD 2006 map legend is shown on the left bottom side, also refer to Table 1

access in a provisional version in February 2011, the USGS NLCD 2006 map was "based primarily on the unsupervised classification" of Landsat-5 Thematic Mapper (TM) and "Landsat-7 Enhanced Thematic Mapper (ETM)+ images acquired in circa 2006" (Xian & Homer, 2010). It has been released as a 30 m resolution raster product in the Albers Equal Area projection, which is the cartographic projection reference standard for continental scale cartography produced by U.S. agencies. Its legend consists of 16 LC classes defined according to the Level II LC classification system; refer to Table 1 (EPA, 2007). Validation of the USGS NLCD 2006 map provided an overall accuracy (OA) of 78%, which increased to 84% when the 16 LC classes were aggregated into 9 LC classes (Stehman, Wickham, Wade, & Smith, 2008; Wickham et al., 2010, 2013). Noteworthy, these 9 LC classes are conceptually equivalent to an "augmented" 3-level 9-class FAO LCCS-DP taxonomy; refer to previous Chapter 1. The validated USGS NLCD 2006 map's OA values of 78% and 84% with, respectively, a 16 and a 9 LC class legend can be considered state of the art. For example, these OA estimates are superior to OAs featured by national-scale maps recently generated by pixel-based random forest classifiers from monthly WELD composites, whose OA is 65%–67% using 22 detailed classes and 72%–74% using 12 aggregated national classes (Wessels et al., 2016). In general, renowned experts in Geographical Information Science (GIScience) suggest that "the widely used target accuracy of 85% may often be inappropriate and that the approach to accuracy assessment adopted commonly in RS can be pessimistically biased" (Foody, 2006, 2016).

Based on these observations, we considered the USGS NLCD 2006 map's official OA estimate of 84% realistic and state of the art at the U.S. national scale when the 3-level 9-class "augmented" FAO LCCS-DP legend is adopted. We concluded that the reference USGS NLCD 2006 map was eligible for use in a GEO-CEOS Stage 4 *Val* of the SIAM application software whose output color map legend had to be reconciled with the "augmented" 3-level 9-class FAO LCCS-DP taxonomy of the reference USGS NLCD 2006 map and of a target ESA EO Level 2 product; refer to previous Chapter 1. When a test SIAM-WELD map and a reference USGS NLCD map share the same 30 m spatial resolution and spatial extent, then they can be compared wall-to-wall without sampling. Because no conventional sampling-theory procedure is employed (Lunetta & Elvidge, 1999), a wall-to-wall OA(Test SIAM-WELD; Reference NLCD) estimate $\in$ [0, 100%] is provided with a confidence interval (degree of uncertainty in measurement), $\pm \delta \geq 0$, considered mandatory by the GEO-CEOS QA4EO *Val* guidelines, equal to $\pm \delta = 0\%$.

From a statistic standpoint, the aforementioned experimental work specifications imply the following. Let us identify with OA(Test SIAM-WELD; "Ultimate" GroundTruth) $\in$ [0, 100%] = 100%—Mismatch(Test SIAM-WELD; "Ultimate" GroundTruth) the OA of an EO data-derived SIAM test map with respect to an "ultimate" (ideal) ground truth and with OA(Test SIAM-WELD; Reference NLCD) $\pm$ 0% = 100%—Mismatch (Test SIAM-WELD; Reference NLCD) $\pm$ 0% the overall degree of agreement provided with its confidence interval of a test SIAM-WELD map compared wall-to-wall without sampling with a reference USGS NLCD map at the same spatial resolution and spatial extent. It is known that (Stehman et al., 2008; Wickham et al., 2010, 2013)

OA(Reference NLCD 2006, "augmented" 3-level 9-class FAO LCCS-DP; "Ultimate" GroundTruth 2006, "augmented" 3-level 9-class FAO LCCS-DP) = 84% = 100% - Mismatch (Reference NLCD 2006, "augmented" 3-level 9-class FAO LCCS-DP; "Ultimate" GroundTruth 2006, "augmented" 3-level 9-class FAO LCCS-DP) = 100% - 16%.

Similarly (Stehman et al., 2008; Wickham et al., 2010, 2013),

OA(Reference NLCD 2006, NLCD 16 classes; "Ultimate" GroundTruth 2006, NLCD 16 classes) = 78% = 100% - Mismatch(Reference NLCD 2006, NLCD 16 classes; "Ultimate" GroundTruth 2006, NLCD 16 classes) = 100% - 22%.

Based on the superposition principle, see Figure 5, it is possible to write

Table 1. Definition of the USGS NLCD 2001/2006/2011 classification taxonomy, Level II.[2]Alaska only, consisting of 16 land cover (LC) class names. For further details, refer to the "National Land Cover Database 2006 (NLCD 2006)," Multi-Resolution Land Characteristics Consortium (MRLC), 2013. The right column instantiates a possible binary relationship R: A ⇒ B ⊆ A × B from set A = 16-class NLCD legend to set B = 2-level 4-class Dichotomous Phase (DP) taxonomy of the Food and Agriculture Organization of the United Nations (FAO)—Land Cover Classification System (LCCS) (Di Gregorio & Jansen, 2000), refer to Figure

NLCD 2001/2006/2011 classification scheme (legend), level II				LCCS-DP, level 1: A = Veg, B = Non-veg, and level 2: 1 = Terrestrial, 2 = Aquatic
Code	ID	Name	Land cover (LC) class definition	ID
11	OW	Open water	OW: Areas of open water, generally with less than 25% cover of vegetation or soil	B4—Non-vegetated aquatic
12	PIS	Perennial Ice/Snow	PIS: Areas characterized by a perennial cover of ice and/or snow, generally greater than 25% of total cover.	B4
21 22 23 24	DOS DLI DMI DHI	• Developed, Open Space • Developed, Low Intensity • Developed, Medium Intensity • Developed, High Intensity	DOS: Includes areas with a mixture of some constructed materials, but mostly vegetation in the form of lawn grasses. Impervious surfaces account for less than 20 percent of total cover. These areas most commonly include large-lot single-family housing units, parks, golf courses, and vegetation planted in developed settings for recreation, erosion control, or aesthetic purposes. DLI, DMI, DHI: refer to the "National Land Cover Database 2006 (NLCD 2006)," Multi-Resolution Land Characteristics Consortium (MRLC), 2013.	B3—Non-vegetated terrestrial/A1— Vegetated terrestrial
31	BL	Barren Land (Rock/ Sand/Clay)	BL: Barren areas of bedrock, desert pavement, scarps, talus, slides, volcanic material, glacial debris, sand dunes, strip mines, gravel pits, and other accumulations of earthen material. Generally, vegetation accounts for less than 15% of total cover. As a consequence of this constraint, class BL covers only 1.21% of the CONUS total surface.	B3
41 42 43	DF EF MF	• Deciduous Forest • Evergreen Forest • Mixed Forest	DF: Areas dominated by trees generally greater than 5 m tall, and greater than 20% of total vegetation cover. More than 75 percent of the tree species shed foliage simultaneously in response to seasonal change. EF: Areas dominated by trees generally greater than 5 m tall, and greater than 20% of total vegetation cover. More than 75 percent of the tree species maintain their leaves all year. Canopy is never without green foliage. MF: Mixed Forest—Areas dominated by trees generally greater than 5 m tall, and greater than 20% of total vegetation cover. Neither deciduous nor evergreen species are greater than 75 percent of total tree cover.	A1

(Continued)

NLCD 2001/2006/2011 classification scheme (legend), level II				LCCS-DP, level 1: A = Veg, B = Non-veg, and level 2: 1 = Terrestrial, 2 = Aquatic
Code	ID	Name	Land cover (LC) class definition	ID
51 52	-SS	• Dwarf Scrub [2] • Scrub/Shrub	SS: Areas dominated by shrubs; less than 5 m tall with shrub canopy typically greater than 20% of total vegetation. This class includes true shrubs, young trees in an early successional stage or trees stunted from environmental conditions. The aforementioned definition of class BL means that class SS may feature a vegetated cover which accounts for 15% of total cover or more.	A1/B3
71 72 73 74	GH-	• Grassland/ Herbaceous • Sedge Herbaceous [2] • Lichens [2] • Moss [2]	GH: Areas dominated by grammanoid or herbaceous vegetation, generally greater than 80% of total vegetation. These areas are not subject to intensive management such as tilling, but can be utilized for grazing. The aforementioned definition of class BL means that class GH may feature a vegetated cover that accounts for 15% of total cover or more.	A1/B3
81 82	PH CC	• Pasture/Hay • Cultivated Crops	PH: Areas of grasses, legumes, or grass-legume mixtures planted for livestock grazing or the production of seed or hay crops, typically on a perennial cycle. Pasture/hay vegetation accounts for greater than 20 percent of total vegetation. CC: Areas used for the production of annual crops, such as corn, soybeans, vegetables, tobacco, and cotton, and also perennial woody crops such as orchards and vineyards. Crop vegetation accounts for greater than 20% of total vegetation. This class also includes all land being actively tilled.	A1
90 95	WW EHW	• Woody Wetlands • Emergent Herbaceous • Wetland	WW: Areas where forest or shrubland vegetation accounts for greater than 20 percent of vegetative cover and the soil or substrate is periodically saturated with or covered with water. EHW: Areas where perennial herbaceous vegetation accounts for greater than 80% of vegetative cover and the soil or substrate is periodically saturated with or covered with water.	A2—Vegetated aquatic

OA(Test SIAM-WELD; "Ultimate" GroundTruth) $\in$ [0, 100%] = {OA(Reference NLCD; "Ultimate" GroundTruth) $\pm$ Mismatch(Test SIAM-WELD; Reference NLCD) $\pm$ 0%} $\in$ [Worst Case, Best Case], where Worst Case = max{0%, Lower Bound} and Best Case = min{100%, Upper Bound}, with Lower Bound $\leq$ Upper Bound $\in$ [0%, 100%].

When the "Ultimate" GroundTruth adopts an "augmented" 9-class LCCS-DP legend, then the aforementioned Lower and Upper Bounds become (Stehman et al., 2008; Wickham et al., 2010, 2013)

Table 2. List of the 19 spectral macro-categories generated from the aggregation by the independent human expert of the SIAM's 48 spectral categories originally detected at the intermediate level of color quantization. The *"Water or Shadow"* (WA) spectral macro-category results from the aggregation of six original SIAM categories, the *"Snow"* (SN) spectral macro-category from two and the spectral macro-category *"Others"* (O) from the aggregation of 24 original spectral categories covering disturbances typically minimized or removed in an annual composite (clouds, smoke plumes, fire fronts, etc.) as well as the original spectral category *"Unknowns."* Hence, (19–3) + 6 + 2 + 24 = 48, which is the SIAM's intermediate color discretization level. In the proposed names of spectral macro-categories, acronym Near Infra-Red (NIR) indicates Landsat TM/ETM+ band 4 (0.9 μm) and acronym Medium Infra-Red (MIR) indicates Landsat TM/ETM+ band 5 (1.6 μm). The right column instantiates a possible binary relationship R: A ⇒ C ⊆ A × C from set A = 19-class SIAM legend to set C = 2-level 4-class Dichotomous Phase (DP) taxonomy of the Food and Agriculture Organization of the United Nations (FAO)—Land Cover Classification System (LCCS) (Di Gregorio & Jansen, 2000), refer to Figure

SIAM, Intermediate discretization level (featuring 48 spectral categories) reassembled into 19 spectral macro-categories				LCCS-DP, level 1: A = Veg, B = Non-veg, and level 2: 1 = Terrestrial, 2 = Aquatic
ID	Abbreviation	OR-Aggregations	Spectral macro-category name	ID
1	sV_HC	1	Strong evidence vegetation, high canopy cover	A1—Vegetated terrestrial
2	aV_HC	1	Average evidence vegetation, high canopy cover	A1
3	wV_HC	1	Weak evidence vegetation, high canopy cover	A1
4	sV_MC	1	Strong evidence vegetation, medium canopy cover	A1
5	aV_MC	1	Average evidence vegetation, medium canopy cover	A1
6	sV_LC	1	Strong evidence vegetation, low canopy cover	A1
7	aV_LC	1	Average evidence vegetation, low canopy cover	A1
8	wbV_MLC	1	Weak evidence bright vegetation, medium or low canopy cover	A1
9	wdV_MLC	1	Weak evidence dark vegetation, medium or low canopy cover	A1/A2—Vegetated aquatic
10	sbS_1	1	Strong evidence bright soil AND NIR ≤ MIR	B3—Non-vegetated terrestrial
11	sbS_2	1	Strong evidence bright soil AND NIR > MIR	B3
12	smS_1	1	Strong evidence medium soil AND NIR ≤ MIR	B3
13	smS_2	1	Strong evidence medium soil AND NIR > MIR	B3
14	sdS	1	Strong evidence dark soil	B3
15	aS	1	Average evidence soil	B3
16	wS	1	Weak evidence soil	B3
17	SN	2	Snow	B4—Non-vegetated aquatic
18	WA	6	Water or Shadow	B4
19	O	24	Others	B—Non-vegetated
	TOT.	48		

Table 3. Spectral category-specific percentage of occurrences in the SIAM-WELD 2006/2007/ 2008/2009 test maps at the intermediate level of color quantization, where 48 basic color names were aggregated into 19 spectral macro-categories by an independent human expert. Adopted acronyms for the SIAM's 19 spectral macro-categories: refer to Table

Spectral category	2006	2007	2008	2009	Mean	Std Dev
sV_HC	33.11%	32.56%	33.79%	34.06%	33.38%	0.68%
aV_HC	19.94%	23.31%	20.02%	20.86%	21.03%	1.57%
wV_HC	0.18%	0.17%	0.17%	0.19%	0.18%	0.01%
sV_MC	0.00%	0.00%	0.00%	0.00%	0.00%	0.00%
aV_MC	20.05%	18.79%	18.07%	17.93%	18.71%	0.97%
sV_LC	0.00%	0.00%	0.00%	0.00%	0.00%	0.00%
aV_LC	0.40%	0.18%	0.31%	0.22%	0.28%	0.10%
wbV_MLC	4.12%	3.60%	3.73%	3.30%	3.69%	0.34%
wdV_MLC	1.48%	1.37%	1.19%	1.53%	1.39%	0.15%
Total vegetation	79.28%	79.98%	77.29%	78.10%	78.66%	1.20%
sbS_1	5.00%	5.44%	6.28%	5.39%	5.53%	0.54%
sbS_2	0.09%	0.13%	0.08%	0.12%	0.11%	0.02%
smS_1	4.65%	3.51%	5.38%	4.78%	4.58%	0.78%
smS_2	0.19%	0.16%	0.24%	0.20%	0.20%	0.03%
sdS	0.25%	0.28%	0.25%	0.29%	0.27%	0.02%
aS	8.04%	8.18%	8.24%	8.70%	8.29%	0.29%
wS	0.02%	0.01%	0.01%	0.01%	0.01%	0.00%
Total soils	18.24%	17.71%	20.48%	19.49%	18.98%	1.25%
SN	0.01%	0.01%	0.02%	0.01%	0.01%	0.01%
WA	1.28%	1.28%	1.25%	1.27%	1.27%	0.02%
O	1.19%	1.02%	0.96%	1.13%	1.07%	0.10%

Lower Bound = [OA(Reference NLCD 2006, "augmented" 3-level 9-class FAO LCCS-DP; "Ultimate" GroundTruth 2006, "augmented" 3-level 9-class FAO LCCS-DP) – Mismatch(Test SIAM-WELD; Reference NLCD 2006, "augmented" 3-level 9-class FAO LCCS-DP) ± 0%] = [100% – Mismatch(Reference NLCD 2006, "augmented" 3-level 9-class FAO LCCS-DP; "Ultimate" GroundTruth 2006, "augmented" 3-level 9-class FAO LCCS-DP) – (100% - OA(Test SIAM-WELD; Reference NLCD 2006, "augmented" 3-level 9-class FAO LCCS-DP) ± 0%)] = [OA (Test SIAM-WELD; Reference NLCD 2006, "augmented" 3-level 9-class FAO LCCS-DP) ± 0% - Mismatch(Reference NLCD 2006, "augmented" 3-level 9-class FAO LCCS-DP; "Ultimate" GroundTruth 2006, "augmented" 3-level 9-class FAO LCCS-DP)] = [OA(Test SIAM-WELD; Reference NLCD 2006, "augmented" 3-level 9-class FAO LCCS-DP) ± 0% - 16%],

and (Stehman et al., 2008; Wickham et al., 2010, 2013)

Upper Bound = [OA(Reference NLCD 2006, "augmented" 3-level 9-class FAO LCCS-DP; "Ultimate" GroundTruth 2006, "augmented" 3-level 9-class FAO LCCS-DP) + Mismatch(Test SIAM-WELD; Reference NLCD 2006, "augmented" 3-level 9-class FAO LCCS-DP) ± 0%] = [84% + (100% - OA(Test SIAM-WELD; Reference NLCD 2006, "augmented" 3-level 9-class FAO LCCS-DP) ± 0%)] = [184% - OA(Test SIAM-WELD; Reference NLCD 2006, "augmented" 3-level 9-class FAO LCCS-DP) ± 0%].

To recapitulate, when the "Ultimate" GroundTruth adopts an "augmented" 3-level 9-class FAO LCCS-DP legend, it is expected that (Stehman et al., 2008; Wickham et al., 2010, 2013)

Table 4. Non-square OAMTRX = FrequencyCount(A × B) instance generated from a wall-to-wall overlap between the annual SIAM-WELD 2006 test map of the CONUS with legend A = 19 spectral macro-categories and the reference USGS NLCD 2006 map with legend B = 16 LC class names. Gray entry-pair cells identify the binary relationship R: A ⇒ B ⊆ A × B chosen by the independent human expert to guide the interpretation process of the OAMTRX = FrequencyCount (A × B). Statistically independent O-Q2Is, to be jointly maximized, are CVPAI2(R: A ⇒ B) ∈ [0, 1] (where value 1 means perfect harmonization between the two input sets A and B) = 0.6689 and OA(OAMTRX = FrequencyCount(A × B)) = 96.88%. Adopted acronyms for reference LC classes and test spectral macro-categories are described in Tables and , respectively

2006 OAMTRX, probabilities (%). Rows: SIAM™-WELD 2006, 19 spectral categories; Columns: NLCD 2006, 16 land cover classes.

NLCD code	11	12	21	22	23	24	31	41	42	43	52	71	81	82	90	95	
NLCD class	OW	PIS	DOS	DLI	DMI	DHI	BL	DF	EF	MF	SS	GH	PH	CC	WW	EHW	
FAO LCCD-DP1&2	B4	B4	B3 -> A1	B3 -> A1	B3 -> A1	B3 -> A1	B3	A1	A1	A1	A1 -> B3	A1 -> B3	A1	A1	A2	A2	
SIAM™ Intermediate Granularity, 19 Spectral Categories.																	
sV_HC	0.07%	0.00%	1.00%	0.13%	0.01%	0.00%	0.02%	9.51%	4.29%	1.73%	1.10%	0.77%	3.04%	8.35%	2.76%	0.31%	33.11%
aV_HC	0.25%	0.00%	1.31%	0.82%	0.20%	0.02%	0.04%	1.38%	4.75%	0.33%	1.67%	1.66%	2.32%	3.71%	0.99%	0.50%	19.94%
wV_HC	0.04%	0.00%	0.00%	0.01%	0.01%	0.00%	0.00%	0.00%	0.04%	0.00%	0.02%	0.01%	0.00%	0.01%	0.00%	0.02%	0.18%
sV_MC	0.00%	0.00%	0.00%	0.00%	0.00%	0.00%	0.00%	0.00%	0.00%	0.00%	0.00%	0.00%	0.00%	0.00%	0.00%	0.00%	0.00%
aV_MC	0.04%	0.00%	0.70%	0.28%	0.09%	0.01%	0.05%	0.48%	2.31%	0.06%	4.93%	6.66%	1.47%	2.55%	0.19%	0.23%	20.05%
sV_LC	0.00%	0.00%	0.00%	0.00%	0.00%	0.00%	0.00%	0.00%	0.00%	0.00%	0.00%	0.00%	0.00%	0.00%	0.00%	0.00%	0.00%
aV_LC	0.00%	0.00%	0.01%	0.00%	0.00%	0.00%	0.00%	0.00%	0.01%	0.00%	0.07%	0.27%	0.02%	0.02%	0.00%	0.00%	0.40%
wbV_MLC	0.01%	0.00%	0.07%	0.07%	0.09%	0.03%	0.04%	0.00%	0.31%	0.00%	2.54%	0.79%	0.02%	0.13%	0.01%	0.01%	4.12%
wdV_MLC	0.02%	0.00%	0.02%	0.03%	0.06%	0.02%	0.03%	0.01%	0.38%	0.00%	0.71%	0.12%	0.01%	0.05%	0.00%	0.03%	1.48%
sbS_1	0.01%	0.00%	0.09%	0.04%	0.04%	0.02%	0.57%	0.00%	0.01%	0.00%	2.88%	0.90%	0.03%	0.38%	0.01%	0.01%	5.00%
sbS_2	0.00%	0.00%	0.00%	0.00%	0.00%	0.01%	0.03%	0.00%	0.00%	0.00%	0.05%	0.00%	0.00%	0.00%	0.00%	0.00%	0.09%
smS_1	0.01%	0.00%	0.05%	0.01%	0.00%	0.00%	0.04%	0.01%	0.04%	0.00%	2.09%	1.94%	0.03%	0.43%	0.00%	0.01%	4.65%
smS_2	0.00%	0.00%	0.00%	0.00%	0.00%	0.00%	0.05%	0.00%	0.00%	0.00%	0.12%	0.00%	0.00%	0.00%	0.00%	0.00%	0.19%
sdS	0.01%	0.00%	0.00%	0.00%	0.01%	0.01%	0.03%	0.00%	0.01%	0.00%	0.16%	0.01%	0.00%	0.00%	0.00%	0.00%	0.25%
aS	0.01%	0.00%	0.08%	0.04%	0.04%	0.05%	0.17%	0.01%	0.12%	0.00%	5.47%	1.76%	0.02%	0.26%	0.01%	0.01%	8.04%
wS	0.00%	0.00%	0.00%	0.00%	0.00%	0.00%	0.00%	0.00%	0.00%	0.00%	0.00%	0.01%	0.00%	0.00%	0.01%	0.00%	0.02%
SN	0.00%	0.00%	0.00%	0.00%	0.00%	0.00%	0.00%	0.00%	0.00%	0.00%	0.00%	0.00%	0.00%	0.00%	0.00%	0.00%	0.01%
WA	1.10%	0.00%	0.00%	0.00%	0.01%	0.01%	0.04%	0.00%	0.00%	0.00%	0.03%	0.01%	0.00%	0.01%	0.00%	0.05%	1.28%
O	0.14%	0.00%	0.04%	0.04%	0.03%	0.01%	0.09%	0.01%	0.08%	0.00%	0.34%	0.11%	0.03%	0.23%	0.01%	0.02%	1.19%
	1.71%	0.02%	3.36%	1.48%	0.59%	0.20%	1.21%	11.41%	12.35%	2.13%	22.19%	15.03%	6.99%	16.12%	3.99%	1.21%	100.00%

OA(Test SIAM-WELD; "Ultimate" GroundTruth 2006, "augmented" 3-level 9-class FAO LCCS-DP) ∈ [max{0%, Lower Bound}, min{100%, Upper Bound}] = [max{0%, OA(Test SIAM-WELD; Reference NLCD 2006, "augmented" 3-level 9-class FAO LCCS-DP) ± 0% - 16%}, min{100%, 184% - OA(Test SIAM-WELD; Reference NLCD 2006, "augmented" 3-level 9-class FAO LCCS-DP) ± 0%}].   (1)

Similarly, when the "Ultimate" GroundTruth adopts a 16-class NLCD legend, then it is expected that (Stehman et al., 2008; Wickham et al., 2010, 2013)

OA(Test SIAM-WELD; "Ultimate" GroundTruth 2006, NLCD 16 classes) ∈ [max{0%, Lower Bound}, min{100%, Upper Bound}] = [max{0%, OA(Test SIAM-WELD; Reference NLCD 2006, NLCD 16 classes) ± 0% - 22%}, min{100%, 178% - OA(Test SIAM-WELD; Reference NLCD 2006, NLCD 16 classes) ± 0%}].   (2)

Equations (1) and (2) are useful because, first, they highlight the undisputable fact that per se the reference USGS NLCD 2006 map is not a "ground truth" for the test SIAM-WELD map, but only a reference baseline for comparison purposes. Second, they support the validity of this experimental project by showing that a summary statistic OA(Test SIAM-WELD; "Ultimate" GroundTruth 2006) can be inferred from an estimated OA(Test SIAM-WELD; Reference NLCD 2006) ± 0% known that OA(Reference NLCD 2006; "Ultimate" GroundTruth 2006) is equal to 84% or 78% when the NLCD 2006 map and the "Ultimate" GroundTruth adopt an "augmented" 3-level 9-class FAO LCCS-DP legend or the 16-class NLCD legend, respectively.

Supported by NASA and distributed by the USGS EDC (WELD, 2016), the annual WELD image composites for years 2006, 2007, 2008, and 2009 were selected as a large-scale EO image time-series radiometrically calibrated into TOARF values as required by a GEO-CEOS stage 4 *Val* of the SIAM application software in comparison with the reference USGS NLCD 2006 map; see Figure 4. Each annual WELD image composite consists of approximately 8,000 Landsat-5/7 image acquisitions per year over the CONUS, starting from year 2003 to year 2012. The current WELD processing workflow requires as input Landsat sensor series L1T images with cloud cover ≤ 20%. The WELD composite of the CONUS encompasses 501 fixed location tiles defined in the Albers Equal Area projection. Each tile is 5000 × 5000 pixels in size, equal to 150 × 150 km (Homer et al., 2004). The Landsat sensor series L1T image geolocation error in the CONUS, including areas with substantial terrain relief, is less than 30 m (< 1 pixel) (Lee, Storey, Choate, & Hayes, 2004). The most recent Landsat data radiometric *Cal* expertise is employed in the WELD workflow to ensure harmonization and interoperability of multi-sensor Landsat image time-series, with a 5% absolute reflective band *Cal* uncertainty (Markham & Helder, 2012), in agreement with the GEO-CEOS QA4EO *Cal/Val* requirements (GEO-CEOS, 2010). Figure 6 shows the annual WELD 2006 image composite over the CONUS, where TOARF values are depicted in true colors and linearly stretched for visualization purposes, with the WELD tiling scheme overlaid in white.

To account for typical non-stationary geospatial statistics, an inter-map statistical comparison on a stratified (masked) basis should be accomplished at a local spatial extent, where strata convey some geospatial criteria of land surface information invariance. The USGS NLCD 2006 reference map was partitioned into Level III ecoregions of North America collected from the EPA (EPA, 2013). There are 86 ecoregions across the CONUS, each ecoregion featuring similar ecological and climatic characteristics (Griffith & Omernik, 2009). Distributed as vector data, the EPA Level III ecoregions were rasterized to 30 m resolution in the Albers Equal Area projection. Figure 4 shows the USGS NLCD 2006 map with boundaries of ecoregions overlaid in black.

3. Methods
A wall-to-wall comparison without sampling between the test SIAM-WELD map time-series and the reference USGS NLCD 2006 map, sharing the same 30 m spatial resolution at the CONUS spatial extent, but whose legends A = VocabularyOfColorNames (see Table 2) and B = LegendOfObjectClassNames

(see Table 1) do not coincide and must be harmonized, was designed and implemented for a GEO-CEOS stage 4 *Val* purpose. These working hypotheses differ from thematic map accuracy assessment protocols adopted by the large majority of the RS community, typically based on an either random (probabilistic) or non-random sampling in combination with a square and sorted confusion matrix, CMTRX. A CMTRX is defined as a special case of a two-way contingency table (bivariate frequency table), BIVRFTAB = FrequencyCount(A × B), where A × B is a 2-fold Cartesian product between two univariate categorical variables A and B of the same population and where A ≠ B in general (Kuzera & Pontius, 2008; Pontius & Connors, 2006; Pontius & Millones, 2011). In particular, a CMTRX is square and sorted because the test and reference categorical variables A and B of the same population are required to be the same, to let the main diagonal guide the interpretation process; refer to the Part 1, Chapter 2.

In Baraldi, Boschetti, and Humber (2014), a crisp thematic map assessment protocol was proposed based on: (i) a probability sampling strategy, (ii) a pair of test and reference thematic map legends A and B that may differ, (iii) a crisp overlapping area matrix, OAMTRX = FrequencyCount(A × B), defined as a BIVRFTAB instantiation estimated from a geospatial population with or without sampling, (Beauchemin & Thomson, 1997; Ortiz & Oliver, 2006), whose spatial unit x is (0D) pixel, (iv) a set of thematic quantitative quality indicators (Q^2Is), TQ^2Is, extracted from the OAMTRX and (v) a set of spatial Q^2Is (SQ^2Is) extracted from sub-symbolic image-objects (image-segments) in the multi-level map domain, where image-objects are either (0D) pixels, (1D) lines or (2D) polygons according to the OGC nomenclature (OGC, 2015). Whereas the construction of an OAMTRX is straightforward and non-controversial when categorical labels of sampling units are crisp (hard), the method to construct an OAMTRX when categorical labels are soft (fuzzy) is not obvious at all; for example, refer to (Kuzera & Pontius, 2008). Hence, those authors focused on crisp OAMTRX instances, exclusively.

To accomplish our present working hypotheses, the crisp thematic map probability sampling protocol proposed in (Baraldi et al., 2014) was modified as follows.

- The original hybrid eight-step guideline proposed in the Part 1, Chapter 4 was adopted to streamline the inherently subjective selection by human experts of a binary relationship R:

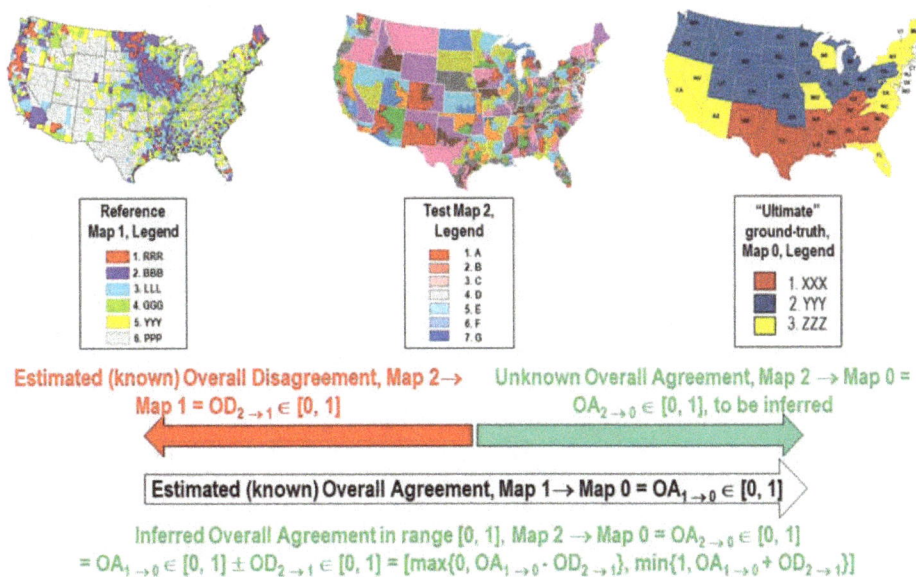

Figure 5. Superposition princi-ple in a sequence of thematic map accuracy estimates.

A = VocabularyOfColorNames ⇒ B = LegendOfObjectClassNames ⊆ A × B that guides the interpretation process of a crisp OAMTRX = FrequencyCount(A × B) = FrequencyCount (VocabularyOfColorNames × LegendOfObjectClassNames); see Table 3 in the Part 1 of this paper.

- Given a binary relationship R: A = VocabularyOfColorNames ⇒ B = LegendOfObjectClassNames to guide the interpretation process of a crisp OAMTRX = FrequencyCount(A × B) ≠ R: A ⇒ B ⊆ A × B, a novel formulation CVPAI2(R: A ⇒ B ⊆ A × B) was adopted as a relaxed version of the CVPAI1 formulation proposed in (Baraldi et al., 2014); refer to the Part 1, Chapter 5 and to Figure 18 in the Part 1 of this paper.

- Traditional 30 m resolution Landsat image classifiers are pixel-based because MS color information tends to dominate spatial information in 30 m resolution MS imagery acquired from space where; for example, individual man-made objects, such as individual buildings, roads, or agricultural fields, are typically hard to distinguish. Hence, in the 30 m resolution WELD composites, the most informative planar entity is (0D) pixel, rather than image-object, either (1D) line or (2D) polygon (OGC, 2015). As a consequence, for the sake of simplicity, in the present thematic map comparison image-object-based SQ^2Is were omitted. Rather, the following pixel-based TQ^2Is were estimated from the crisp OAMTRX = FrequencyCount(A × B) estimated wall-to-wall with spatial unit x equal to pixel.

  o An OA(OAMTRX = FrequencyCount(A × B)) ± 0% was computed in line with (Baraldi et al., 2014; Pontius & Millones, 2011; Stehman & Czaplewski, 1998). This OA estimate is guided by the binary relationship R: A = VocabularyOfColorNames ⇒ B = LegendOfObjectClassNames identified and community-agreed upon in advance; refer to this text above. In an OAMTRX = FrequencyCount(A × B) estimated from a wall-to-wall inter-map comparison, where no sample data is investigated, any adopted TQ^2I features a degree of uncertainty in measurement equal to ± 0%; for example, see Equation (1).

  o User's and producer's accuracies, computed in (Baraldi et al., 2014; Pontius & Millones, 2011; Stehman & Czaplewski, 1998), were replaced by class-conditional probabilities, $p(r \mid t)$ of reference class r given test class t and, vice versa, $p(t \mid r)$ of test class t given reference class r, with r = 1, …, RC, and t = 1, …, TC, where RC = |B| = b = ObjectClassLegendCardinality and TC = |A| = a = ColorVocabularyCardinality are the total numbers of reference and test classes, respectively.

The proposed ensemble of TQ^2I summary statistics, specifically, CVPAI2(R: A ⇒ B ⊆ A × B), OA (OAMTRX = FrequencyCount(A × B)) and class-conditional probabilities(OAMTRX), is an original minimally dependent and maximally informative (mDMI) set (Si Liu, Hairong Liu, Latecki, Xu, & Lu, 2011; Peng, Long, & Ding, 2005) of outcome Q^2Is (O-Q^2Is), to be jointly maximized according to the Pareto formal analysis of multi-objective optimization problems (Boschetti, Flasse, & Brivio, 2004); refer to the Part 1, Chapter 1.

4. Validation session
According to the GEO-CEOS *Val* guidelines (GEO-CEOS, 2010; GEO-CEOS WGCV, 2015), *Val* is the process of assessing, by independent means, the quality of an information processing system by means of an mDMI set (Si Liu et al., 2011; Peng et al., 2005) of community-agreed outcome and process (OP) Q^2Is (OP- Q^2Is), each one provided with a degree of uncertainty in measurement, ± δ, with δ ≥ 0%.

In the present study, the following definition is adopted: an information processing system can be considered in operating mode (ready-to-go) if it scores "high" in all of its OP-Q^2I estimates; refer to the Part 1, Chapter 1.

To comply with the GEO-CEOS stage 4 *Val* requirements (GEO-CEOS WGCV, 2015), refer to previous Chapter 1, the SIAM-WELD data mapping process and outcome were validated by a human expert independent of the present authors (refer to Acknowledgments). This independent human expert accomplished the following tasks. (I) Run without user interaction an off-the-shelf

SIAM application upon the 30 m resolution annual WELD 2006 to 2009 image composites of the CONUS. (II) Overlap wall-to-wall the test SIAM-WELD annual map time-series with the reference USGS NLCD 2006 map to generate instances of an OAMTRX = FrequencyCount(A × B) (Baraldi et al., 2014). (III) Estimate an mDMI set of OP-Q^{2}Is, defined as follows, in agreement with the Part 1, Chapter 1 (Baraldi & Boschetti, 2012a, 2012b; Baraldi & Humber, 2015; Baraldi et al., 2013).

(i) Product effectiveness. Proposed outcome Q^{2}Is (O-Q^{2}Is) were the TQ^{2}Is presented in Chapter 3: (a) CVPAI2(R: A ⇒ B ⊆ A × B; (b) OA(OAMTRX = FrequencyCount(A × B)), and (c) class-conditional probabilities $p(r \mid t)$ and p(t $\mid$ r) with test class $t = 1, ..., TC = |A| = $ ColorVocabularyCardinality and reference class $r = 1, ..., RC = |B| = $ ObjectClassLegendCardinality.

(ii) Process efficiency. Proposed process Q^{2}Is (P-Q^{2}Is) were: (a) computation time and (b) run-time memory occupation.

(iii) Process degree of automation, monotonically decreasing with the number of system's free-parameters to be user defined.

(iv) Process robustness to changes in the input dataset. For post-classification change/no-change detection (Lunetta & Elvidge, 1999), the SIAM-WELD 2006 to 2009 maps were compared with one another when one year apart.

(v) Process robustness to changes in input parameters, if any.

(vi) Process scalability, to keep up with changes in users' needs and sensor specifications.

(vii) Product timeliness, defined as the time between data acquisition and product generation.

(viii) Product costs in manpower and computer power.

For the sake of paper simplicity, the following decisions were undertaken.

• Two-of-three SIAM-WELD 2006 output color maps, specifically, the one featuring 96 color names, equivalent to a fine color granularity, and the one featuring 48 color names, equivalent to an intermediate color granularity, were compared with the 16-class NLCD 2006 map, while the SIAM-WELD 2006 color map featuring 18 color names, equivalent to a coarse color granularity, was ignored, see Table 1 in the Part 1. This implied the following.

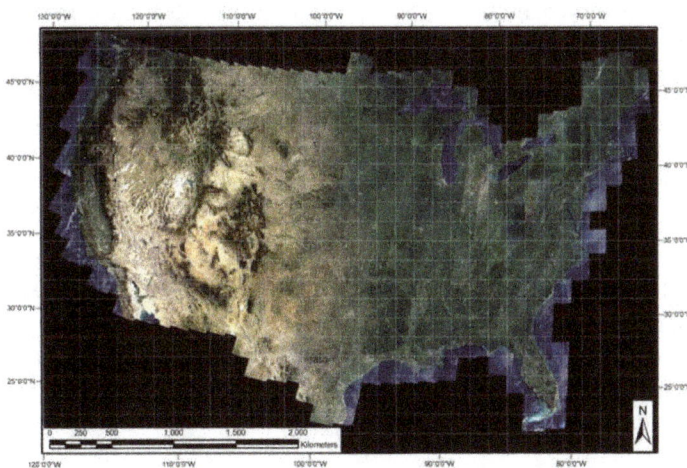

Figure 6. 30 m resolution annual Web-Enabled Landsat Data (WELD) image composite for the year 2006 (December 2005 to November 2006) of the conterminous U.S. (CONUS), radiometrically calibrated into top-of-atmosphere reflectance (TOARF) values. Depicted in true colors (red: Band 3, 0.63–0.69 µm; green: Band 2, 0.53–0.61 µm, and blue: Band 1, 0.45–0.52 µm), linearly stretched for visualization pur-poses. The white grid shows locations of the 501 WELD tiles of the CONUS. Each tile is 5000 × 5000 pixels in size, cov-ering a surface area of 150 × 150 km. Pixels are geo-graphically projected in the Albers Equal Area projection.

o In the case of a SIAM-WELD 2006 map featuring 96 color names at the SIAM fine color
   discretization level, an OAMTRX = FrequencyCount(A × B) instance consisted of a test set
   A = 96 spectral categories as rows and a reference set B = 16 NLCD classes as columns.
   Because of its excessive size, equal to 96 × 16 cells, this OAMTRX instance cannot be shown
   in a technical paper. However, it is made available on an anonymous ftp site (SIAM-WELD-
   NLCD FTP, 2016), and its TQ^2I summary statistics are reported in the present paper.

o In the case of a SIAM-WELD 2006 map featuring 48 color names at the SIAM intermediate
   color discretization level, an OAMTRX = FrequencyCount(A × B) instance consisted of a test
   set A = 48 spectral categories as rows and a reference set B = 16 NLCD classes as columns.
   Because of its excessive size, equal to 48 × 16 cells, this OAMTRX instance cannot be shown
   in a technical paper. Hence, the SIAM ensemble of 48 basic color (BC) names at intermediate color granularity was perceptually ("subjectively") reassembled into 19 spectral macro-
   categories, refer to Table 2 (Benavente, Vanrell, & Baldrich, 2008; Berlin & Kay, 1969; Gevers,
   Gijsenij, Van De Weijer, & Geusebroek, 2012; Griffin, 2006), by the independent human
   expert who adopted in support the hybrid guideline for binary relationship detection proposed in the Part 1, Chapter 4. This reduced set of 19 spectral macro-categories was
   constrained to be mutually exclusive and totally exhaustive, in line with the Congalton
   and Green's requirements of a classification scheme (Congalton & Green, 1999). Such a
   grouping of BC names into parent spectral macro-categories scrutinized and agreed upon
   by a human expert pertains to the inherently equivocal (subjective) domain of *information-
   as-data-interpretation*; refer to the Part 1, Chapter 4 (Capurro & Hjørland, 2003). Among the
   19 spectral macro-categories reassembled by the independent human expert, 16 macro-
   categories coincided exactly with one BC name in the SIAM set of 48 BC names at intermediate granularity. One-of-19 spectral macro-category, named "*Others*" by the independent human expert, grouped 25-of-48 BC names detected by SIAM at intermediate color
   granularity. Among these 25-of-48 BC names, one is identified by SIAM as category
   "*Unknowns*" (outliers), while the remaining 24 BC names are all related to spectral signatures equivalent to "noisy" data (no terrain data), such as spectral signatures typical of LC
   classes cloud, smoke plume, active fire, and so on, which are typically minimized in an
   annual WELD image composite according to the known WELD multi-temporal pixel selection policies; refer to previous Chapter 2. As a consequence of grouping the SIAM's 48 BC
   names into 19 spectral macro-categories, a simplified OAMTRX instance of reduced size was
   generated as OAMTRX = FrequencyCount(A^ × B), where a test set A^ = 19 spectral macro-
   categories was adopted as rows and a reference set B = 16 NLCD classes was adopted as
   columns. Thanks to its size, reduced to 19 × 16 cells, this OAMTRX instance can be shown, in
   combination with its estimated TQ^2I values, in the present paper.

• In agreement with the previous paragraph, the annual SIAM-WELD 2006 to 2009 maps at the
  SIAM intermediate color discretization level of 48 color names were all reassembled into 19
  spectral macro-categories; see Table 2.

4.1. *Verification of the co-registration requirements for pixel-based inter-map comparison*

In the requirements specification of RS projects dealing with per-pixel post-classification change/
no-change detection, the required RS image co-registration error is typically < 1 pixel. For example,
in (Lunetta & Elvidge, 1999), it is recommended that the root-mean-square (RMS) co-registration
error between any pair of two-date imagery should not exceed 0.5 pixels.

In (Dai & Khorram, 1998), simulated misregistration effects are investigated upon multi-temporal Landsat images of North Carolina across four study areas representative of land cover types:
forest land, agricultural land, bare soil, and urban/residential area. In these experiments, a
registration accuracy < 1/5 of a pixel is considered necessary to achieve a land cover change
detection error < 10%. This conclusion is more severe than the one-pixel co-registration constraint
typically adopted in most change detection applications.

The annual WELD composites and the USGS NLCD 2006 thematic map were derived from the same sensory dataset of Landsat L1T images acquired by the USGS EDC. It means that the SIAM-WELD 2006 pre-classification maps and the USGS NLCD 2006 reference map were derived from the same sensory dataset. Hence, it is reasonable to assume that the co-registration error between these data-derived maps is negligible.

4.2. Inter-annual SIAM-WELD map comparisons for years 2006 to 2009

The consistency across time and space of the annual SIAM-WELD 2006 to 2009 map time-series featuring a map legend of 19 spectral macro-categories was investigated. Based on a priori knowledge of the multi-temporal pixel-based selection criteria adopted by the USGS EDC for the generation of annual WELD composites (refer to Chapter 2) and of the LC/LC change (LCC) dynamics in the real-world CONUS, a small percentage of LCC counts was expected to be detected one year apart at the CONUS spatial extent.

Table 3 shows class-conditional percentages collected at the CONUS spatial extent across the annual WELD image composite time-series for each of the 19 SIAM-WELD spectral macro-categories. The green-as-"*Vegetation*" spectral macro-categories are predominant (refer to the total vegetation statistic reported in Table 3), with an average 79% of the CONUS pixels, followed by MS color names such as brown-as-"*Bare soils or built-up*" (19% on average), followed by the remaining spectral macro-categories that, altogether, account for about 2%. The standard deviation through time of the occurrence of each SIAM-WELD spectral macro-category at the CONUS spatial extent is lower than 1%, with the exception of two vegetation spectral macro-categories (specifically, aV_HC and aV_MC) where a larger variance can be attributed mostly to phenology. If a vegetation-through-time spectral variability due to changes in phenology affects the annual WELD composites then the data-derived SIAM-WELD color quantization maps will be affected by changes in phenology too. This diagnosis was verified as follows. Because of the limited availability of cloud-free Landsat observations at a generic pixel location per year, the Julian day of the year of the observation selected at a given location (pixel) of the annual WELD image composite changes through years (Roy et al., 2010). This is illustrated in Figure 7 where, at any fixed location across a target "ground-truth" area of deciduous forest in a pair of monthly August-November WELD composites, the SIAM spectral labels change significantly, but consistently with the phenological season. The same consideration holds when changes in phenology affect the annual WELD composites. This can explain the "high" intra-vegetation spectral variability observed by the SIAM vegetation-related spectral macro-categories aV_HC and aV_MC in the tested time-series of annual WELD composites for years 2006 to 2009.

Non-stationary spatial phenomena occurring at the CONUS spatial extent in the geospatial physical world can be oversighted by global statistics. To be better captured, spatial non-stationarities require more local statistics, such as class-conditional global statistics described in Table 3.

According to previous Chapter 3, for every pair of one annual SIAM-WELD test map with legend A = 19 spectral macro-categories for year 2006 to year 2009 overlapped at the CONUS spatial extent with a reference USGS NLCD 2006 map with legend B = NLCD 16 classes, the pair of summary statistics CVPAI2(R: A $\Rightarrow$ B $\subseteq$ A $\times$ B) $\in$ [0, 1.0] and OA(OAMTRX = FrequencyCount (A $\times$ B)) $\in$ [0, 1.0] should be maximized jointly. Shown as gray entry-pair cells in Table 4, a binary relationship R: A $\Rightarrow$ B was identified by the independent human expert, who adopted the hybrid eight-step guideline for identification of a categorical variable-pair relationship proposed in the Part 1, Chapter 4. The binary relationship R: A $\Rightarrow$ B, selected by the human expert and shown in Table 4, provides a CVPAI2(R: A $\Rightarrow$ B) = 0.6689, while the OA(OAMTRX = FrequencyCount (A $\times$ B) = Table 4) = OA(Test SIAM-WELD 2006, 19 spectral macro-categories; Reference NLCD 2006, NLCD 16 classes) = 96.88% ± 0%. With a binary relationship R: A $\Rightarrow$ B kept fixed, where CVPAI2(R: A $\Rightarrow$ B) = 0.6689, the OA(OAMTRX) estimate became equal to 97.02%, 96.69%, and 96.75% for the annual SIAM-WELD map of year 2007 to year 2009 compared with the reference USGS NLCD 2006 map.

4.3. Comparison of the SIAM-WELD 2006 and NLCD 2006 thematic maps

The pair of SIAM-WELD 2006 test maps at intermediate and fine color quantization levels, featuring 48 and 96 BC names respectively, were compared wall-to-wall with the USGS NLCD 2006 reference map as described below.

4.3.1. Test case A

The SIAM-WELD 2006 map of the CONUS at the intermediate color discretization level reassembled into 19 spectral macro-categories is shown in Figure 8. The OAMTRX = FrequencyCount(A × B) instance generated from the overlap between the test SIAM-WELD 2006 map with legend A = 19 spectral macro-categories at the CONUS spatial extent with a reference USGS NLCD 2006 map with legend B = NLCD 16 classes is shown in Table 4, where each cell reports a joint probability value p (SIAM-WELD$_t$, NLCD$_r$), $r = 1, ..., RC = |B| = 16$, refer to Table 2, and $t = 1, ..., TC = |A| = 19$, refer to Table 1. Gray entry-pair cells identify the binary relationship R: A ⇒ B ⊆ A × B ≠ OAMTRX = FrequencyCount(A × B) chosen by the independent human expert to guide the OAMTRX interpretation process. The distribution of these "correct" entry-pairs shows that every NLCD class overlaps with several discrete color types, with the exceptions of two SIAM-NLCD entry-pairs, specifically, entry-pair [SIAM spectral macro-category, NLCD class] = [MS white-as-"Snow" (SN, see Table 2), NLCD class "Perennial ice/snow" (PIS, see Table 1)] and entry-pair [SIAM spectral macro-category, NLCD class] = [MS blue-as-"Water or Shadow" (WA, see Table 2), NLCD class "Open water" (OW, see Table 1)], which are both characterized by a 1–1 matching relation. According to their specific definitions in natural language (refer to Table 1), anthropic NLCD classes, such as "Developed, Open Space" (DOS), "Developed, Low Intensity" (DLI), "Developed, Medium Intensity" (DMI) and "Developed, High intensity" (DHI), are a mixture of vegetated surfaces, impervious surfaces and bare soil, in agreement with the popular vegetation-impervious surface-soil model for urban ecosystem analysis (Ridd, 1995). In agreement with their definitions in human language, these NLCD classes overlap exclusively with the SIAM spectral macro-categories related to vegetation or bare soil. The USGS NLCD class "Barren Land" (BL, see Table 1) overlaps with all of the SIAM spectral macro-categories related to bare soil. Noteworthy, according to Table 4, the USGS NLCD class BL covers only 1.21% of the CONUS total surface. This is due to the USGS NLCD 2006 definition of class BL (Rock/Sand/Clay), very restrictive with regard to the presence of vegetation, which has to account for less than 15% of total cover. The USGS NLCD definition of class BL means that the USGS NLCD classes "Shrub/Scrub" (SS) and "Grassland/Herbaceous" (GH, refer to Table 1) may feature a vegetated cover which accounts for 15% of total cover or more. The USGS NLCD forest classes "Deciduous forest" (DF), "Evergreen Forest" (EF) and "Mixed forest" (MF, refer to Table 1) overlap with the SIAM's high and medium canopy cover-related spectral macro-categories. The USGS NLCD vegetation classes "Shrub/Scrub" (SS) and "Grassland/Herbaceous" (GH, refer to Table 1) overlap with the SIAM-WELD 2006 medium and low canopy cover-related spectral macro-categories, but, in case of dry or sparse vegetation, also with some of the SIAM-WELD 2006 spectral macro-categories related to bare soil, namely, sbS_1, SmS_1, and aS (refer to Table 2). The overlap between the reference USGS NLCD 2006 vegetation classes SS and GH and the test SIAM-WELD 2006 bare soil spectral macro-categories sbS_1, SmS_1, and aS is the only case of comprehensive (systematic) "semantic mismatch" recorded across the wall-to-wall SIAM-WELD 2006 and NLCD 2006 thematic map pair comparison. Hence, it is worth a deeper analysis in comparison with an "ultimate" ground truth. Reported above in this chapter, the USGS NLCD 2006 definition of class "Barren Land" (BL, see Table 1) means that the USGS NLCD vegetation classes SS and GH may feature a vegetated cover that accounts for 15% of total cover or more. Two consequence of these three NLCD class definitions are that, whereas the USGS NLCD class BL covers only 1.21% of the CONUS total surface, the USGS NLCD vegetation classes SS and GH map the near totality of desert areas across the CONUS. Hence, there is a systematic "semantic mismatch" between the USGS NLCD 2006 vegetation classes SS and GH and the SIAM-WELD 2006 bare soil spectral macro-categories across nearly all desert areas of the CONUS. Figure 9 shows real-world examples of geographic locations mapped as vegetation classes "Scrub/Shrub" (SS) or "Grassland/Herbaceous" (GH) in the USGS NLCD 2006 map (refer to Table 1), while they are mapped predominantly as the bare soil spectral categories sbS_1, SmS_1, and aS in the SIAM-WELD 2006 map (refer to Table 2).

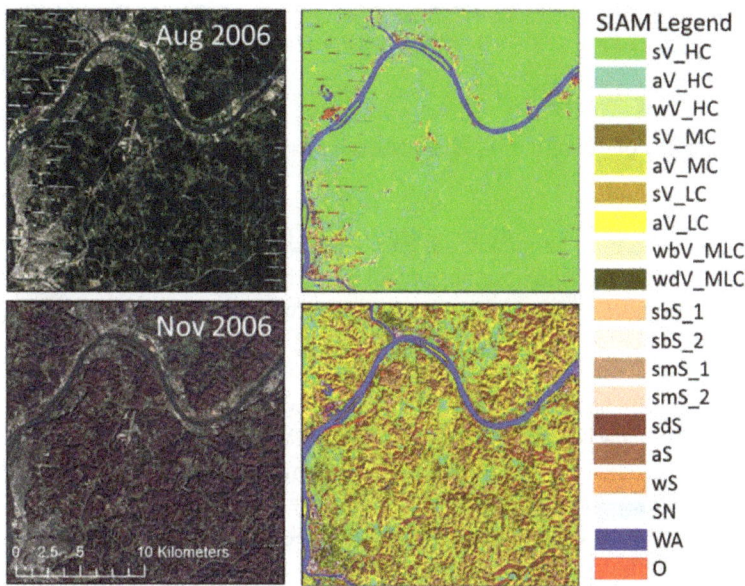

Figure 7. Changes through time of the 19-class SIAM spectral macro-category labels due to vegetation phenology affecting the monthly WELD composite. *Left side*:30mresolution monthly WELD composites, radiometrically calibrated into top-of-atmosphere reflectance (TOARF) values, for August and November 2006, showing an area predominantly covered by broadleaf forest in the Mid-Western United States (Ohio). Depicted in true colors (red: Band 3, 0.63–0.69 μm; green: Band 2, 0.53–0.61 μm, and blue: Band 1, 0.45–0.52 μm). To allow inter-imagecomparison,thetwo images are displayed with an identical contrast stretch. *Right side*: SIAM-WELD color maps generated from the two WELD images shownonthe leftside. The SIAM map legend, consisting of 19 spectral macro-categories, isshownonthe rightside, also refer to Table 2.

For more comments about this systematic case of "conceptual mismatch" between the test SIAM-WELD and reference USGS NLCD 2006 maps, refer to Figure 12.

Additional inter-map overlaps highlighted by Table 4 reveal that the USGS NLCD class *"Pasture/Hay"* (PH, see Table 1) occurs together with high and medium canopy cover-related BC names in the SIAM-WELD map. The USGS NLCD class *"Cultivated crops"* (CC, see Table 1) matches with both SIAM's spectral macro-categories MS green-as-*"Vegetation"* and MS brown-as-*"Bare soil or built-up."* Finally, the USGS NLCD classes of wetland, specifically, *"Woody Wetlands"*, WW, and *"Emergent Herbaceous Wetland,"* EHW, see Table 1, overlap with the SIAM's vegetated spectral macro-categories or with spectral macro-category MS blue-as-*"Water or Shadow"* (WA, refer to Table 2).

As reported in previous Chapter 4.2, in the OAMTRX = FrequencyCount(A × B) instance shown in Table 4, gray entry-pair cells were identified as "correct" by the independent human expert, based on the hybrid eight-step guideline proposed in the Part 1, Chapter 4. They identify the binary relationship, R: A ⇒ B ⊆ A × B ≠ OAMTRX = FrequencyCount(A × B), suitable for guiding the interpretation process in the OAMTRX at hand. In the OMATRX instance shown in Table 4, the mDMI set of O-Q^{2}Is to be jointly maximized comprises summary statistics OA (OAMTRX = FrequencyCount(A × B)) = OA(Test SIAM-WELD 2006, 19 spectral macro-categories; Reference NLCD 2006, NLCD 16 classes) = 96.88% ± 0% with CVPAI2(R: A ⇒ B ⊆ A × C) = 0.6689. As a consequence, according to Equation (2),

OA(OAMTRX = FrequencyCount(A × B)) = OA(Test SIAM-WELD 2006, 19 spectral macro-categories; "Ultimate" GroundTruth 2006, NLCD 16 classes) ∈ [max{0%, Lower Bound}, min {100%, Upper Bound}] = [max{0%, OA(Test SIAM-WELD 2006, 19 spectral macro-categories; Reference NLCD 2006, NLCD 16 classes) ± 0% - 22%}, min{100%, 178% - OA(Test SIAM-WELD

2006, 19 spectral macro-categories; Reference NLCD 2006, NLCD 16 classes) ± 0%}] = [max {0%, 96.88% ± 0% - 22%}, min{100%, 178% - 96.88% ± 0%}] = [74.88%, 81.12%],

with

CVPAI2(R: A = SIAM vocabulary of 19 BC names ⇒ B = NLCD legend of 16 LC class names) =0.6689.   (3)

Hence, the semantic information gap, to be minimized, from input sub-symbolic sensory data to output symbolic NLCD classes, left to be filled (disambiguated) by further stages in the hierarchical EO-IUS pipeline, where spatial information is masked by first-stage color names, see Figure 3, is equal to (1—CVPAI2) = 0.3311; refer to the Part 1, Chapter 5.

When disagreements between the two reference and test maps were back-projected onto the WELD 2006 image domain, these specific WELD sites were photointerpreted by the independent human expert to provide an additional independent source of thematic evidence for GEO-CEOS stage 4 *Val* of the annual SIAM-WELD 2006 test map in BC names. The large majority of the CONUS areas where the USGS NLCD vegetation classes overlap with the SIAM spectral macro-category MS blue-as-"*Water or Shadow*" (WA, refer to Table 2) or, vice versa, where the SIAM vegetation spectral macro-categories overlap with the USGS NLCD reference class "*Open Water*" (OW, refer to Table 1) were identified by the independent human photointerpreter as riparian zones. In practice, these riparian zones were labeled by the annual SIAM-WELD 2006 and NLCD 2006 maps in two different conditions of their annual surface status. Also in this case, the SIAM-WELD labeling appears consistent with the human photointerpretation of the annual WELD composite, irrespective of the semantic disagreement between this SIAM-WELD labeling and the reference USGS NLCD 2006 map.

Based on evidence collected by the independent photointerpreter with regard to systematic "conceptual mismatches" between the test SIAM-WELD 2006 map and the reference USGS NLCD 2006 map across nearly all desert areas and nearly all riparian zones of the CONUS, validated conclusions were twofold. First, according to Equation (1), where the reference USGS NLCD 2006 map is acknowledged to be no "ground truth" for the annual SIAM-WELD 2006 test map, but only a reference baseline for comparison purposes, "conceptual mismatches" between the test SIAM-

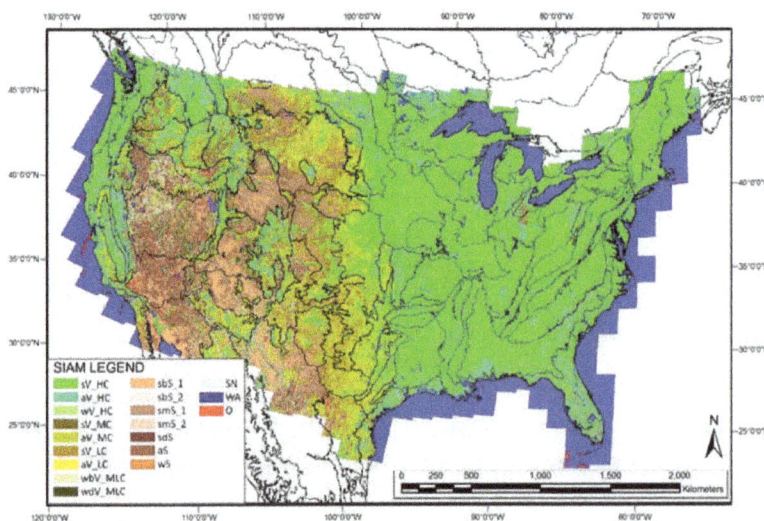

Figure 8. Automatically gener-ated SIAM-WELD 2006 color map depicted at an intermedi-ate discretization level of 48 color names, reassembled into 19 spectral macro-categories by an independent human expert. Black lines across the SIAM-WELD 2006 map represent the boundaries of the 86 EPA Level III ecoregions of the CONUS. The reassembled 19-class SIAM map legend is depicted at bottom left, also refer to Table 2.

Figure 9. See note.

WELD 2006 map and the reference USGS NLCD 2006 map should not be misconceived as mapping errors by the test SIAM-WELD 2006 map with respect to an "ultimate" ground truth. Second, according to Equation (3), summary statistic OA(OAMTRX = FrequencyCount(A × B) = OA(Test SIAM-WELD 2006, 19 spectral macro-categories; "Ultimate" GroundTruth 2006, NLCD 16 classes) was inferred to belong to range [74.88%, 81.12%], to be assessed in combination with a CVPAI2(R: A ⇒ B) value in range [0, 1], estimated equal to 0.6689.

It is important to recall here that for any given two-way frequency table OAMTRX = FrequencyCount (A × B) generated from two categorical variables A and B of the same population, where A ≠ B in general, the OAMTRX's pair of O-Q^2Is forming an mDMI set of quality indexes to be jointly maximized is OA(OAMTRX = FrequencyCount(A × B)) ∈ [0, 1] and CVPAI2(R: A ⇒ B ⊆ A × B ≠ OAMTRX = FrequencyCount(A × B)) ∈ [0, 1], where the latter, estimated from the binary relationship guiding the interpretation process of the OAMTRX at hand, is independent of the OA(OAMTRX) estimated value. Only if A = B then OAMTRX = (square and sorted) CMTRX whose main diagonal guides the interpretation process, CVPAI2(R: A ⇒ B) = 1 and OA(OAMTRX) ∈ [0, 1] becomes the sole O-Q^2I informative per se about the degree of match between bivariate occurrences A and B.

4.3.2. Test case B

To reveal the inherent ill-posedness of "conceptual matching" between two categorical variables A and B investigated by Ahlqvist (2005) (refer to the Part 1, Chapter 4), one co-author of this paper, different from the independent human expert (refer to Acknowledgments), conducted a second inherently equivocal selection of "correct" entry-pairs in a binary relationship, R: A $\Rightarrow$ B $\subseteq$ A $\times$ B, to guide the interpretation process of the OAMTRX = FrequencyCount(A $\times$ B) instance shown in Table 4. This second experiment provided an mDMI pair of O-Q^2Is equal to OA(OAMTRX = FrequencyCount(A $\times$ B)) = 97.28% $\pm$ 0% and CVPAI2(R: A $\Rightarrow$ B) = 0.6731. They are both superior to (better than) the pair of summary statistics OA(OAMTRX = FrequencyCount(A $\times$ B)) = 96.88% $\pm$ 0% and CVPAI2(R: A $\Rightarrow$ B) = 0.6689 provided by the independent human expert in the test case A. This alternative binary relationship R: A $\Rightarrow$ B looks sparser and therefore less intuitive to understand than that shown as gray entry-pair cells in Table 4. Hence, it is not shown in this paper, although it is made available via anonymous ftp (SIAM-WELD-NLCD FTP, 2016). When these summary statistics replace variables in Equation (2), we obtain

OA(OAMTRX = FrequencyCount(A $\times$ B)) = OA(Test SIAM-WELD 2006, 19 spectral macro-categories; "Ultimate" GroundTruth 2006, NLCD 16 classes) $\in$ [max{0%, Lower Bound}, min{100%, Upper Bound}] = [max{0%, OA(Test SIAM-WELD 2006, 19 spectral macro-categories; Reference NLCD 2006, NLCD 16 classes) $\pm$ 0% - 22%}, min{100%, 178% - OA(Test SIAM-WELD 2006, 19 spectral macro-categories; Reference NLCD 2006, NLCD 16 classes) $\pm$ 0%}] =[max{0%, 97.28% $\pm$ 0% - 22%}, min{100%, 178% - 97.28% $\pm$ 0%}] =[75.28%, 80.72%],

with

CVPAI2(R: A = SIAM vocabulary of 19 BC names $\Rightarrow$ B = NLCD legend of 16 LC class names) = 0.6731.   (4)

Hence, the semantic information gap, to be minimized, from input sub-symbolic sensory data to output symbolic NLCD classes, left to be filled (disambiguated) by further stages in the hierarchical EO-IUS pipeline, where spatial information is masked by first-stage color names, see Figure 3, is equal to (1—CVPAI2) = 0. 3269; also refer to the Part 1, Chapter 5.

4.3.3. Test case C

The wall-to-wall overlap between the test SIAM-WELD 2006 map, whose legend A = 96 BC names belonging to the SIAM fine color discretization level (refer to Table 1 in the Part 1 of this paper), and the reference USGS NLCD 2006 map, with legend B = 16 LC classes, generated another OAMTRX = FrequencyCount(A $\times$ B) instance, 96 $\times$ 16 cells in size, too large to be shown in a technical paper, but made available via anonymous ftp (SIAM-WELD-NLCD FTP, 2016). Once again, the hybrid inference procedure described in the Part 1, Chapter 4 was employed by the independent human expert to select "correct" entry-pairs in the binary relationship R: A $\Rightarrow$ B $\subseteq$ A $\times$ B eligible for guiding the interpretation process of this OAMTRX instance. Estimated mDMI set of O-Q^2Is became OA(OAMTRX = FrequencyCount(A $\times$ B)) = 95.41% $\pm$ 0% and CVPAI2(R: A $\Rightarrow$ B $\subseteq$ A $\times$ B) = 0.5809. When these summary statistics replace variables in Equation (2), we obtained

OA(OAMTRX = FrequencyCount(A $\times$ B)) = OA(Test SIAM-WELD 2006, 96 BC names; "Ultimate" GroundTruth 2006, NLCD 16 classes) $\in$ [max{0%, Lower Bound}, min{100%, Upper Bound}] = [max{0%, OA(Test SIAM-WELD 2006, 96 spectral categories; Reference NLCD 2006, NLCD 16 classes) $\pm$ 0% - 22%}, min{100%, 178% - OA(Test SIAM-WELD 2006, 96 spectral macro-categories; Reference NLCD 2006, NLCD 16 classes) $\pm$ 0%}] = [max{0%, 95.41% $\pm$ 0% - 22%}, min{100%, 178% - 95.41% $\pm$ 0%}] = [73.41%, 82.59%],

with

CVPAI2(R: A = SIAM vocabulary of 96 BC names $\Rightarrow$ B = NLCD legend of 16 LC class names) = 0. 5809.   (5)

Hence, the semantic information gap, to be minimized, from input sub-symbolic sensory data to output symbolic NLCD classes, left to be filled (disambiguated) by further stages in the hierarchical EO-IUS pipeline, where spatial information is masked by first-stage color names, see Figure 3, is equal to (1—CVPAI2 = 0. 4191); also refer to the Part 1, Chapter 5.

4.3.4. Test case D

When the USGS NLCD 2006 classification taxonomy became less discriminative (coarser) because reassembled from its original 16 LC class names into either 9 LC class names (Stehman et al., 2008; Wickham et al., 2010, 2013), refer to previous Chapter 2, or 4 LC class names, see Table 1, constrained by inter-level parent-child relationships in agreement with the FAO LCCS-DP taxonomy, see Figure 1, then the following inequality holds true.

OA(Reference NLCD 2006, NLCD 16 classes; "Ultimate" GroundTruth 2006, NLCD 16 classes) = 78% (Wickham et al. 2010; Wickham et al. 2013; Stehman et al. 2008) ≤ OA(Reference NLCD 2006, "augmented" 3-level 9-class FAO LCCS-DP; "Ultimate" GroundTruth 2006, "augmented" 3-level 9-class FAO LCCS-DP) = 84% (Wickham et al. 2010; Wickham et al. 2013; Stehman et al. 2008) ≤OA(Reference NLCD 2006, 2-level 4-class FAO LCCS-DP taxonomy; "Ultimate" GroundTruth 2006, 2-level 4-class FAO LCCS-DP taxonomy) = XX%, hence, XX% ≥ 84%,   (6)

where XX% is an unknown variable expected to remain unspecified because we have no chance to access the "Ultimate" GroundTruth 2006 dataset, adopted in past works to validate the USGS NLCD 2006 map (Stehman et al., 2008; Wickham et al., 2010, 2013), to reassemble its original "augmented" 3-level 9-class FAO LCCS-DP taxonomy into a 2-level 4-class FAO LCCS-DP taxonomy consisting of LC classes (see Figure 1):

- A1 = Primarily Vegetated Terrestrial Areas = Cultivated Areas (A11) or (Semi) Natural Vegetation (A12).
- A2 = Primarily Vegetated Aquatic or Regularly Flooded Areas = Cultivated Aquatic Areas (A23) or (Semi) Natural Aquatic Vegetation (A24).
- B3 = Primarily Non-vegetated Terrestrial Areas = Artificial Surfaces (B35) or Bare Areas (B36).
- B4 = Primarily Non-vegetated Aquatic or Regularly Flooded Areas = Artificial (B47) or Natural Waterbodies, Snow and Ice (B48).

In agreement with Equations (1), (2), and (6), we could write

OA(Test SIAM-WELD 2006, 19 spectral macro-categories; "Ultimate" GroundTruth 2006, 2-level 4-class FAO LCCS-DP taxonomy) ∈ [max{0%, Lower Bound}, min{100%, Upper Bound}] = [max{0%, OA(Test SIAM-WELD 2006, 19 spectral macro-categories; Reference NLCD 2006, 2-level 4-class FAO LCCS-DP taxonomy) ± 0% - (100% - XX%)}, min{100%, 100% + XX% - OA (Test SIAM-WELD 2006, 19 spectral macro-categories; Reference NLCD 2006, 2-level 4-class FAO LCCS-DP taxonomy) ± 0%}], where unknown variable XX% = Equation (6) = OA(Reference NLCD 2006, 2-level 4-class FAO LCCS-DP taxonomy; "Ultimate" GroundTruth 2006, 2-level 4-class FAO LCCS-DP taxonomy) ≥ 84%.   (7)

As shown in the test case A, according to the USGS NLCD taxonomy definitions (refer to Table 1), LC classes "Developed, Open Space" (DOS), "Developed, Low Intensity" (DLI), "Developed, Medium Intensity" (DMI) and "Developed, High intensity" (DHI) describe a spatial mixture of vegetated surfaces, impervious surfaces and bare soil types, in agreement with the popular vegetation-impervious surface-soil model for urban ecosystem analysis (Ridd, 1995). It means that a logical OR combination of the USGS NLCD classes DOS or DLI or DMI or DHI mainly matches with 2-of-4 FAO LCCS-DP 2nd-level classes, either B3 or A1. Experts in the domain of world ontologies and in the harmonization of LC class taxonomies, the present authors concluded it is not possible to define without ambiguity each of the four LC classes in the 2-level 4-class FAO LCCS-DP taxonomy as a mutually exclusive and totally exhaustive parent-child relationship starting from the 16 LC classes in the USGS NLCD taxonomy. Nevertheless, to harmonize these two thematic map legends, we arbitrarily selected a sub-optimal binary relationship from set A = NLCD 16 LC class taxonomy to set B = 2-level 4-class FAO LCCS-DP taxonomy, constrained as a mutually exclusive and totally exhaustive parent-child relationship, reported in Table 5. Implemented grouping rules are summarized below.

- A1 = Primarily Vegetated Terrestrial Areas = Cultivated Areas or (Semi) Natural Vegetation ≈ NLCD 16-classes DF (41) or EF (42) or MF (43) or SS (52) or GH (71) or PH (81) or CC (82). Actually, this OR-combination of NLCD classes is an expected mixture of the FAO LCCS-DP 2nd-level classes A1 and B3 as first- and second-best match, respectively.
- A2 = Primarily Vegetated Aquatic or Regularly Flooded Areas = Cultivated Aquatic Areas or (Semi) Natural Aquatic Vegetation ≈ NLCD 16-classes WV (90) or EHW (95).
- B3 = Primarily Non-vegetated Terrestrial Areas = Artificial Surfaces or Bare Areas ≈ NLCD 16-classes DOS (21) or DLI (22) or DMI (23) or DHI (24) or BL (31). Actually, this OR-combination of NLCD classes is an expected mixture of the FAO LCCS-DP 2nd-level classes B3 and A1 as first- and second-best match, respectively.
- B4 = Primarily Non-vegetated Aquatic or Regularly Flooded Areas = Artificial or Natural Waterbodies, Snow and Ice ≈ NLCD 16-classes OW (11) or PIS (12).

Following the aforementioned "arbitrary" (subjective) aggregation of an NLCD 16-class legend into a 2-level 4-class FAO LCCS-DP legend, equivalent to an inherently equivocal (qualitative) *information-as-data-interpretation* task, we accomplished the following.

- A binary relationship R: A ⇒ C ⊆ A × C from set A = SIAM legend of 19 spectral macro-categories, identified by the independent human expert, to set C = 2-level 4-class FAO LCCS-DP legend, identified by the present authors, was (subjectively) identified by the present authors according to the hybrid eight-step strategy for categorical variable-pair relationship identification proposed in the Part 1, Chapter 4. Depicted as gray entry-pair cells in Table 5, this binary relationship is eligible for guiding the interpretation process of an OAMTRX = FrequencyCount(A × C).
- An OAMTRX = FrequencyCount(A × C) was generated by the wall-to-wall overlap between the test SIAM-WELD map with legend A = 19 spectral macro-categories as rows and the reference USGS NLCD map whose original 16-class legend was grouped into legend C = 2-level 4-class FAO LCCS-DP taxonomy as columns, see Table 5.

The mDMI set of O-Q^2Is estimated from Table 5 were OA(OAMTRX = FrequencyCount (A × C)) = 93.09% ± 0% and CVPAI2(R: A ⇒ C ⊆ A × C) = 0.7486. When these summary statistics replace variables in Equation (7), we obtained

OA(Test SIAM-WELD 2006, 19 spectral macro-categories; "Ultimate" GroundTruth 2006, 2-level 4-class FAO LCCS-DP taxonomy) ∈ [max{0%, Lower Bound}, min{100%, Upper Bound}] = [max{0%, 93.09% ± 0% - (100% - XX%)}, min{100%, 100% + XX% - 93.09% ± 0%}] = [XX% - 6.91%, XX% + 6.91%],   (8)

where

unknown variable XX% = Equation (6) = OA(Reference NLCD 2006, 2-level 4-class FAO LCCS-DP taxonomy; "Ultimate" GroundTruth 2006, 2-level 4-class FAO LCCS-DP taxonomy) ≥ 84%,

with

CVPAI2(R: A = SIAM vocabulary of 19 spectral macro-categories ⇒ C = 2-level 4-class FAO LCCS-DP taxonomy) = 0.7486.

Hence, the semantic information gap, to be minimized, from input sub-symbolic sensory data to output symbolic classes belonging to the 2-level 4-class FAO LCCS-DP legend, left to be filled (disambiguated) by further stages in the hierarchical EO-IUS pipeline, where spatial information is masked by first-stage color names, see Figure 3, is equal to (1—CVPAI2) = 0. 2514; refer to the Part 1, Chapter 5.

4.4. Probabilities of the SIAM-WELD test labels conditioned by the USGS NLCD reference labels and vice versa

Table 4 shows the OAMTRX instance generated from the wall-to-wall overlap of the annual SIAM-WELD 2006 test map featuring 19 spectral macro-categories, see Table 2 and Figure 8, with the reference USGS NLCD 2006 map featuring 16 LC classes, see Table 1 and Figure 4. The division of each probability cell of Table 4 by its column-sum generates the class-conditional probability p (SIAM-WELD$_t$ / NLCD$_r$) of the SIAM-WELD 2006 test spectral category t, with $t = 1, \ldots, TC = 19$, given the USGS NLCD 2006 reference class r, with $r = 1, \ldots, RC = 16$, refer to Figure 10 and Table 6, where Table 6 is a summarized text version of Figure 10. To prove their plausibility, conditional probabilities p(SIAM-WELD$_t$ / NLCD$_r$), $t = 1, \ldots, TC, r = 1, \ldots, RC$, should agree with theoretical expectations stemming from human experience. For instance, it was expected that the USGS NLCD 2006 reference classes "*Deciduous Forest*" (DF), "*Evergreen Forest*" (EF) and "*Mixed Forest*" (MF), refer to Table 1, overlap with vegetated spectral categories in the test SIAM-WELD 2006 map, while the USGS NLCD reference class "*Developed, High Intensity*" (DHI, see Table 1) was expected to be mostly matched by bare soil-related spectral macro-categories in the test SIAM-WELD 2006 map. Overall, these prior knowledge-based expectations about specific class-conditional probabilities appear satisfied by both Figure 10 and Table 6.

In the RS common practice, once a generic user has generated at no cost in manpower and computer power, that is, in near real time and without user-machine interaction, a SIAM color map from an unknown EO image, what this user wishes to do is to infer from the EO image a set of LC classes (say, "*Forest*"), conditioned by the detected SIAM's BC names (say, MS green-as-"*Vegetation*"). To accomplish this spectral category-conditional inference, class-conditional probabilities p(SIAM-WELD$_t$ / NLCD$_r$), $t = 1, \ldots, TC, r = 1, \ldots, RC$, shown in Table 6, are not useful. Rather, this generic user can found helpful to know the conditional probabilities of an NLCD 2006 reference class r, with $r = 1, \ldots, RC = 16$, given the SIAM-WELD 2006 spectral category t, with $t = 1, \ldots, TC = 19$. These are the class-conditional probabilities p(NLCD$_r$ | SIAM-WELD$_t$), $t = 1, \ldots, TC, r = 1, \ldots, RC$, generated by dividing each probability cell of Table 4 by its row-sum. They are shown in Figure 11 and summarized in text form in Table 7. Very intuitive to understand, Table 7 clearly highlights the two main semantic inconsistencies found between the reference USGS NLCD 2006 map and the test SIAM-WELD 2006 map already reported in previous Chapter 4.3. First, the SIAM vegetation-related spectral macro-category wV_HC ("*Weak evidence vegetation with high canopy cover*", refer to Table 2) is best matched by the reference NLDC class "*Open Water*" (OW, refer to Table 2). Because this semantic mismatch occurs almost exclusively in the CONUS areas recognized by the independent human expert as riparian zones typically depicted as mixed pixels at 30 m resolution, then the 30 m resolution SIAM labeling can be considered reasonable, if we consider that the crisp SIAM implementation is not expected to accomplish pixel unmixing. Second, the USGS NLCD 2006 reference class "*Shrub/Scrub*" (SS, refer to Table 2) appears to be the best match for several of the SIAM bare soil-related spectral macro-categories. Figure 9 shows examples of geographic locations where this semantic mismatch occurs. In these locations, 30 m resolution pixels are typically affected by mixed spectral contributions the crisp SIAM implementation is not expected to unmix.

4.5. Stratification by ecoregions

According to a simplistic interpretation of the central limit theorem, the sum of a large number of independent random variables tends to form a Gaussian distribution, where independent "local" data distributions (like basis functions) become indistinguishable from the whole. For example, in human vision, the neural computations are inherently spatially local in the (2D) image-domain; next, a global spatial average is superimposed on the local computational processes. In general, non-stationary local features do not survive the averaging process, that is, the precise position of each local contribution is no longer perceived after the averaging process (Victor, 1994). Because the WELD composite of the CONUS is about 10 billion pixels in size, summary statistics of the SIAM mapping quality at the CONUS spatial extent are inadequate to demonstrate the local-scale capability of the SIAM expert system to correctly map EO images, characterized by non-stationary local statistics. To investigate the SIAM mapping capability at local spatial extent, the test SIAM-WELD 2006 map with legend A of 19 BC names and the reference USGS NLCD 2006 map with legend B of 16 LC class names, where set A ≠ set B in cardinality and semantics, were stratified using the 86 EPA Level III ecoregions of the CONUS (see Figure 4) and an individual OAMTRX = FrequencyCount(A × B) was generated per ecoregion. All 86 ecoregion-specific OAMTRX instances are available as supplemental online material (SIAM-WELD-NLCD FTP, 2016). As one example of an inter-map comparison at the ecoregion spatial scale of analysis, let us consider the SIAM-WELD 2006 and NLCD 2006 maps of the Wyoming Basin ecoregion, which is predominantly desert, see Figure 12 where the ecoregion boundary is highlighted in red. Table 8 reports the corresponding OAMTRX instance. Table 8 shows that the predominantly desertic Wyoming Basin ecoregion is predominantly classified as the LC classes "*Scrub/Shrub*" (SS) and "*Grassland/Herbaceous*" (GH) in the reference USGS NLCD 2006 map (refer to Table 1) and as bare soil-related spectral categories (sbS_1, SmS_1, aS) in the test SIAM-WELD 2006 map (refer to Table 2). This semantic disagreement was already observed in previous Chapter 4.3; also refer to Figure 9.

Figure 13 provides a synthetic representation of the full dataset of 86 ecoregion-specific OAMTRX instances available as supplemental online material (SIAM-WELD-NLCD FTP, 2016). It shows for each of the 16 reference NLDC classes, with index $r = 1, \ldots, RC = 16$, the box-and-whisker diagram of the USGS NLCD-class-conditional probabilities $p(\text{SIAM-WELD}_{er,\ t} \mid \text{NLCD}_{er,\ r})$, with $t = 1, \ldots,$ $TC = 19$, collected across the 86 ecoregions, each ecoregion identified with an index $er = 1, \ldots,$ $ER = 86$. In each of the $TC = 19$ boxes of an NLCD class-specific boxplot, the median (shown as a horizontal line within the box) represents the general trend of the distribution and the dispersion around it describes the distribution variability across ecosystems. A small dispersion around the median value indicates a reference-to-test class mapping whose occurrence is nearly constant across ecosystems, while a large dispersion around the median indicates that occurrences of this inter-map relationship change significantly across ecosystems.

4.6. mDMI set of OP-Q^2I values estimated by independent means in a GEO-CEOS stage 4 Val of the SIAM process and product

Described in the introduction to Chapter 4 (Duke, 2016), an mDMI set of OP-Q^2Is was estimated by the independent human expert (refer to Acknowledgments) in compliance with a GEO-CEOS stage 4 *Val* of the SIAM product and process, input with a 30 m resolution annual WELD 2006 to 2009 image composite time-series of the CONUS. These *Val* results are summarized below.

(i) Process degree of automation. In line with theoretical expectations about expert systems (refer to previous Chapter 1), the SIAM computer program required neither user-defined parameters nor reference samples to run. Hence, its ease of use cannot be surpassed by any alternative inference approach.

(ii) Outcome effectiveness. An mDMI set of O-Q^2Is (Si Liu et al., 2011; Peng et al., 2005), comprising OA(OAMTRX = FrequencyCount(A × B)), CVPAI2(R: A ⇒ B ⊆ A × B), class-conditional probabilities $p(r \mid t)$ of reference class $r = 1, \ldots, RC = |B|$, given test class $t = 1, \ldots,$ $TC = |A|$, and class-conditional probabilities $p(t \mid r)$, with $r = 1, \ldots, RC = |B|$, $t = 1, \ldots, TC = |A|$, was estimated in the four test cases described in previous Chapter 4.3 to Chapter 4.5.

Table 5. Non-square OAMTRX = FrequencyCount(A × C) instance generated from a wall-to-wall overlap between the annual SIAM-WELD 2006 test map of the CONUS with legend A = 19 spectral macro-categories and the reference USGS NLCD 2006 map with legend C = 16 LC class names grouped by the present authors into a 2-level 4-class FAO LCCS-DP taxonomy, see Figure , as proposed in Table 1. Gray entry-pair cells identify the binary relationship R: A ⇒ C ⊆ A × C chosen by the present authors to guide the interpretation process of the OAMTRX = FrequencyCount(A × C). Statistically independent O-Q$_2$Is, to be jointly maximized, are OA(OAMTRX = FrequencyCount(A × C)) = 93.09% with CVPAI2(R: A ⇒ C ⊆ A × C) ∈ [0, 1] (where value 1 means perfect harmonization between the two input sets A and C) = 0.7486. Adopted acronyms for reference LC classes and test spectral macro-categories are described in Tables 1 and 2, respectively

	NLCD CODE (class acronym), 16 classes	41 (DF), 42 (EF), 43 (MF), 52 (SS) <-> B3, 71 (GH) <-> B3, 81 (PH), 82 (CC)	90 (WV), 95 (EHW)	21 (DOS) <-> A1, 22 (DLI) <-> A1, 23 (DMI) <-> A1, 24 (DHI) <-> A1, 31 (BL)	11 (OW), 12 (PIS)	
	FAO LCCS-DP1&2 Code, 4 classes	≈ A1	≈ A2	≈ B3	≈ B4	
	FAO LCCS-DP1&2 Class name, 4 classes	Veg terstrl	Veg aqutc	Non-veg terstrl	Non-veg aqutc	Sum per row
SIAM™ Intermediate Granularity, 19 Spectral Categories.	sV_HC	28.80%	3.07%	1.17%	0.07%	33.11%
	aV_HC	15.82%	1.49%	2.38%	0.25%	19.94%
	wV_HC	0.08%	0.03%	0.03%	0.04%	0.18%
	sV_MC	0.00%	0.00%	0.00%	0.00%	0.00%
	aV_MC	18.45%	0.42%	1.13%	0.05%	20.05%
	sV_LC	0.00%	0.00%	0.00%	0.00%	0.00%
	aV_LC	0.39%	0.00%	0.01%	0.00%	0.40%
	wbV_MLC	3.80%	0.02%	0.30%	0.01%	4.12%
	wdV_MLC	1.27%	0.04%	0.15%	0.02%	1.48%
	sbS_1	4.21%	0.01%	0.76%	0.01%	5.00%
	sbS_2	0.06%	0.00%	0.03%	0.00%	0.09%
	smS_1	4.53%	0.01%	0.10%	0.01%	4.65%
	smS_2	0.13%	0.00%	0.06%	0.00%	0.19%
	sdS	0.18%	0.00%	0.06%	0.01%	0.25%
	aS	7.63%	0.02%	0.38%	0.01%	8.04%
	wS	0.02%	0.00%	0.00%	0.00%	0.02%
	SN	0.00%	0.00%	0.00%	0.00%	0.01%
	WA	0.06%	0.05%	0.07%	1.11%	1.28%
	O	0.80%	0.03%	0.21%	0.14%	1.19%
	Sum per column	**86.23%**	**5.20%**	**6.84%**	**1.73%**	

(iii) Process efficiency: run-time memory occupation and computation time. About run-time memory occupation, the SIAM computer program adopts a tile streaming implementation, where the dynamic memory (random access memory, RAM) maximum occupation is a known function of the tile size to be fixed in advance, irrespective of the image size. In these experiments the RAM maximum occupation was set equal to 800 MB, which can be considered a "small" RAM value. About computation time: when run on a Dell Power Edge 710 server with dual Intel Xeon @ 2.70 GHz processor with 64 GB of RAM and a 64-bit Linux operating system, the SIAM software application required less than 45 s to generate its

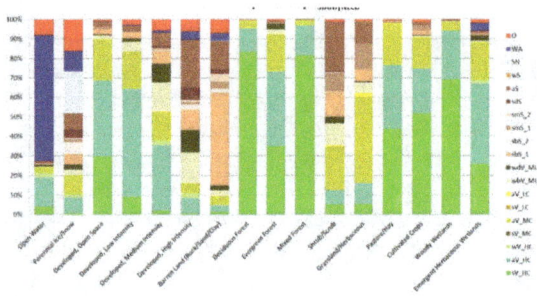

Figure 10. Histogram of the conditional probabilities of the 19 SIAM-WELD 2006 spectral macro-categories (shown as the right column of acronyms, refer to Table 2) at the SIAM intermediate color discretiza-tion level, conditioned by one-of-16 NLCD 2006 classes, listed along the horizontal axis. These class-conditional probabilities are derived from Table 4 by normalizing each cell of Table 4 by its column-sum. The same class-conditional probabilities are summarized in text form in Table 6. In this histogram, pseudo-colors associated with the SIAM color types make the interpretation of the histogram columns more intuitive. Green pseudo-colors are associated with the SIAM vegetation-related spectral categories (identified by acronyms of type xV_y on the right column of labels), brown pseudo-colors are selected for the SIAM bare soil-related spectral categories (identified by acronyms of type xS_y on the right column of labels), the pseudo-color blue is chosen for the SIAM spectral category named "Water or Shadow" (WA), the light blue pseudo-color is linked to the SIAM spectral category named snow (SN), etc. As a conse-quence, the column of the USGS NLCD class "Open Water" is expected to look blue, columns of the USGS NLCD vegetation-related classes are expected to look green, etc.

complete set of per-image output products from a 7-band Landsat-7 ETM+ WELD tile of 5000 × 5000 pixels, which means about 8 h to map an annual WELD composite of the CONUS. In our data mapping workflow, such an output rate was not inferior to the input rate of an annual WELD composite being implemented or delivered to end-users. Hence, the SIAM computation time was considered equivalent to near real time, where the SIAM computational complexity increases linearly with image size.

(iv) Process robustness to changes in the input dataset. The SIAM mapping consistency of the annual WELD composites from year 2006 to 2009 was estimated to be "high" at the CONUS spatial extent; refer to Chapter 4.2 to Chapter 4.5.

(v) Process robustness to changes in input parameters, if any. Because SIAM requires no user-defined parameter to run, its robustness to changes in input parameters cannot be surpassed by alternative approaches.

(vi) Process maintainability/scalability/re-usability, to keep up with changes in users' needs and sensor specifications. The multi-source SIAM physical model can be applied to any existing or future planned spaceborne/airborne MS imaging sensor provided with a radio-metric calibration metadata file; refer to the existing literature (Baraldi et al., 2010a, 2010b, 2010c; Baraldi & Humber, 2015) and to previous Chapter 2.

(vii) Outcome timeliness, defined as the time span between data acquisition and product generation. Because it is prior knowledge based and near real-time, the SIAM application reduces timeliness from image acquisition to color map generation to almost zero, equal to computation time, which increaseas linearly with image size.

(viii) Outcome costs, monotonically increasing with manpower and computer power. Because it is prior knowledge based, therefore automated, and near real time in a standard laptop computer, the SIAM costs are almost negligible.

5. Discussion

Table 3 shows that the 30 m resolution annual SIAM-WELD map time-series for years 2006 to 2009 at the CONUS spatial extent featuring a SIAM intermediate color discretization legend of 48 BC names reassembled into 19 spectral macro-categories by the independent human expert (refer to previous Chapter 3) is characterized by a standard deviation of the annual frequency counts collected for each spectral macro-category lower than 1%, with the exception of two vegeta-tion-related spectral macro-categories, specifically, aV_HC and aV_MC (see Table 2). These two larger variations in spectral category-specific annual frequency counts at the CONUS spatial extent

Table 6. Class-conditional probability p(SIAM-WELD | NLCD), t = 1, ..., $TC = |A| = 19$ color names, r = 1, ..., $RC = |B| = 16$ LC class names. For each NLCD 2006 reference map's LC class, the five best-matching SIAM-WELD 2006 test map's spectral macro-categories, belonging to the finite set A of 19 spectral macro-categories, are shown as SIAM1 to SIAM5

NLCD class	SIAM1	p(SIAM1 | NLCD)	SIAM2	p(SIAM2 | NLCD)	SIAM3	p(SIAM3 | NLCD)	SIAM4	p(SIAM4 | NLCD)	SIAM5	p(SIAM5 | NLCD)
Open water	WA	0.64	aV_HC	0.15	O	0.08	sV_HC	0.04	wV_HC	0.03
Ice/Snow	SN	0.22	O	0.16	WA	0.10	aV_MC	0.10	aS	0.08
Developed, Open	aV_HC	0.39	sV_HC	0.30	aV_MC	0.21	sbS_1	0.03	aS	0.02
Developed, Low	aV_HC	0.55	aV_MC	0.19	sV_HC	0.09	wbV_MLC	0.05	sbS_1	0.03
Developed, Medium	aV_HC	0.33	wbV_MLC	0.15	aV_MC	0.15	wdV_MLC	0.10	aS	0.07
Developed, High	aS	0.24	wbV_MLC	0.16	wdV_MLC	0.11	sbS_1	0.11	aV_HC	0.08
Rock/Sand/Clay	sbS_1	0.48	aS	0.14	O	0.07	aV_MC	0.04	smS_2	0.04
Deciduous Forest	sV_HC	0.83	aV_HC	0.12	aV_MC	0.04	O	0.00	wdV_MLC	0.00
Evergreen Forest	aV_HC	0.38	sV_HC	0.35	aV_MC	0.19	wdV_MLC	0.03	wbV_MLC	0.03
Mixed Forest	sV_HC	0.81	aV_HC	0.16	aV_MC	0.03	O	0.00	wbV_MLC	0.00
Scrub/Shrub	aS	0.25	aV_MC	0.22	sbS_1	0.13	wbV_MLC	0.11	smS_1	0.09
Grassland/ Herbaceous	aV_MC	0.44	smS_1	0.13	aS	0.12	aV_HC	0.11	sbS_1	0.06
Pasture/Hay	sV_HC	0.43	aV_HC	0.33	aV_MC	0.21	sbS_1	0.00	smS_1	0.00
Cultivated Crops	sV_HC	0.52	aV_HC	0.23	aV_MC	0.16	smS_1	0.03	sbS_1	0.02
Woody Wetlands	sV_HC	0.69	aV_HC	0.25	aV_MC	0.05	wbV_MLC	0.00	O	0.00
Herbaceous Wetland	aV_HC	0.41	sV_HC	0.26	aV_MC	0.19	WA	0.04	wdV_MLC	0.03

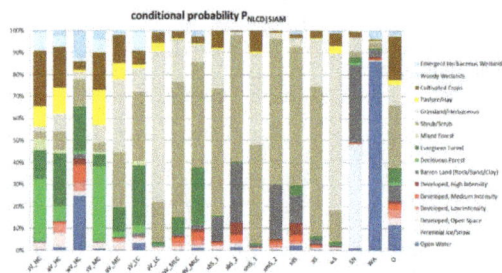

Figure 11. Histogram of the conditional probabilities of the USGS NLCD 2006 map's 1 6 L C   classes (shown as the right column of class names) condi-tioned by one-of-19 SIAM-WELD 2006 spectral macro-categories, listed along the horizontal axis as acronyms, refer to Table 2, at the SIAM intermediate color discretiza-tion level. This histogram is derived from Table 4 by nor-malizing each cell of Table 4 by its row-sum. The same class-conditional probabilities are summarized in text form in Table 7. In this histogram, pseudo-colors associated with the USGS NLCD classes make the interpretation of the histo-gram columns more intuitive. Green pseudo-colors are asso-ciated with vegetation NLCD classes, brown pseudo-colors are selected for bare soil NLCD classes, the pseudo-color blue is chosen for the USGS NLCD class "*Open Water,*" the light blue pseudo-color is linked to the USGS NLCD class "*Perennial Ice/Snow,*" etc. As a conse-quence, the column corre-sponding to the SIAM spectral category "*Water or Shadow*" (WA) is expected to look blue, column corresponding to the SIAM vegetation-related spec-tral categories, identified by acronyms of type xV_y located along the horizontal axis, are expected to look green, etc.

can be attributed mostly to vegetation phenology. This was proved in previous Chapter 4.2: changes in phenology affect the monthly WELD and annual WELD image composites and, as a consequence, the data-derived SIAM-WELD maps. These numerical results agree with the a priori knowledge of RS experts about the CONUS surface dynamics, whose inter-annual LCC summary statistics are expected to score low. The conclusion is that observations stemming from the annual SIAM-WELD map time-series with a legend of 19 spectral macro-categories comply with the domain knowledge of RS experts about the LC and LCC dynamics in the geophysical domain of the CONUS.

The interpretation process of the OAMTRX = FrequencyCount(A × B) shown in Table 4, generated from the wall-to-wall overlap between the test SIAM-WELD 2006 map featuring a set A = VocabularyOfColorNames = 19 spectral macro-categories and the reference USGS NLCD 2006 map with a set B = LegendOfObjectClassNames = 16 LC classes, is guided by the inter-dictionary binary relationship R: A ⇒ B ⊆ A × B, whose entry-pair cells, shown in gray, were selected as "correct" by the independent human expert (refer to Acknowledgments) according to the hybrid eight-step guideline for identification of a categorical variable-pair relationship proposed in the Part 1, Chapter 4. Table 4 reveals one single systematic case of "conceptual mismatch" between the USGS NLCD 2006 reference vegetation classes "*Scrub/Shrub*" (SS) or "*Grassland/Herbaceous*" (GH, refer to Table 1) and the SIAM-WELD 2006 bare soil-related spectral macro-categories sbS_1, SmS_1, and aS (refer to Table 2). These inter-map semantic mismatches occur in geographical locations where the CONUS landscapes look like those shown in Figure 9. When these land surface types are observed from space with a Landsat-like spatial resolution of 30 m, a one-pixel surface area of 900 m^2 becomes a spectral mixture of sparse vegetation, rangeland, cheatgrass, dry long grass and/or short grass as foreground, with a background of sand, clay and/or rocks, especially if the percentage of vegetation cover can be slightly above the 15% of total cover required by the USGS NLCD definitions of classes SS and GH (refer to Table 2). In these mixed pixels at 30 m resolution, the spectral detection of the vegetated component is impossible for a hard (crisp) classifier, while it would be more manageable by a fuzzy classifier (Baraldi, 2011). In these experiments, since the SIAM expert system is run in crisp mode (refer to Chapter 3), then no pixel unmixing strategy can be applied to diminish or avoid the observed case of "semantic mismatch." The conclusion is that the "conceptual mismatch" between the USGS NLCD 2006 reference vegetation classes SS and GH and the SIAM-WELD 2006 bare soil-related spectral categories is a possible example of systematic disagreement between the test and reference thematic maps featuring the same spatial resolution whose occurrence should be carefully scru-tinized by RS experts in comparison with an "ultimate" ground truth; see Figure 9.

Table 7. Class-conditional probability p(NLCD | SIAM-WELD), t = 1, ..., TC = |A| = 19 color names, r = 1, ..., RC = |B| = 16 LC class names. For each SIAM-WELD 2006 test map's spectral macro-category, the five best-matching NLCD 2006 reference map's LC classes, belonging to the finite set B of 16 LC classes, are shown as NLCD1 to NLCD5

SIAM	NLCD1	p(NLCD1 | SIAM)	NLCD2	p(NLCD2 | SIAM)	NLCD3	p(NLCD3 | SIAM)	NLCD4	p(NLCD4 | SIAM)	NLCD5	p(NLCD15 | SIAM)
sV_HC	Deciduous Forest	0.29	Cultivated Crops	0.25	Evergreen Forest	0.13	Pasture/Hay	0.09	Woody Wetlands	0.08
aV_HC	Evergreen Forest	0.24	Cultivated Crops	0.19	Pasture/Hay	0.12	Shrub/Scrub	0.08	Grassland/Herbaceous	0.08
wV_HC	Open Water	0.25	Evergreen Forest	0.20	Herbaceous Wetlands	0.12	Shrub/Scrub	0.12	Developed, Medium	0.08
sV_MC	Deciduous Forest	0.33	Cultivated Crops	0.17	Pasture/Hay	0.15	Grassland/Herbaceous	0.08	Woody Wetlands	0.06
aV_MC	Grassland/Herbaceous	0.33	Shrub/Scrub	0.25	Cultivated Crops	0.13	Evergreen Forest	0.12	Pasture/Hay	0.07
sV_LC	Shrub/Scrub	0.04	Evergreen Forest	0.03	Grassland/Herbaceous	0.01	Woody Wetlands	0.01	Cultivated Crops	0.01
aV_LC	Grassland/Herbaceous	0.68	Shrub/Scrub	0.18	Cultivated Crops	0.05	Pasture/Hay	0.04	Evergreen Forest	0.02
wbV_MLC	Shrub/Scrub	0.62	Grassland/Herbaceous	0.19	Evergreen Forest	0.08	Cultivated Crops	0.03	Developed, Medium	0.02
wdV_MLC	Shrub/Scrub	0.48	Evergreen Forest	0.26	Grassland/Herbaceous	0.08	Developed, Medium	0.04	Cultivated Crops	0.03
sbS_1	Shrub/Scrub	0.58	Grassland/Herbaceous	0.18	Rock/Sandy/Clay	0.11	Cultivated Crops	0.08	Developed, Open	0.02
sbS_2	Shrub/Scrub	0.58	Rock/Sandy/Clay	0.27	Developed, High	0.06	Developed, Medium	0.03	Grassland/Herbaceous	0.02
smS_1	Shrub/Scrub	0.45	Grassland/Herbaceous	0.42	Cultivated Crops	0.09	Developed, Open	0.01	Rock/Sandy/Clay	0.01

(Continued)

Table 7. (Continued)

SIAM	NLCD1	p(NLCD1 | SIAM)	NLCD2	p(NLCD2 | SIAM)	NLCD3	p(NLCD3 | SIAM)	NLCD4	p(NLCD4 | SIAM)	NLCD5	p(NLCD15| SIAM)
smS_2	Shrub/Scrub	0.66	Rock/Sandy/Clay	0.25	Grassland/Herbaceous	0.03	Developed, Low	0.01	Developed, Medium	0.01
sdS	Shrub/Scrub	0.63	Rock/Sandy/Clay	0.13	Developed, High	0.05	Evergreen Forest	0.04	Grassland/Herbaceous	0.04
aS	Shrub/Scrub	0.68	Grassland/Herbaceous	0.22	Cultivated Crops	0.03	Rock/Sandy/Clay	0.02	Evergreen Forest	0.01
wS	Grassland/Herbaceous	0.71	Shrub/Scrub	0.14	Cultivated Crops	0.06	Pasture/Hay	0.03	Evergreen Forest	0.02
SN	Ice/Snow	0.47	Rock/Sandy/Clay	0.35	Grassland/Herbaceous	0.06	Shrub/Scrub	0.03	Cultivated Crops	0.03
WA	Open Water	0.86	Herbaceous Wetlands	0.04	Rock/Sandy/Clay	0.03	Shrub/Scrub	0.03	Grassland/Herbaceous	0.01
O	Shrub/Scrub	0.28	Cultivated Crops	0.20	Open Water	0.12	Grassland/Herbaceous	0.09	Rock/Sandy/Clay	0.07

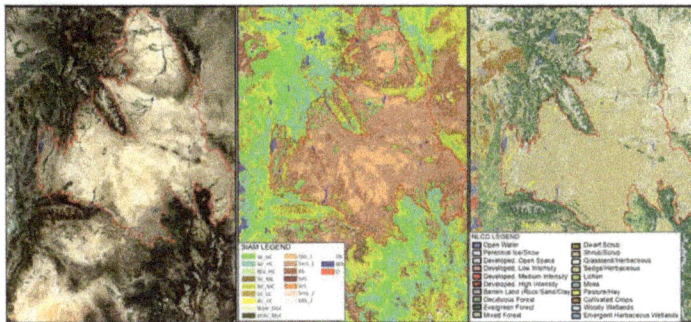

Figure 12. Wyoming Basin ecoregion, as part of the "North American deserts" level 1 ecor-egion 10.1.4. *Left*: WELD 2006 tile (true color). *Middle*: SIAM test map of the WELD 2006 tile shown at left, with 19 spectral macro-categories at the intermediate color discretization level. *Right*: NLCD 2006 reference map, featuring 16 LC classes. In these three images, the boundary of the Wyoming Basin ecoregion is overlaid in red. The desertic Wyoming Basin ecoregion is classified as predominantly *"Scrub/Shrub"* (SS) and *"Grassland/Herbaceous"* (GH) in the USGS NLCD 2006 reference map (refer to Table 1), and predo-minantly as bare soil (sbS_1, SmS_1, aS) in the SIAM-WELD 2006 test map (refer toTable 2). This phenomenon of comprehensive "semantic mis-match" between the USGS NLCD 2006 and SIAM-WELD 2006 thematic maps is explained thoroughly in Chapter 4.3.

A different strategy to aesthetically (rather than formally) remove the aforementioned inter-diction-ary "conceptual mismatch" would be to change color names in the SIAM color map legend, without changing the SIAM decision tree for MS reflectance space hyperpolyhedralization. In other words, based on thematic evidence collected on an a posteriori basis from the USGS NLCD 2006 reference map, it would be possible to change color names attached to the SIAM-WELD 2006 map legend and consider that, at the Landsat spectral and spatial resolution of an annual WELD composite of the CONUS, the SIAM spectral categories sbS_1, SmS_1, and aS are more likely to map the USGS NLCD reference vegetation classes *"Shrub/Scrub"* (SS) or *"Grassland/herbaceous"* (GH) than bare soil surface types.

Starting from the same OAMTRX = FrequencyCount(A × B) shown in Table 4, two independent selections by two different RS experts of a binary relationship R: A ⇒ B ⊆ A × B suitable for guiding the interpretation process of the OAMTRX instance at hand provided two alternative mDMI pairs of O-Q^{2}I values to be jointly maximized, namely, an OA_1(OAMTRX = FrequencyCount (A × B)) = 96.88% ± 0% with a CVPSI2_1(R: A ⇒ B ⊆ A × B) = 0.6689 in test case A and an OA_2 (OAMTRX = FrequencyCount(A × B)) = 97.28% ± 0% with a CVPSI2_2(R: A ⇒ B ⊆ A × B) = 0.6731 in test case B. These alternative O-Q^{2}I pairs highlight the inherent ill-posedness of any inter-dic-tionary conceptual harmonization, although a specific protocol to reduce heuristic decisions by human experts in the identification of a binary relationship R: A ⇒ B ⊆ A × B was proposed in the Part 1, Chapter 4. According to a Pareto multi-objective optimization principle, the latter O-Q^{2}I value pair should be preferred to the former. This choice proves that the OA of the test SIAM-WELD 2006 map compared with the reference NLDC 2006 map scores "very high," with a semantic information gap from sub-symbolic sensory data to symbolic NLCD classes left to be filled (dis-ambiguated) by further stages in the hierarchical EO-IUS pipeline, where spatial information is masked by first-stage color names, see Figure 3, equal to (1—CVPSI2) = 0.3269.

At the fine discretization level of the SIAM-WELD 2006 test map, featuring a legend A = 96 BC names, another inter-map wall-to-wall overlap with the USGS NLCD 2006 reference map, whose legend B = 16 LC classes, provided an mDMI pair of O-Q^{2}I values equal to OA (OAMTRX = FrequencyCount(A × B)) = 95.41% ± 0% and CVPAI2(R: A ⇒ B ⊆ A × B) = 0.5809 in test case C. When compared to the two pairs of O-Q^{2}I values collected from the test case A and the test case B, this third O-Q^{2}I value pair proves that a finer hyperpolyhedralization of the MS reflectance space for color naming is not necessarily more convenient to cope with by human experts in the stratification of an LC classification problem according to a spectral and spatial convergence-of-evidence approach, refer to Equation (3) in the Part 1 (Hunt & Tyrrell, 2012).

When an approximated binary relationship R: B ⇒ C ⊆ B × C was identified from set B = NLCD 16-class legend to set C = 2-level 4-class FAO LCCS-DP legend, see Figure 1, a binary relationship R: A ⇒

Table 8. OAMTRX instance generated from a wall-to-wall overlap over the Wyoming Basin Ecoregion between the test SIAM-WELD 2006 map with legend A = 19 spectral macro-categories and the reference USGS NLCD 2006 map with legend B = 16 LC class names. Gray squares identify the binary relationship R: A → B ⊆ A × B chosen by the independent human expert to guide the interpretation process of the OAMTRX = FrequencyCount(A × B), same as in Table 4. The Wyoming Basin Ecoregion is predominantly desertic. It is classified as LC class "Scrub/Shrub" (SS) or LC class "Grassland/Herbaceous" (GH) in the USGS NLCD 2006 reference map (refer to Table 1), and predominantly as spectral macro-categories of bare soil (sbS_1, SmS_1, aS) in the SIAM-WELD 2006 testmap (refer to Table 2). This phenomenon of large-scale "conceptual mismatch" between the USGS NLCD 2006 and SIAM-WELD 2006 thematic map pair is discussed thoroughly in Chapter 4.3

Wyoming Basin Ecoregion, 2006 OAMTRX. Probabilities (%). Rows: SIAM™-WELD 2006, 19 spectral categories; Columns: NLCD 2006, 16 land cover classes.

NLCD code	11	12	21	22	23	24	31	41	42	43	52	71	81	82	90	95	
NLCD class	OW	PIS	DOS	DLI	DMI	DHI	BL	DF	EF	MF	SS	GH	PH	CC	WW	EHW	
FAO LCCD-DP1&2	B4	B4	B3 -> A1	B3 -> A1	B3 -> A1	B3 -> A1	B3	A1	A1	A1	A1 -> B3	A1 -> B3	A1	A1	A2	A2	
sV_HC	0.00%	0.00%	0.03%	0.01%	0.00%	0.00%	0.00%	0.03%	0.02%	0.01%	0.06%	0.02%	1.04%	0.23%	0.09%	0.11%	1.65%
aV_HC	0.02%	0.00%	0.05%	0.03%	0.01%	0.00%	0.00%	0.07%	0.43%	0.02%	0.41%	0.13%	1.05%	0.16%	0.38%	0.35%	3.11%
wV_HC	0.01%	0.00%	0.00%	0.00%	0.00%	0.00%	0.00%	0.00%	0.01%	0.00%	0.01%	0.00%	0.00%	0.00%	0.01%	0.00%	0.05%
sV_MC	0.00%	0.00%	0.00%	0.00%	0.00%	0.00%	0.00%	0.00%	0.00%	0.00%	0.00%	0.00%	0.00%	0.00%	0.00%	0.00%	0.00%
aV_MC	0.01%	0.00%	0.06%	0.02%	0.00%	0.00%	0.00%	0.08%	0.68%	0.01%	3.47%	1.21%	0.60%	0.06%	0.23%	0.42%	6.85%
sV_LC	0.00%	0.00%	0.00%	0.00%	0.00%	0.00%	0.00%	0.00%	0.00%	0.00%	0.00%	0.00%	0.00%	0.00%	0.00%	0.00%	0.00%
aV_LC	0.00%	0.00%	0.00%	0.00%	0.00%	0.00%	0.00%	0.00%	0.00%	0.00%	0.00%	0.01%	0.00%	0.00%	0.00%	0.00%	0.07%
wbV_MLC	0.01%	0.00%	0.04%	0.03%	0.01%	0.00%	0.00%	0.00%	0.18%	0.00%	2.35%	0.55%	0.04%	0.02%	0.05%	0.07%	3.35%
wdV_MLC	0.01%	0.00%	0.01%	0.00%	0.00%	0.00%	0.00%	0.00%	0.14%	0.00%	0.57%	0.05%	0.01%	0.00%	0.01%	0.01%	0.81%
sbS_1	0.01%	0.00%	0.12%	0.04%	0.01%	0.00%	0.69%	0.00%	0.01%	0.00%	16.90%	5.66%	0.09%	0.04%	0.04%	0.11%	23.72%
sbS_2	0.00%	0.00%	0.00%	0.00%	0.00%	0.00%	0.00%	0.00%	0.00%	0.00%	0.02%	0.01%	0.00%	0.00%	0.00%	0.00%	0.03%
smS_1	0.00%	0.00%	0.12%	0.01%	0.00%	0.00%	0.03%	0.00%	0.06%	0.00%	32.71%	4.11%	0.04%	0.01%	0.02%	0.09%	37.20%
smS_2	0.00%	0.00%	0.00%	0.00%	0.00%	0.00%	0.01%	0.00%	0.00%	0.00%	0.04%	0.02%	0.00%	0.00%	0.00%	0.00%	0.07%
sdS	0.01%	0.00%	0.00%	0.00%	0.00%	0.00%	0.08%	0.00%	0.00%	0.00%	0.08%	0.01%	0.00%	0.00%	0.00%	0.00%	0.10%
aS	0.02%	0.00%	0.17%	0.07%	0.01%	0.00%	0.08%	0.00%	0.13%	0.00%	17.86%	3.25%	0.05%	0.02%	0.03%	0.10%	21.79%
wS	0.00%	0.00%	0.00%	0.00%	0.00%	0.00%	0.00%	0.00%	0.00%	0.00%	0.00%	0.00%	0.00%	0.00%	0.00%	0.00%	0.00%
SN	0.00%	0.00%	0.00%	0.00%	0.00%	0.00%	0.00%	0.00%	0.00%	0.00%	0.00%	0.00%	0.00%	0.00%	0.00%	0.00%	0.00%
WA	0.54%	0.00%	0.00%	0.00%	0.00%	0.00%	0.01%	0.00%	0.01%	0.00%	0.06%	0.01%	0.00%	0.00%	0.01%	0.01%	0.64%
O	0.02%	0.00%	0.01%	0.01%	0.00%	0.00%	0.01%	0.00%	0.01%	0.00%	0.31%	0.06%	0.05%	0.03%	0.02%	0.03%	0.56%
	0.65%	0.00%	0.61%	0.22%	0.04%	0.01%	0.83%	0.18%	1.67%	0.04%	74.91%	15.10%	2.97%	0.57%	0.89%	1.30%	100.00%

SIAM™ Intermediate Granularity, 19 Spectral Categories.

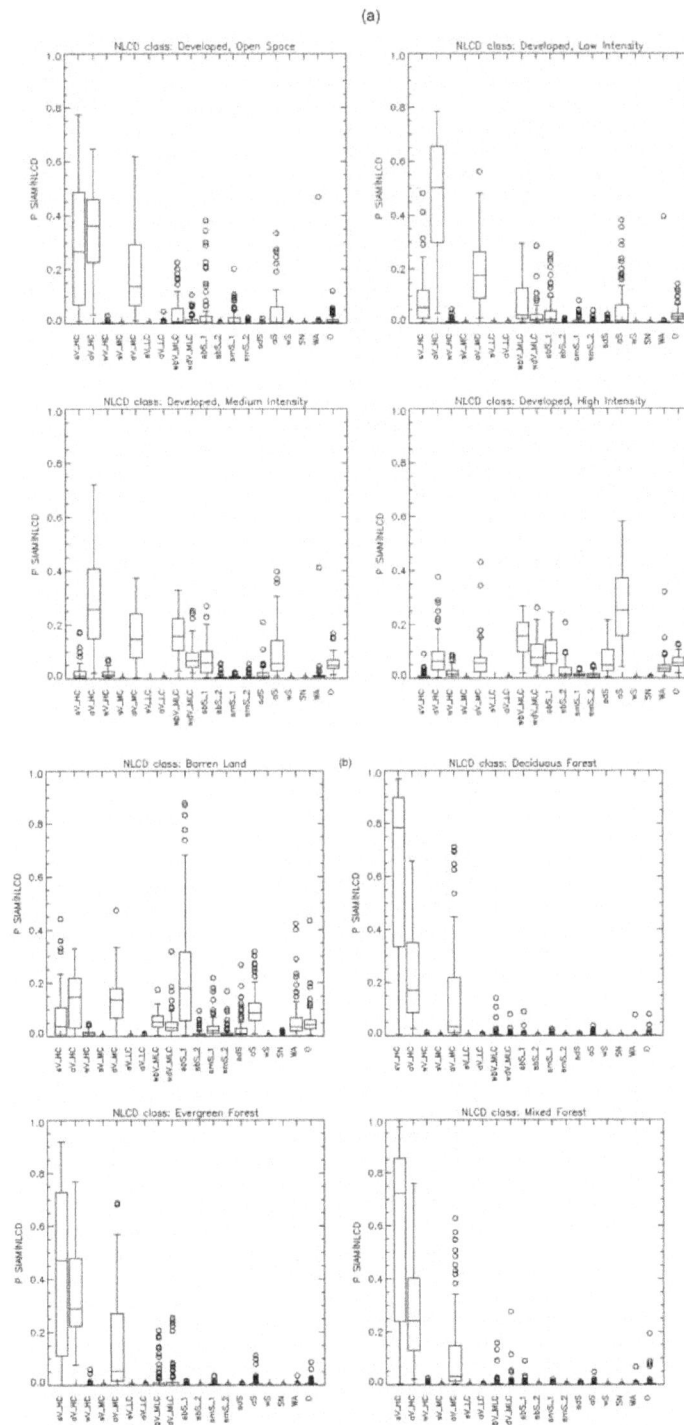

Figure 13. (a)—(d). Reference USGS NLCD class-specific box-and-whisker diagrams, identi-fied by index $r = 1 , \ldots , \quad RC = 16$, of the USGS NLCD class-condi-tional probabilities $p(\text{SIAM-WELD}_{er, t} | \text{NLCD}_{er, r})$, with $t = 1 , \quad \ldots, \quad TC = 19$, collected across ecoregions $er = 1, \ldots, ER = 86$. The 19 spectral categories of the SIAM-WELD test map, identified by their acronyms (refer to Table 2), are distribu-ted along the x axis of each NLCD class-specific diagram. Each of the 19 boxes in a box-and-whisker diagram extends from the 25th to the 75th per-centile, with a horizontal line to represent the median (50th percentile) of the distribution. The whiskers extend to the maximum or minimum value of the data set, or to 1.5 times the interquantile range, whichever comes first. If there is data beyond this range, it is repre-sented by open circles.

Figure 13. Continued.

$C \subseteq A \times C$ was defined from set A = SIAM 19-class legend as rows to set C = 2-level 4-class FAO LCCS-DP legend as columns (reassembled from original columns of the 16-class legend B) and an OAMTRX = FrequencyCount(A × C) was generated by the wall-to-wall overlap between the test SIAM-WELD map with legend A and the reference USGS NLCD map with 16-class legend grouped into the 4-class legend C as reported in Table 5 (test case D), then $O\text{-}Q^2I$ values were OA (OAMTRX = FrequencyCount(A × C)) = 93.09% ± 0% with CVPAI2(R: A ⇒ C) = 0.7486. From these results, we could infer the following.

OA(OAMTRX = FrequencyCount(A × C)) = OA(Test SIAM-WELD 2006, 19 spectral macro-categories; "Ultimate" GroundTruth 2006, 2-level 4-class FAO LCCS-DP taxonomy) ∈ [XX% - 6.91%, XX% + 6.91%], where

unknown variable XX% = Equation (6) = OA(Reference NLCD 2006, 2-level 4-class FAO LCCS-DP taxonomy; "Ultimate" GroundTruth 2006, 2-level 4-class FAO LCCS-DP taxonomy) ≥ 84%,

with a semantic information gap, to be minimized, from input sub-symbolic sensory data to output symbolic classes belonging to the 2-level 4-class FAO LCCS-DP legend, left to be filled (disambiguated) by further stages in the hierarchical EO-IUS pipeline, where spatial information is masked by first-stage color names, see Figure 3, equal to (1 – CVPAI2) = 0. 2514, refer to the Part 1, Chapter 5.

This inference supports the thesis investigated by the present experimental work, where the off-the-shelf SIAM lightweight computer program for prior knowledge-based MS reflectance space hyperpolyhedralization into BC names was considered eligible for systematic ESA EO Level 2 SCM product generation, with an SCM legend consistent with the "augmented" 9-class 3-level FAO LCCS-DP taxonomy; more specifically, the SIAM color maps are consistent with the 2-level 4-class FAO LCCS-DP taxonomy; see Figure 1 (Baraldi, Tiede, Sudmanns, Belgiu, & Lang, 2016, 2017).

To complete the interpretation of the OAMTRX instance shown in Table 4, two histograms of class-conditional probabilities, shown in Figures 10 and 11 respectively, together with their summarized text versions, shown as Tables 6 and 7 respectively, were generated from the OAMTRX of interest. Figure 10 and Table 6 reveal that any test SIAM-WELD 2006 spectral category conditioned by one NLCD 2006 reference class appears consistent with the USGS NLCD class definition (refer to Table 1) and with an a priori domain knowledge of RS experts about the geophysical CONUS domain, spatially sampled at 30 m resolution. Analogously, Figure 11 and Table 7 show that any NLCD 2006 reference class conditioned by one SIAM-WELD 2006 spectral category appears consistent with the spectral properties of the SIAM color type and with an a priori domain knowledge of RS experts about the geophysical CONUS domain, depicted at 30 m resolution. To conclude, class-conditional probabilities generated from Table 4 appear reasonable and confirm the statistical plausibility of the OAMTRX instance shown in Table 4 as a whole.

Figure 13 shows that if, for example, the boxplot of the USGS NLCD 2006 reference class "Deciduous Forest" (DF), "Evergreen Forest" (EF) and "Mixed Forest" (MF, refer to Table 1) is compared to that of reference class "Developed, Low Intensity" (DLI), "Developed, Medium Intensity" (DMI) and "Developed, High Intensity" (DHI, refer to Table 1), then a monotonic decrease of the class-conditional probability of the SIAM-WELD vegetation-related spectral categories conditioned by the USGS NLCD reference class and collected at a regional spatial extent corresponding to a population of 86 ecoregions is observed in parallel with a monotonic increase of the class-conditional probability of the SIAM-WELD bare soil-related spectral categories. This is perfectly consistent with the a priori domain knowledge of RS experts about the spatial and spectral properties of urban and industrial areas in the CONUS, in agreement with the popular vegetation-impervious surface-soil model for urban ecosystem analysis (Ridd, 1995). In addition, these boxplots confirm that, at the local spatial extent of individual ecoregions, the USGS NLCD 2006 reference classes "Deciduous Forest" (DF), "Evergreen Forest" (EF) and "Mixed Forest" (MF, refer to Table 1) are almost entirely (> 90%) covered by the SIAM-WELD vegetation-related spectral categories, in agreement with theoretical expectations about the SIAM-WELD test map. In line with preliminary outcomes discussed in previous Chapter 4.3 and in Figure 12, boxplots shown in Figure 13 confirm that the USGS NLCD 2006 reference classes *"Scrub/Shrub"* (SS) and *"Grassland/Herbaceous"* (GH) have a strong heterogeneity of matches with the SIAM-WELD 2006 spectral categories collected at the ecoregion spatial extent. This is tantamount to saying that spectral signatures of these NLCD classes feature a strong variability when collected at regional scale; also refer to Figure 9. More properties of the USGS NLCD 2006 class-specific box diagrams collected at the local spatial extent of ecoregions appear reasonable, based on a priori human knowledge of the geophysical CONUS domain at the ecoregion spatial extent. For example, first, the USGS NLCD 2006 reference classes *"Pasture/Hay"* (PH) and *"Cultivated Crops"* (CC, refer to Table 1) are largely matched across ecoregions by the SIAM-WELD vegetation-related spectral categories. Second, the USGS NLCD reference class *"Perennial Ice/Snow"* (PIS, refer to Table 1) is best matched across ecoregions by the SIAM-WELD spectral category MS white-as-*"Snow"* (SN, refer to Table 2). Third, across ecoregions, the USGS NLCD 2006 reference class *"Open water"* (OW, refer to Table 1) is best matched by the SIAM-WELD spectral category MS blue-as-*"Water or Shadow"* (WA, refer to Table 2). To summarize, collected at the local extent of ecoregions to account for non-stationary spatial properties, boxplots shown in Figure 13a are considered statistically and semantically consistent with the definitions of the two legends adopted by the test and reference maps, they agree with a priori domain knowledge of RS experts about the LC and LCC dynamics in the geophysical CONUS domain and appear consistent with global (non-stratified by ecoregions) statistics collected at the CONUS spatial extent, as reported in previous Chapter 4.2 to Chapter 4.4.

6. Conclusions

To pursue the GEO-CEOS visionary goal of a GEOSS implementation plan for years 2005–2015 not yet accomplished by the RS community, this interdisciplinary work aimed at filling an analytic and pragmatic information gap from EO big data to systematic ESA EO Level 2 product generation at the ground segment, never achieved to date by any EO data provider and postulated as necessary not sufficient precondition to GEOSS development. For the sake of readability, this paper was split into two, the preliminary Part 1 - Theory and the present Part 2 - Validation.

Provided with a relevant survey value, the Part 1 of this paper reviewed a long history of prior knowledge-based MS reflectance space partitioners for static color naming developed by the RS community for use in hybrid (combined deductive and inductive) EO-IUSs for EO image enhancement and classification tasks in operating mode, but never validated in compliance with the GEO-CEOS QA4EO *Cal/Val* requirements. Original contributions of the Part 1 include, first, an analytic expression of a "naïve" Bayes classifier proposed as a biologically plausible hybrid CV system suitable for convergence of color and spatial evidence. Second, a hybrid eight-step protocol was proposed to infer a binary relationship, R: A ⇒ B ⊆ A × B, from categorical variable A to categorical variable B estimated from the same population. This eight-step protocol is of practical use because identification of a binary relationship R: A ⇒ B is mandatory to guide the interpretation process of a bivariate frequency table, BIVRFTAB = FrequencyCount(A × B), where A ≠ B in general. Third, in compliance with the GEO-CEOS QA4EO *Val* guidelines, two original and alternative formulations, CVPAI2(R: A ⇒ B) ∈ [0, 1] and CVPAI3 (R: A ⇒ B) ∈ [0, 1], were proposed as a categorical variable-pair normalized degree of association (harmonization, matching) in a binary relationship, R: A ⇒ B ⊆ A × B, from categorical variable A to categorical variable B estimated from the same population, where A ≠ B in general. When CVPAI2 or CVPAI3 is maximum, equal to 1, then the two categorical variables A and B are considered fully harmonized (reconciled). If A ≠ B, an mDMI set of O-Q^2Is for a two-way frequency table BIVRFTAB = FrequencyCount(A × B) comprises OA(BIVRFTAB = FrequencyCount(A × B)) ∈ [0, 1] and CVPAI2(R: A ⇒ B) ∈ [0, 1] to be jointly maximized. Only if A = B then BIVRFTAB is equal to a well-known square and sorted CMTRX, intuitive to use because its main diagonal guides the interpretation process and where CVPAI2(R: A ⇒ B) = CVPAI3(R: A ⇒ B) = 1.

In the present Part 2, from an ensemble of expert systems for color naming found in the RS literature, an off-the-shelf SIAM lightweight computer program was selected as potential candidate for systematic ESA EO Level 2 SCM product generation in operating mode, to be submitted to a GEO-CEOS stage 4 *Val* by independent means, at large spatial extent and multiple time samples, in agreement with the GEO-CEOS QA4EO *Cal/Val* guidelines. The selected input data set was the 30 m resolution annual WELD image composite time-series in TOARF values from year 2006 to 2009 at the CONUS spatial extent. The selected reference map was the 30 m resolution USGS NLCD 2006 map of the U.S., whose legend consists of 16 LC classes. The original methodological contribution of the present Part 2 consists of a novel protocol for wall-to-wall inter-map comparison without sampling, where the test and reference thematic maps feature the same spatial resolution and spatial extent, but whose legends A and B are not the same and must be harmonized. These working hypotheses are fully complementary to those of traditional protocols for map accuracy assessment based on random sampling and a pair of coincident test and reference map legends.

Conclusions of the present Part 2 are twofold. First, in agreement with the definition of an information processing system in operating mode proposed in this work, the off-the-shelf SIAM software executable submitted to a GEO-CEOS stage 4 *Val* can be considered in operating mode because its whole set of OP-Q^2I estimates scored "high." Second, the off-the-shelf SIAM lightweight computer program in operating mode can be considered suitable for systematic generation of an ESA EO Level 2 SCM product instantiation whose legend agrees with the standard 2-level 4-class FAO LCCS-DP taxonomy (first DP level: vegetation vs non-vegetation, second DP level: aquatic vs terrestrial), preliminary to an "augmented" 3-level 9-class FAO LCCS-DP taxonomy, defined as a standard 3-level 8-class FAO LCCD-DP legend, see Figure 1, augmented with class "Others," which includes quality layers cloud and cloud-shadow. At the CONUS spatial extent, the experimental portion of the present Part 2 inferred that OA(OAMTRX = FrequencyCount(A × C)) = OA(Test SIAM-WELD 2006, A = 19 spectral macro-categories; "Ultimate" GroundTruth 2006, C = 2-level 4-class FAO LCCS-DP taxonomy) ∈ [XX%—6.91%, XX% + 6.91%,], where unknown variable XX% = OA (Reference NLCD 2006, 2-level 4-class FAO LCCS-DP taxonomy; "Ultimate" GroundTruth 2006, 2-level 4-class FAO LCCS-DP taxonomy) ≥ 84%, with CVPAI2(R: A ⇒ C) = 0.7486. Hence, the semantic information gap, to be minimized, from input sensory data to the output 2-level 4-class FAO LCCS-DP legend left to be filled (disambiguated) by further stages in the EO-IUS pipeline following the SIAM first stage, see Figure 3, is equal to (1—CVPAI2) = 0.2514 ∈ [0, 1]. This is tantamount to saying that, although spatial information dominates color information in vision (Baraldi, 2017; Matsuyama & Hwang, 1990), the spatial context-insensitive SIAM expert system for color naming, whose legend A was a vocabulary of 19 color names equivalent to hyperpolyhedra in a MS reflectance space, when input with the annual WELD 2006 image composite of the CONUS and when compared with the reference USGS NLCD 2006 map, whose original legend B consisting of 16 LC class names was reassembled into a dictionary C of four LC classes belonging to a standard 2-level 4-class FAO LCCS-DP taxonomy, carried out a CVPAI2(R: A ⇒ C) estimate equal to 74.86% of the classification (discrimination) work required to disentangle the four LC classes of a standard 2-level 4-class FAO LCCS-DP taxonomy, with an OA(OAMTRX = FrequencyCount(A × C)) = [XX%—6.91%, XX% + 6.91%], where variable XX% ≥ 84%.

Ongoing developments of this work aim at systematic generation from multi-source MS imagery of an ESA EO Level 2 SCM product whose legend is the "augmented" 3-level 9-class FAO LCCS-DP taxonomy, which includes quality layers cloud and cloud-shadow, as a viable alternative to the non-standard ESA EO Level 2 SCM legend adopted by the SEN2COR software toolbox distributed by ESA to be run on user side. Preliminary OP-Q^2Is collected from a prototypical implementation of the hybrid (combined physical model-based and statistical model-based) feedback EO-IUS architecture sketched in Figure 3, where convergence of color and spatial evidence is pursued, were considered encouraging (Baraldi, 2015, 2017; Baraldi et al., 2016, 2017; Tiede, Baraldi, Sudmanns, Belgiu, & Lang, 2016). Degrees of novelty of the hybrid EO-IUS under development for systematic ESA EO Level 2 product generation include, first, a novel 2D wavelet-based spatial filter bank for automated image-contour detection and image segmentation (raw primal sketch, in the Marr terminology), consistent with human visual perception, such as the Mach bands illusion (Baraldi, 2017). Second, a "universal" hybrid spatial context-sensitive cloud/cloud-shadow detector in single-date MS imagery, eligible for use with past, present, and future optical imaging sensors provided with metadata *Cal* parameters (Baraldi, 2015, 2017 2015), alternative to existing cloud/ cloud-shadow detectors, including that implemented in the SEN2COR software toolbox. Third, an original mDMI set of scale-invariant planar shape indexes (Baraldi, 2017; Baraldi & Soares, 2017), which includes a novel implementation of a straightness-of-boundaries indicator (Nagao & Matsuyama, 1980), particularly useful to discriminate managed (man-made, anthropic) LC classes from natural or semi-natural surface types, such as in the 3rd-level FAO LCCS-DP taxonomy, where LC class A11 (Cultivated and Managed Terrestrial Vegetated Areas) must be discriminated from LC class A12 (Natural and Semi-Natural Terrestrial Vegetation) and where LC class B35 (Artificial Surfaces and Associated Areas) must be separated from LC class B36 (Bare Areas); see Figure 1.

Abbreviation

AI:	Artificial Intelligence
ATCOR:	Atmospheric/Topographic Correction commercial softare product
AVHRR:	Advanced Very High Resolution Radiometer
BC:	Basic Color
BIVRFTAB:	Bivariate Frequency Table
Cal:	Calibration
Cal/Val:	Calibration and Validation
CCRS:	Canada Centre for Remote Sensing
CEOS:	Committee on Earth Observation Satellites
CLC:	CORINE Land Cover (taxonomy)
CMTRX:	Confusion Matrix
CNES:	Centre national d'études spatiales
CONUS:	Conterminous U.S.
CORINE:	Coordination of Information on the Environment
CV:	Computer Vision
CVPAI:	Categorical Variable Pair Association (matching) Index (in range [0, 1])
DCNN:	Deep Convolutional Neural Network
DLR:	Deutsches Zentrum für Luft- und Raumfahrt (German Aerospace Center)
DN:	Digital Number
DP:	Dichotomous Phase (in the FAO LCCS taxonomy)
DRIP:	Data-Rich, Information-Poor (scenario)
EDC:	EROS Data Center
EO-IU:	EO Image Understanding
EO-IUS:	EO-IU System
EPA:	Environmental Protection Agency
EROS:	Earth Resources Observation Systems
ESA:	European Space Agency
FAO:	Food and Agriculture Organization
GEO:	Intergovernmental Group on Earth Observations
GEOSS:	Global EO System of Systems
GIGO:	Garbage In, Garbage Out principle of error propagation
GIS:	Geographic Information System

GIScience:	Geographic Information Science
GUI:	Graphic User Interface
HS:	Hyper-Spectral
IGBP:	International Global Biosphere Programme
IUS:	Image Understanding System
LC:	Land Cover
LCC:	Land Cover Change
LCCS:	Land Cover Classification System (taxonomy)
LCLU:	Land Cover Land Use
LEDAPS:	Landsat Ecosystem Disturbance Adaptive Processing System
mDMI:	Minimally Dependent and Maximally Informative (set of quality indicators)
MHP:	Modular Hierarchical Phase (in the FAO LCCS taxonomy)
MIR:	Medium InfraRed
MODIS:	Moderate Resolution Imaging Spectroradiometer
MS:	Multi-Spectral
NASA:	National Aeronautics and Space Administration
NIR:	Near InfraRed
NLCD:	National Land Cover Data
NOAA:	National Oceanic and Atmospheric Administration
OA:	Overall Accuracy
OAMTRX:	Overlapping Area Matrix
OBIA:	Object-Based Image Analysis
ODGIS:	Ontology-Driven GIS
OGC:	Open Geospatial Consortium
OP:	Outcome (product) and Process
O-Q2Is:	Outcome Quantitative Quality Index
OP-Q^2I:	Outcome and Process Quantitative Quality Index
P-Q2Is:	Product Quantitative Quality Index
QA:	Quality Assurance
QA4EO:	Quality Accuracy Framework for Earth Observation
Q^2I:	Quantitative Quality Indicator
R&D:	Research and Development
RGB:	monitor-typical Red-Green-Blue data cube
RMSE:	Root Mean Square Error
RS:	Remote Sensing
SCM:	Scene Classification Map
SDC:	Single-Date Classification
SEN2COR:	Sentinel 2 (atmospheric) Correction Prototype Processor
SQ^2I:	Spatial Quantitative Quality Index
SIAM™:	Satellite Image Automatic Mapper™
STRATCOR:	Stratified Topographic Correction
SS:	Super-Spectral

SURF:	Surface Reflectance
TIR:	Thermal InfraRed
TM (superscript):	(non-registered) Trademark
TOA:	Top-Of-Atmosphere
TOARF:	TOA Reflectance
TQ^{2}I:	Thematic Quantitative Quality Index
UML:	Unified Modeling Language
USGS:	US Geological Survey
Val:	Validation
VQ:	Vector Quantization
WELD:	Web-Enabled Landsat Data set
WGCV:	Working Group on Calibration and Validation

Acknowledgments

Prof. Ralph Maughan, Idaho State University, is kindly acknowledged for his contribution as active conservationist and for his willingness to share his invaluable photo archive with the scientific community as well as the general public. Andrea Baraldi thanks Prof. Raphael Capurro, Hochschule der Medien, Germany, and Prof. Christopher Justice, Chair of the Department of Geographical Sciences, University of Maryland, College Park, MD, for their support. Above all, the authors acknowledge the fundamental contribution of Prof. Luigi Boschetti, currently at the Department of Forest, Rangeland and Fire Sciences, University of Idaho, Moscow, Idaho, who conducted by independent means all experiments whose results are proposed in this validation paper. The authors also wish to thank the Editor-in-Chief, Associate Editor, and reviewers for their competence, patience, and willingness to help.

Funding

To accomplish this work, Andrea Baraldi was supported in part by the National Aeronautics and Space Administration (NASA) under Grant No. NNX07AV19G issued through the Earth Science Division of the Science Mission Directorate. Andrea Baraldi, Dirk Tiede, and Stefan Lang were supported in part by the Austrian Science Fund (FWF) through the Doctoral College GIScience (DK W1237- N23).

Competing interests

As a source of potential competing interests of professional or financial nature, co-author Andrea Baraldi reports he is the sole developer and intellectual property right (IPR) owner of the Satellite Image Automatic Mapper™ (non-registered trademark) computer program validated by an independent third party in this research and licensed by the one-man-company Baraldi Consultancy in Remote Sensing (siam.andreabaraldi.com) to academia, public institutions, and private companies, eventually free of charge. Throughout his scientific career, Andrea Baraldi has put in place approved plans for managing any potential conflict arising from his IPR ownership of the SIAM™ computer software.

Author details

Andrea Baraldi[1,2,3,4]
E-mail: andrea6311@gmail.com
ORCID ID: http://orcid.org/0000-0001-5196-9944

Michael Laurence Humber[2]
E-mail: mhumber@umd.edu
Dirk Tiede[3]

E-mail: dirk.tiede@sbg.ac.at
ORCID ID: http://orcid.org/0000-0002-5473-3344
Stefan Lang[3]
E-mail: stefan.lang@sbg.ac.at
[1] Department of Agricultural Sciences, University of Naples Federico II, Portici, Italy.
[2] Department of Geographical Sciences, University of Maryland, College Park, MD, USA.
[3] Department of Geoinformatics – Z_GIS, University of Salzburg, Salzburg, Austria.
[4] Italian Space Agency (ASI), Rome, Italy.

Notes

Figure 3. Six-stage hybrid (combined deductive and inductive) feedback EO image understanding system (EO-IUS) design, based on a convergence-of-evidence approach consistent with Bayesian naïve classification (Baraldi, 2017). Alternative to inductive feedforward EO-IUS architectures adopted by the RS mainstream, it supports a hierarchical approach to low-level (preliminary, general-purpose, sensor-, application- and user independent) EO image understanding followed by high-level (sensor-, application- and user-specific) EO image understanding (classification), consistent with the standard fully nested Land Cover Classification System (LCCS) taxonomy promoted by the Food and Agriculture Organization (FAO) of the United Nations (Di Gregorio & Jansen, 2000). For the sake of visualization each of the six data processing stages plus stage-zero for EO data pre-processing is depicted as a rectangle with a different color fill. Visual evidence stems from multiple information sources, specifically, numeric color values quantized into categorical color names, local shape, texture and inter-object spatial relationships, either topological or non-topological. An example of preliminary (low-level) general-purpose, user- and application-independent EO image classification taxonomy required by an ESA EO Level 2 Scene Classification Map (SCM) product is the 3-level 8-class FAO LCCS Dichotomous Phase (DP) legend, in addition to quality layers such as cloud and cloud-shadow. High-level EO image classification is user- and application-specific, where a thematic map product of Level 3 or superior is provided with a legend consistent with the FAO LCCS Modular Hierarchical Phase (MHP) taxonomy (Di Gregorio & Jansen, 2000); refer to Figure 3 in the Part 1 of this paper. Acronym SIAM stays for Satellite Image Automatic Mapper (SIAM), a lightweight computer program for MS reflectance space hyperpolyhedralization into a static vocabulary of MS color names, superpixel detection and vector quantization (VQ) quality assessment (Baraldi, 2017; Baraldi et al., 2006, 2010a, 2010b, 2010c; Baraldi, 2011; Baraldi & Boschetti, 2012a, 2012b; Baraldi et al., 2013; 2016; Baraldi & Humber, 2015).

Figure 9. Examples of geographic locations mapped as vegetation classes *"Scrub/Shrub"* (SS) or *"Grassland/ Herbaceous"* (GH) in the USGS NLCD 2006 reference map (refer to Table 1) and predominantly as bare soil spectral categories (sbS_1, SmS_1, aS) in the SIAM-WELD 2006 test map (refer to Table 2), as pointed out in Table 8. The SIAM's color names sbS_1, SmS_1, and aS mean that, from space, with a pixel size of 30 m × 30 m = 900 m^2, the contribution of sparse vegetation, rangeland, cheatgrass, dry long grass or short grass as foreground, mixed with a background of sand, clay, or rocks, like those shown in these pictures, becomes extremely difficult to detect, especially if a hard (crisp, defuzzified) label rather than a set of fuzzy class labels must be provided as the output product. (a) Sublette, WY, Rangeland, 42° 51' 37" N, 109° 43' 7" W. Copyright Ralph Maughan, Idaho State Univ. Reproduced with permission of the author. Acquisition date: 6/16/2011. [Online]. Available: http://www.panor amio.com (accessed on 24 February 2013). (b) Twin Falls, ID, Ripening cheatgrass infestation, 42° 23' 52" N, 114° 21' 9" W. Copyright Ralph Maughan, Idaho State Univ. Reproduced with permission of the author. Acquisition date: April 2010? [Online]. Available: http:// www.panoramio.com (accessed on 24 February 2013). (c) Overton, NV, 36° 25' 42" N, 114° 27' 21" W. Copyright Ralph Maughan, Idaho State Univ. Reproduced with permission of the author. Acquisition date: 2/11/2009. [Online]. Available: http://www.panor amio.com (accessed on 24 February 2013). (d) San Juan, UT, 37° 16' 43" N, 109° 40' 27" W. Copyright Ralph Maughan, Idaho State Univ. Reproduced with permission of the author. Acquisition date: 3/4/2009. [Online]. Available: http://www.panoramio.com (accessed on 24 February 2013). (e) Springerville, AZ, Volcanic Field, 34° 15' 6" N, 109° 21' 9" W. Copyright Ralph Maughan, Idaho State Univ. Reproduced with permission of the author. Acquisition date: 3/3/2009. [Online]. Available: http://www.panoramio.com (accessed on 24 February 2013). (f) Esmeralda, NV, 38° 1' 40" N, 117° 43' 21" W. Copyright Ralph Maughan, Idaho State Univ. Reproduced with permission of the author. Acquisition date: 4/22/2010. [Online]. Available: http://www.panoramio.com (accessed on 24 February 2013).

References

Ackerman, S. A., Strabala, K. I., Menzel, W. P., Frey, R. A., Moeller, C. C., & Gumley, L. E. (1998). Discriminating clear sky from clouds with MODIS. *Journal of Geophysical Research, 103*(32), 141–157. doi:10.1029/1998JD200032

Ahlqvist, O. (2005). Using uncertain conceptual spaces to translate between land cover categories. *International Journal of Geographical Information Science, 19*, 831–857.

Arvor, D., Madiela, B. D., & Corpetti, T. (2016). Semantic pre-classification of vegetation gradient based on linearly unmixed Landsat time series. *Geoscience and Remote Sensing Symposium (IGARSS), IEEE International* (pp. 4422–4425).

Baraldi, A. (2011). Fuzzification of a crisp near-real-time operational automatic spectral-rule-based decision-tree preliminary classifier of multisource multispectral remotely sensed images. *IEEE Transactions on Geoscience and Remote Sensing, 49*, 2113–2134. doi:10.1109/TGRS.2010.2091137

Baraldi, A. (2015). "Automatic Spatial Context-Sensitive Cloud/Cloud-Shadow Detection in Multi-Source Multi-Spectral Earth Observation Images – AutoCloud+," Invitation to tender ESA/AO/1-8373/15/I-NB – "VAE: Next Generation EO-based Information Services", DOI: 10.13140/RG.2.2.34162.71363. arXiv: 1701.04256. Retrieved Jan., 8, 2017 from https:// arxiv.org/ftp/arxiv/papers/1701/1701.04256.pdf

Baraldi, A. (2017). Pre-processing, classification and semantic querying of large-scale earth observation spaceborne/airborne/terrestrial image databases: Process and product innovations". Ph.D. dissertation in Agricultural and Food Sciences, University of Naples "Federico II". *Department of Agricultural Sciences, Italy. Ph.D. defense: 16 May 2017.* doi:10.13140/RG.2.2.25510.52808. Retrieved January, 30 2018, from https://www.researchgate. net/publication/317333100_Pre-processing_classifi cation_and_semantic_querying_of_large-scale_ Earth_observation_spaceborneairborneterrestrial_ image_databases_Process_and_product_ innovations

Baraldi, A., & Boschetti, L. (2012a). Operational automatic remote sensing image understanding systems: Beyond Geographic Object-Based and Object-Oriented Image Analysis (GEOBIA/GEOOIA) – part 1: Introduction. *Remote Sensing, 4*, 2694–2735. doi:10.3390/rs4092694

Baraldi, A., & Boschetti, L. (2012b). Operational automatic remote sensing image understanding systems: Beyond Geographic Object-Based and Object-Oriented Image Analysis (GEOBIA/GEOOIA) – part 2: Novel system architecture, information/knowledge representation, algorithm design and implementation. *Remote Sensing, 4*, 2768–2817.

Baraldi, A., Boschetti, L., & Humber, M. (2014). Probability sampling protocol for thematic and spatial quality assessments of classification maps generated from spaceborne/airborne very high resolution images. *IEEE Transactions on Geoscience and Remote Sensing, 52*(1), 701–760. doi:10.1109/ TGRS.2013.2243739

Baraldi, A., Durieux, L., Simonetti, D., Conchedda, G., Holecz, F., & Blonda, P. (2010a). Automatic spectral rule-based preliminary classification of radiometrically calibrated SPOT-4/-5/IRS, AVHRR/ MSG,AATSR, IKONOS/QuickBird/OrbView/GeoEye and DMC/SPOT-1/-2 imagery – part I: System design and implementation. *IEEE Transactions on Geoscience and Remote Sensing, 48*, 1299–1325. doi:10.1109/ TGRS.2009.2032457

Baraldi, A., Durieux, L., Simonetti, D., Conchedda, G., Holecz, F., & Blonda, P. (2010b). Automatic spectral rule-based preliminary classification of radiometrically calibrated SPOT-4/-5/IRS, AVHRR/ MSG, AATSR, IKONOS/QuickBird/OrbView/GeoEye and DMC/SPOT-

1/-2 imagery – part II: Classification accuracy assessment. *IEEE Transactions on Geoscience and Remote Sensing, 48,* 1326–1354. doi:10.1109/TGRS.2009.2032064

Baraldi, A., Gironda, M., & Simonetti, D. (2010c). Operational three-stage stratified topographic correction of spaceborne multi-spectral imagery employing an automatic spectral rule-based decision-tree preliminary classifier. *IEEE Transactions Geosci Remote Sensing, 48*(1), 112–146. doi:10.1109/TGRS.2009.2028017

Baraldi, A., & Humber, M. (2015). Quality assessment of pre-classification maps generated from spaceborne/airborne multi-spectral images by the Satellite Image Automatic Mapper™ and Atmospheric/Topographic Correction™-Spectral Classification software products: Part 1 – theory. *IEEE Journal of Selected Topics in Applied Earth Observations and Remote Sensing, 8*(3), 1307–1329.

Baraldi, A., Humber, M., & Boschetti, L. (2013). Quality assessment of pre-classification maps generated from spaceborne/airborne multi-spectral images by the Satellite Image Automatic Mapper™ and Atmospheric/Topographic Correction™-Spectral Classification software products: Part 2 – experimental results. *Remote Sensing, 5,* 5209–5264. doi:10.3390/rs5105209

Baraldi, A., Puzzolo, V., Blonda, P., Bruzzone, L., & Tarantino, C. (2006). Automatic spectral rule-based preliminary mapping of calibrated Landsat TM and ETM+ images. *IEEE Transactions on Geoscience and Remote Sensing, 44,* 2563–2586. doi:10.1109/TGRS.2006.874140

Baraldi, A., & Soares, J. V. B. (2017). Multi-objective software suite of two-dimensional shape descriptors for object-based image analysis. *Subjects: Computer Vision and Pattern Recognition (cs.CV), arXiv:1701.01941.* Retrieved January 8, 2017, from [Online] https://arxiv.org/ftp/arxiv/papers/1701/1701.01941.pdf

Baraldi, A., Tiede, D., Sudmanns, M., Belgiu, M., & Lang, S. (2016). *Automated near real-time earth observation level 2 product generation for semantic querying.* Enschede, The Netherlands: GEOBIA 2016, 14–16 Sept., University of Twente Faculty of Geo-Information and Earth Observation (ITC).

Baraldi, A., Tiede, D., Sudmanns, M., Belgiu, M., & Lang, S. (2017, March 28–30). *Systematic ESA EO level 2 product generation as pre-condition to semantic content-based image retrieval and information/knowledge discovery in EO image databases.* 2017 Conf. on Big Data From Space, BiDS'17, Toulouse, France.

Baraldi, A., Wassenaar, T., & Kay, S. (2010d). Operational performance of an automatic preliminary spectral rule-based decision-tree classifier of spaceborne very high resolution optical images. *IEEE Transactions on Geoscience and Remote Sensing, 48,* 3482–3502. doi:10.1109/TGRS.2010.2046741

Beauchemin, M., & Thomson, K. (1997). The evaluation of segmentation results and the overlapping area matrix. *International Journal of Remote Sens, 18,* 3895–3899. doi:10.1080/014311697216720

Benavente, R., Vanrell, M., & Baldrich, R. (2008). Parametric fuzzy sets for automatic color naming. *Journal of the Optical Society of America, 25,* 2582–2593. doi:10.1364/JOSAA.25.002582

Berlin, B., & Kay, P. (1969). *Basic color terms: Their universality and evolution.* Berkeley, CA: University of California.

Bernus, P., & Noran, O. (2017). Data rich – but information poor. In L. Camarinha-Matos, H. Afsarmanesh, & R. Fornasiero (Eds.), *Collaboration in a data-rich world: PRO-VE 2017. IFIP advances in*

information and communication technology (Vol. 506, pp. 206–214).

Bishop, C. M. (1995). *Neural networks for pattern recognition.* Oxford, UK: Clarendon.

Bishop, M. P., & Colby, J. D. (2002). Anisotropic reflectance correction of SPOT-3 HRV imagery. *International Journal of Remote Sens, 23*(10), 2125–2131. doi:10.1080/01431160110097231

Bishop, M. P., Shroder, J. F., & Colby, J. D. (2003). Remote sensing and geomorphometry for studying relief production in high mountains. *Geomorphology, 55*(1–4), 345–361. doi:10.1016/S0169-555X(03)00149-1

Boschetti, L., Flasse, S. P., & Brivio, P. A. (2004). Analysis of the conflict between omission and commission in low spatial resolution dichotomic thematic products: The Pareto boundary. *Remote Sensing of Environment, 91,* 280–292. doi:10.1016/j.rse.2004.02.015

Boschetti, L., Roy, D. P., Justice, C. O., & Humber, M. L. (2015). MODIS–Landsat fusion for large area 30 m burned area mapping. *Remote Sensing of Environment, 161,* 27–42. doi:10.1016/j.rse.2015.01.022

Capurro, R., & Hjørland, B. (2003). The concept of information. *Annual Review of Information Science and Technology, 37,* 343–411. doi:10.1002/aris.1440370109

Cherkassky, V., & Mulier, F. (1998). *Learning from data: Concepts, theory, and methods.* New York, NY: Wiley.

CNES. (2015). *Venμs satellite sensor level 2 product.* Retrieved January 5, 2016, from https://venus.cnes.fr/en/VENUS/prod_l2.htm

Congalton, R. G, & Green, K. (1999). Assessing the accuracy of remotely sensed data. In *Boca raton, fl.* USA: Lewis Publishers.

Dai, X., & Khorram, S. (1998). The effects of image misregistration on the accuracy of remotely sensed change detection. *IEEE Transactions on Geoscience and Remote Sensing, 36,* 1566–1577. doi:10.1109/36.718860

Despini, F., Teggi, S., & Baraldi, A. (2014, September 22). Methods and metrics for the assessment of pansharpening algorithms. In L. Bruzzone, J. A. Benediktsson, & F. Bovolo (Eds.), *SPIE proceedings, Vol. 9244: Image and signal processing for remote sensing XX.* Amsterdam, Netherlands.

Deutsches Zentrum für Luft- und Raumfahrt e.V. (DLR) and VEGA Technologies. (2011). *Sentinel-2 MSI – Level 2A products algorithm theoretical basis document. Document S2PAD-ATBD-0001.* European Space Agency.

Di Gregorio, A., & Jansen, L. (2000). Land cover classification system (LCCS): Classification concepts and user manual. *FAO: Rome, Italy, FAO Corporate Document Repository.* Retrieved February 10, 2015, from http://www.fao.org/DOCREP/003/X0596E/X0596e00.htm

Dillencourt, M. B., Samet, H., & Tamminen, M. (1992). A general approach to connected component labeling for arbitrary image representations. *Journal of Association for Computing Machinery, 39,* 253–280. doi:10.1145/128749.128750

Dorigo, W., Richter, R., Baret, F., Bamler, R., & Wagner, W. (2009). Enhanced automated canopy characterization from hyperspectral data by a novel two step radiative transfer model inversion approach. *Remote Sensing, 1,* 1139–1170. doi:10.3390/rs1041139

Duke Center for Instructional Technology. (2016). *Measurement: Process and outcome indicators.* Retrieved June 20, 2016, from http://patientsafetyed.

duhs.duke.edu/module_a/measurement/measure
ment.html

Environmental Protection Agency (EPA). (2007).
*"Definitions" in multi-resolution land characteristics
consortium (MRLC).* Retrieved November 13, 2013,
from http://www.epa.gov/mrlc/definitions.html#2001

Environmental Protection Agency (EPA). (2013). *Western
ecology division.* Retrieved November 13, 2013, from
http://www.epa.gov/wed/pages/ecoregions.htm

European Space Agency (ESA). (2015). *Sentinel-2 User
Handbook, Standard Document, Issue 1 Rev 2.*

Foody, G. (2016). *"Observations on accuracy assessments
of object-based image classifications," GEOBIA 2016,
14-16 sept.* University of Twente, Faculty of Geo-
Information and Earth Observation (ITC), Enschede,
The Netherlands.

Foody, G. (2006). The evaluation and comparison of thematic maps derived from remote sensing.In Proc. of
the 7th International Symposium on spatial accuracy
in natural resources and environmental sciences M.
Caetano (& M. Painho, eds), 7-9 July, 2006, Lisbon.
Instituto Geográfico Português, Lisbon, 18-31.

GeoTerraImage. (2015). *Provincial and national land cover
30 m.* Retrieved September 22, 2015, from. http://
www.geoterraimage.com/productslandcover.php

Gevers, T., Gijsenij, A., Van De Weijer, J., & Geusebroek, J. M.
(2012). *Color in computer vision.* Hoboken, NJ: Wiley.

Griffin, L. D. (2006). Optimality of the basic color categories for classification. *Journal of the Royal Society,
Interface, 3,* 71–85. doi:10.1098/rsif.2005.0076

Griffith, G. E., & Omernik, J. M. (2009). Ecoregions of the
United States-Level III (EPA). In C. J. Cleveland (Ed.),
Encyclopedia of earth. Washington, D.C.:
Environmental Information Coalition, National
Council for Science and the Environment.

Group on Earth Observation (GEO). (2005). The global
earth observation system of systems (GEOSS) 10-
year implementation plan, adopted 16 February
2005. Retrieved January 10, 2012, from http://www.
earthobservations.org/docs/10-Year%
20Implementation%20Plan.pdf

Group on Earth Observation/Committee on Earth
Observation Satellites (GEO/CEOS). (2010). A quality
assurance framework for earth observation, version
4.0. Retrieved November 15, 2012, from. http://
qa4eo.org/docs/QA4EO_Principles_v4.0.pdf

Group on Earth Observation/Committee on Earth
Observation Satellites (GEO-CEOS) - Working Group
on Calibration and Validation (WGCV). (2015). *Land
product validation (LPV).* Retrieved March 20, 2015,
from http://lpvs.gsfc.nasa.gov/

Hadamard, J. (1902). Sur les problemes aux deriveespar-
tielles et leur signification physique. *Princet University
Bulletin, 13,* 49–52.

Homer, C., Huang, C. Q., Yang, L. M., Wylie, B., & Coan, M.
(2004). Development of a 2001 National Land-Cover
Database for the United States. *Photogrammetric
Engineering & Remote Sensing, 70,* 829–840.
doi:10.14358/PERS.70.7.829

Hunt, N., & Tyrrell, S. (2012). *Stratified sampling.
Coventry University.* Retrieved February 7,
2012from http://www.coventry.ac.uk/ec/~nhunt/
meths/strati.html

IBM. (2016). *The four V's of big data, IBM big data & analytics
hub.*Retrieved January 30, 2018, from http://www.ibm
bigdatahub.com/infographic/four-vs-big-data

Kuzera, K., & Pontius, R. G. (2008). Importance of matrix
construction for multiple-resolution categorical map
comparison. *GIScience & Remote Sensing, 45,* 249–
274. doi:10.2747/1548-1603.45.3.249

Lee, D. S., Storey, J. C., Choate, M. J., & Hayes, R. (2004).
Four years of Landsat-7 on-orbit geometric calibra-

tion and performance. *IEEE Transactions on
Geoscience and Remote Sensing, 42,* 2786–2795.
doi:10.1109/TGRS.2004.836769

Liang, S. (2004). *Quantitative remote sensing of land surfaces.* Hoboken, NJ: Wiley.

Lillesand, T., & Kiefer, R. (1979). *Remote sensing and
image interpretation.* New York, NY: Wiley.

Lipson, H. (2007). Principles of modularity, regularity, and
hierarchy for scalable systems. *Journal of Biological
Physics and Chemistry, 7,* 125–128. doi:10.4024/40701.
jbpc.07.04

Liu, S., Hairong Liu, L. J., Latecki, S. Y., Xu, C., & Lu, H.
(2011). Size adaptive selection of most informative
features. *Association for the Advancement of Artificial
Intelligence·*

Lück, W., & Van Niekerk, A. (2016). Evaluation of a rule-
based compositing technique for Landsat-5 TM and
Landsat-7 ETM+ images. *International Journal of
Applied Earth Observation and Geoinformation, 47,* 1–
14. doi:10.1016/j.jag.2015.11.019

Lunetta, R., & Elvidge, D. (1999). *Remote sensing change
detection: Environmental monitoring methods and
applications.* London, UK: Taylor & Francis.

Luo, Y., Trishchenko, A. P., & Khlopenkov, K. V. (2008).
Developing clear-sky, cloud and cloud shadow mask
for producing clear-sky composites at 250-meter
spatial resolution for the seven MODIS land bands
over Canada and North America. *Remote Sensing of
Environment, 112,* 4167–4185. doi:10.1016/j.
rse.2008.06.010

Markham, B., & Helder, D. (2012). *Forty-year calibrated
record of earth-reflected radiance from Landsat: A
review.* Paper 70. NASA Publications.

Marr, D. (1982). *Vision.* New York, NY: Freeman and C.

Matsuyama, T., & Hwang, V. S. (1990). *SIGMA – a knowledge-based aerial image understanding system.* New
York, NY: Plenum Press.

Muirhead, K., & Malkawi, O. (1989, September 5–8).
Proceedings Fourth AVHRR Data Users Meeting (pp.
31–34). Rottenburg, Germany.

Nagao, M., & Matsuyama, T. (1980). *A structural analysis
of complex aerial photographs.* New York, NY: Plenum.

National Aeronautics and Space Administration (NASA).
(2016). *Getting petabytes to people: How the EOSDIS
facilitates earth observing data discovery and use.
[Online].* Retrieved December 29, 2016, from https://
earthdata.nasa.gov/getting-petabytes-to-people-
how-the-eosdis-facilitates-earth-observing-data-dis
covery-and-use

Open Geospatial Consortium (OGC) Inc. (2015). *OpenGIS®
implementation standard for geographic information
– simple feature access – part 1: Common architecture.* Retrieved March 8, 2015, from http://www.
opengeospatial.org/standards/is

Ortiz, A., & Oliver, G. (2006). On the use of the overlapping
area matrix for image segmentation evaluation: A
survey and new performance measures. *Pattern
Recognition Letters, 27,* 1916–1926. doi:10.1016/j.
patrec.2006.05.002

Peng, H., Long, F., & Ding, C. (2005). Feature selection
based on mutual information: Criteria of max-
dependency, max-relevance, and min-redundancy.
*IEEE Transactions Pattern Analysis Machine
Intelligent, 27,* 1226–1238. doi:10.1109/
TPAMI.2005.159

Pontius, R. G., Jr., & Connors, J. (2006, July 5–7).
Expanding the conceptual, mathematical and practical methods for map comparison. In M. Caetano &
M. Painho (Eds). *Proceedings of the 7th international
symposium on spatial accuracy assessment in natural
resources and environmental sciences* (pp. 64–79).
Lisbon: Instituto Geográfico Português.

Pontius, R. G., Jr., & Millones, M. (2011). Death to Kappa: Birth of quantity disagreement and allocation disagreement for accuracy assessment. *International Journal of Remote Sensing, 32*(15), 4407–4429. doi:10.1080/01431161.2011.552923

Riano, D., Chuvieco, E., Salas, J., & Aguado, I. (2003). Assessment of different topographic corrections in landsat-TM data for mapping vegetation types. *IEEE Transactions on Geoscience and Remote Sensing, 41*(5), 1056–1061. doi:10.1109/TGRS.2003.811693

Richter, R., & Schläpfer, D. (2012a). *Atmospheric/topographic correction for satellite imagery – ATCOR-2/3 User Guide, Version 8.2 BETA.* Retrieved April 12, 2013, from http://www.dlr.de/eoc/Portaldata/60/Resources/dokumente/5_tech_mod/atcor3_manual_2012.pdf

Richter, R., & Schläpfer, D. (2012b). *Atmospheric/Topographic correction for airborne imagery – ATCOR-4 User Guide, Version 6.2 BETA, 2012.* Retrieved April 12, 2013, from http://www.dlr.de/eoc/Portaldata/60/Resources/dokumente/5_tech_mod/atcor4_manual_2012.pdf

Ridd, M. K. (1995). Exploring a V-I-S (vegetation-impervious surface-soil) model for urban ecosystem analysis through remote sensing: Comparative anatomy for cities. *International Journal of Remote Sens, 16*(12), 2165–2185. doi:10.1080/01431169508954549

Roy, D., Ju, J., Kline, K., Scaramuzza, P. L., Kovalskyy, V., Hansen, M., ... Zhang, C. S. (2010). Web-enabled Landsat data (WELD): Landsat ETM plus composited mosaics of the conterminous United States. *Remote Sensing of Environment, 114*, 35–49. doi:10.1016/j.rse.2009.08.011

SIAM-WELD-NLCD FTP. (2016). Retrieved December 13, 2016, from http://tinyurl.com/j4nzwzl

Simonetti, D., Simonetti, E., Szantoi, Z., Lupi, A., & Eva, H. D. (2015). First results from the phenology based synthesis classifier using Landsat-8 imagery. *IEEE Geoscience and Remote Sensing Letters, 12*(7), 1496–1500. doi:10.1109/LGRS.2015.2409982

Sonka, M., Hlavac, V., & Boyle, R. (1994). *Image processing, analysis and machine vision.* London, UK: Chapman & Hall.

Stehman, S. V., & Czaplewski, R. L. (1998). Design and analysis for thematic map accuracy assessment: Fundamental principles. *Remote Sensing of Environment, 64*, 331–344. doi:10.1016/S0034-4257(98)00010-8

Stehman, S. V., Wickham, J. D., Wade, T. G., & Smith, J. H. (2008). Designing a multi-objective, multi-support accuracy assessment of the 2001 National Land Cover Data (NLCD 2001) of the conterminous United States. *Photogrammetric Engineering & Remote Sensing, 74*, 1561–1571. doi:10.14358/PERS.74.12.1561

Tiede, D., Baraldi, A., Sudmanns, M., Belgiu, M., & Lang, S. (2016, March 15–17). *ImageQuerying (IQ) – earth observation image content extraction & querying across time and space, submitted (oral presentation and poster session).* ESA 2016 Conf. on Big Data From Space, BIDS '16, Santa Cruz de Tenerife, Spain.

Vermote, E., & Saleous, N. (2007). *LEDAPS surface reflectance product description – Version 2.0.* University of Maryland at College Park/Dept Geography and NASA/GSFC Code 614.5

Victor, J. (1994). Images, statistics, and textures: Implications of triple correlation uniqueness for texture statistics and the Julesz conjecture: Comment. *Journal of Optical Social American A, 11*(5), 1680–1684. doi:10.1364/JOSAA.11.001680

Vogelmann, J. E., Howard, S. M., Yang, L., Larson, C. R., Wylie, B. K., & Van Driel, N. (2001). Completion of the 1990s National Land Cover Data set for the conterminous United States from Landsat Thematic Mapper data and ancillary data sources. *Photogrammetric Engineering & Remote Sensing, 67*, 650–662.

Vogelmann, J. E., Sohl, T. L., Campbell, P. V., & Shaw, D. M. (1998). Regional land cover characterization using Landsat Thematic Mapper data and ancillary data sources. *Environmental Monitoring and Assessment, 51*, 415–428. doi:10.1023/A:1005996900217

Web-Enabled Landsat Data (WELD) Tile FTP. Retrieved December 12, 2016, from https://weld.cr.usgs.gov/

Wessels, K., Van Den Bergh, F., Roy, D., Salmon, B., Steenkamp, K., MacAlister, B., ... Jewitt, D. (2016). Rapid land cover map updates using change detection and robust random forest classifiers. *Remote Sensing, 8*(888), 1–24. doi:10.3390/rs8110888

Wickham, J. D., Stehman, S. V., Fry, J. A., Smith, J. H., & Homer, C. G. (2010). Thematic accuracy of the USGS NLCD 2001 land cover for the conterminous United States. *Remote Sensing of Environment, 114*, 1286–1296. doi:10.1016/j.rse.2010.01.018

Wickham, J. D., Stehman, S. V., Gass, L., Dewitz, J., Fry, J. A., & Wade, T. G. (2013). Accuracy assessment of NLCD 2006 land cover and impervious surface. *Remote Sensing of Environment, 130*, 294–304. doi:10.1016/j.rse.2012.12.001

Xian, G., & Homer, C. (2010). Updating the 2001 National Land Cover Database impervious surface products to 2006 using Landsat imagery change detection methods. *Remote Sensing of Environment, 114*, 1676–1686. doi:10.1016/j.rse.2010.02.018

Yang, C., Huang, Q., Li, Z., Liu, K., & Hu, F. (2017). Big data and cloud computing: Innovation opportunities and challenges. *International Journal of Digital Earth, 10*(1), 13–53. doi:10.1080/17538947.2016.1239771

Permissions

The contributors of this book come from diverse backgrounds, making this book a truly international effort. This book will bring forth new frontiers with its revolutionizing research information and detailed analysis of the nascent developments around the world.

We would like to thank all the contributing authors for lending their expertise to make the book truly unique. They have played a crucial role in the development of this book. Without their invaluable contributions this book wouldn't have been possible. They have made vital efforts to compile up to date information on the varied aspects of this subject to make this book a valuable addition to the collection of many professionals and students.

This book was conceptualized with the vision of imparting up-to-date information and advanced data in this field. To ensure the same, a matchless editorial board was set up. Every individual on the board went through rigorous rounds of assessment to prove their worth. After which they invested a large part of their time researching and compiling the most relevant data for our readers.

The editorial board has been involved in producing this book since its inception. They have spent rigorous hours researching and exploring the diverse topics which have resulted in the successful publishing of this book. They have passed on their knowledge of decades through this book. To expedite this challenging task, the publisher supported the team at every step. A small team of assistant editors was also appointed to further simplify the editing procedure and attain best results for the readers.

Apart from the editorial board, the designing team has also invested a significant amount of their time in understanding the subject and creating the most relevant covers. They scrutinized every image to scout for the most suitable representation of the subject and create an appropriate cover for the book.

The publishing team has been an ardent support to the editorial, designing and production team. Their endless efforts to recruit the best for this project, has resulted in the accomplishment of this book. They are a veteran in the field of academics and their pool of knowledge is as vast as their experience in printing. Their expertise and guidance has proved useful at every step. Their uncompromising quality standards have made this book an exceptional effort. Their encouragement from time to time has been an inspiration for everyone.

The publisher and the editorial board hope that this book will prove to be a valuable piece of knowledge for researchers, students, practitioners and scholars across the globe.

List of Contributors

Friday Uchenna Ochege
Laboratory for Cartography and GIS, Department of Geography & Environmental Management, University of Port Harcourt, Choba, Rivers State, Nigeria
Faculty of Environmental Studies, Department of Surveying and Geoinformatics, University of Nigeria, Enugu, Nigeria

Chukwunonyelum Okpala-Okaka
Faculty of Environmental Studies, Department of Surveying and Geoinformatics, University of Nigeria, Enugu, Nigeria

R. Hewson
Faculty of Geo-Information Science & Earth Observation (ITC), University of Twente, 7500 AE, Enschede, The Netherlands

D. Robson, A. Carlton and P. Gilmore
Geological Survey of NSW, NSW Trade & Investment, Division of Resources & Energy, Hunter Region Mail Centre, 2310, Sydney, NSW, Australia

Mary Antwi
Department of Crop and Soil Sciences, Kwame Nkrumah University of Science and Technology, Kumasi, Ghana

Alfred Allan Duker
Department of Geomatic Engineering, Kwame Nkrumah University of Science and Technology, Kumasi, Ghana

Mathias Fosu
Council for Scientific and Industrial Research, Savanna Agriculture Research Institute, Tamale, Ghana

Robert Clement Abaidoo
College of Agriculture and Natural Resources, Kwame Nkrumah University of Science and Technology, Kumasi, Ghana
International Institute of Tropical Agriculture, Ibadan, Nigeria

Tatsuya Ishige
Department of Industrial and System Engineering, Hosei University, Tokyo, Japan

Zdena Dobesova
Faculty of Science, Department of Geoinformatics, Palacký University, 17 listopadu 50, 771 46 Olomouc, Czech Republic

Paweł B. Dąbek and Romuald Żmuda
Institute of Environmental Development and Protection, Wrocław University of Environmental and Life Sciences, pl. Grunwaldzki 24a, Wrocław 50-363, Poland

Ciechosław Patrzałek
Department of Landscape Management, Wrocław University of Environmental and Life Sciences, ul. Grunwaldzka 53, Wrocław 50-363, Poland

Bartłomiej Ćmielewski
Institute of History of Architecture, Wrocław University of Technology, LabScan3D, ul. Bolesława Prusa 53/55, Wrocław 50-317, Poland

Bambo Dubula, Solomon Gebremariam Tesfamichael and Isaac Tebogo Rampedi
Department of Geography, Environmental Management and Energy Studies, University of Johannesburg, Johannesburg, South Africa

S. Minu and Amba Shetty
Department of Applied Mechanics and Hydraulics, National Institute of Technology Karnataka, Surathkal, Mangalore 575025, Karnataka, India

Binny Gopal
Department of Agronomy, University of Agricultural and Horticultural Sciences, Navile, Shimoga, Karnataka 577225, India

Andrea Baraldi
Department of Agricultural Sciences, University of Naples Federico II, Portici, Italy
Department of Geoinformatics – Z_GIS, University of Salzburg, Salzburg, Austria
Italian Space Agency (ASI), Rome, Italy

Michael Laurence Humber
Department of Geographical Sciences, University of Maryland, College Park, MD, USA

Dirk Tiede and Stefan Lang
Department of Geoinformatics – Z_GIS, University of Salzburg, Salzburg, Austria

Index